VOLUME ONE HUNDRED AND TEN

SEMICONDUCTORS AND SEMIMETALS

Brillouin Scattering Part 2

SERIES EDITORS

VOLUME ONE HUNDRED AND TEN

SEMICONDUCTORS AND SEMIMETALS

Brillouin Scattering Part 2

Edited by

BENJAMIN J. EGGLETON
School of Physics,
University of Sydney Nano Institute (Sydney Nano),
University of Sydney, NSW, Australia

MICHAEL J. STEEL
School of Mathematical and Physical Sciences,
Macquarie University, Sydney, NSW, Australia

CHRISTOPHER G. POULTON
School of Mathematical and Physical Sciences,
University of Technology Sydney, Sydney, NSW, Australia

ELSEVIER

ACADEMIC PRESS
An imprint of Elsevier

Academic Press is an imprint of Elsevier
50 Hampshire Street, 5th Floor, Cambridge, MA 02139, United States
525 B Street, Suite 1650, San Diego, CA 92101, United States
The Boulevard, Langford Lane, Kidlington, Oxford OX5 1GB, United Kingdom
125 London Wall, London, EC2Y 5AS, United Kingdom

First edition 2022

Notices
Knowledge and best practice in this field are constantly changing. As new research and experience broaden our understanding, changes in research methods, professional practices, or medical treatment may become necessary.

Practitioners and researchers must always rely on their own experience and knowledge in evaluating and using any information, methods, compounds, or experiments described herein. In using such information or methods they should be mindful of their own safety and the safety of others, including parties for whom they have a professional responsibility.

To the fullest extent of the law, neither the Publisher nor the authors, contributors, or editors, assume any liability for any injury and/or damage to persons or property as a matter of products liability, negligence or otherwise, or from any use or operation of any methods, products, instructions, or ideas contained in the material herein.

ISBN: 978-0-323-98931-2
ISSN: 0080-8784

Publisher: Zoe Kruze
Acquisitions Editor: Jason Mitchell
Developmental Editor: Jhon Michael Peñano
Production Project Manager: Abdulla Sait
Cover Designer: Greg Harris

Typeset by STRAIVE, India

Contents

Contributors for Volume 2

Gaurav Bahl
Mechanical Science and Engineering, University of Illinois at Urbana-Champaign, Champaign, IL, United States

X. Bao
Department of Physics, University of Ottawa, Ottawa, ON, Canada

Daniel J. Blumenthal
Department of Electrical and Computer Engineering, University of California Santa Barbara, Santa Barbara, CA, United States

Robert Boyd
University of Ottawa, Ottawa, ON, Canada; University of Rochester, Rochester, NY, United States

A. Choudhary
Ultrafast Optical Communications and High-performance Integrated Photonics (UFO-CHIP) Group, Department of Electrical Engineering, Indian Institute of Technology, Delhi, India

Bill Corcoran
Department of Electrical and Computer Systems Engineering, Monash University, Melbourne, Australia

Omar Florez
Catalan Institute of Nanoscience and Nanotechnology (ICN2), CSIC and BIST, Bellaterra, Spain

Irina Kabakova
School of Mathematical and Physical Sciences, University of Technology Sydney, Broadway, NSW, Australia

Eric A. Kittlaus
Jet Propulsion Laboratory, California Institute of Technology, Pasadena, CA, United States

Huan Li
State Key Laboratory for Modern Optical Instrumentation, College of Optical Science and Engineering, International Research Center for Advanced Photonics, Zhejiang University, Hangzhou, China

Mo Li
University of Washington, Seattle, WA, United States

Yang Liu
Institute of Physics, Swiss Federal Institute of Technology Lausanne (EPFL), Lausanne, Switzerland

Guilhem Madiot
Catalan Institute of Nanoscience and Nanotechnology (ICN2), CSIC and BIST, Bellaterra, Spain

David Marpaung
Nonlinear Nanophotonics Group, MESA+ Institute, University of Twente Enschede, Enschede, The Netherlands

Moritz Merklein
The University of Sydney Nano Institute (Sydney Nano); Institute of Photonics and Optical Science (IPOS); School of Physics, The University of Sydney, Sydney, NSW, Australia

Nils T. Otterstrom
Photonic and Phononic Microsystems, Sandia National Laboratories, Albuquerque, NM, United States

Bingcheng Pan
State Key Laboratory for Modern Optical Instrumentation, College of Optical Science and Engineering, International Research Center for Advanced Photonics, Zhejiang University, Hangzhou, China

Peter T. Rakich
Department of Applied Physics, Yale University, New Haven, CT, United States

Giuliano Scarcelli
Fischell Department of Bioengineering, University of Maryland, College Park, MD, United States

Thomas Schneider
THz Photonics Group, Technische Universität Braunschweig, Braunschweig, Germany

Donggyu B. Sohn
Mechanical Science and Engineering, University of Illinois at Urbana-Champaign, Champaign, IL, United States

Clivia M. Sotomayor Torres
Catalan Institute of Nanoscience and Nanotechnology (ICN2), CSIC and BIST, Bellaterra; ICREA, Pg. Lluís Companys 23, Barcelona, Spain

Birgit Stiller
Max Planck Institute for the Science of Light, Erlangen, Germany

L. Thevenaz
Swiss Federal Institute of Technology (EPFL), Institute of Electrical Engineering, SCI STI LT, Lausanne, Switzerland

Kerry Vahala
Watson Laboratory of Applied Physics, California Institute of Technology, Pasadena, CA, United States

Kerry John Vahala
T.J. Watson Laboratory of Applied Physics, California Institute of Technology, Pasadena, CA, United States

Herbert Winful
University of Michigan, Ann Arbor, MI, United States

Z. Yang
State Key Laboratory of Information Photonics & Optical Communications, Beijing University of Posts and Telecommunications, Beijing, China

Seok-Hyun Yun
Harvard Medical School and Wellman Center for Photomedicine, Massachusetts General Hospital, Boston; Harvard-MIT Health Sciences and Technology, Cambridge, MA, United States

A. Zadok
Faculty of Engineering and Institute for Nano-Technology and Applied Materials, Bar-Ilan University, Ramat-Gan, Israel

Preface

The year of this book's release marks the centenary of Léon Brillouin's seminal 1922 paper (Brillouin, 1922), which, for the first time, provided a detailed theory of the effect that now bears his name. As one might expect, the history of the prediction (and subsequent observation) of Brillouin scattering is more nuanced than the publication of a single paper, and the contemporary view is that both Brillouin and Leonid Mandelstam played vital roles in the formulation of Brillouin scattering theory. One hundred years later, our understanding of this effect plays a critical role in many areas of modern science and technology, particularly lasers, communications, fiber optics, sensing, and imaging applications as well as in experimental physics.

The history of Brillouin scattering is rich and multifaceted and is linked to other significant developments in optical science, such as the invention of the laser, fiber optics, and the rise of integrated photonics. Brillouin scattering, first observed in the 1930s, with early demonstrations focused on characterizing liquids and gases, became an accessible and powerful tool following the invention of the laser, which enabled the transition to stimulated Brillouin scattering (SBS). Following the invention of optical fibers, the detrimental effect of SBS on transport of narrow linewidth signals was quickly recognized, motivating significant research efforts focused on suppressing SBS. During this period, many underlying principles for manipulating SBS in waveguides were developed, offering new insights for enhancement and application of SBS. For example, it was realized that SBS can form the basis of a novel form of optical gain, without the requirement for "lasing materials," an effect now simply referred to as Brillouin lasing. SBS lasers and oscillators have been extensively studied for decades in the form of optical fibers and, more recently, in integrated waveguides and resonators. SBS can also form the basis of sensor modalities, particularly when incorporated into optical fibers, which can extend over long distances and be integrated into infrastructure; SBS-based distributed sensors have been extensively deployed and are the most successful commercially of all the SBS applications.

The SBS research community has recently seen a renaissance in basic research and applications, enabled by advances in nanotechnology, particularly through the developments of complex photonic crystal fibers (PCFs) and photonic integrated circuits, and the simultaneous development of new nonlinear materials, such as chalcogenide glass, silicon, and, very recently, thin film lithium niobate. This new wave of activity is opening possibilities for observing and harnessing SBS on length scales of centimeters and even millimeters, which is compatible with photonic integrated circuits. Furthermore, an increasing number of materials platforms have been developed that allow for strong confinement of both optical waves and acoustic waves, providing completely new degrees of freedom to exploit the interaction of optical waves and acoustic waves. In these new geometries, there is a fascinating convergence of SBS with optomechanics—the nonlinear interaction between light and stationary discrete acoustic modes. The theoretical descriptions of these seemingly disconnected physical phenomena have strong parallels, suggesting new directions for investigations in both fields.

On the more technological side, there has been a paradigm shift in photonic signal processing based on using SBS effects for manipulating both digital and analog information signals with unprecedented resolution and bandwidth. The narrow linewidth Brillouin gain or loss that is usually a nuisance in lasers thus becomes the basis of a very powerful signal processing filter with unparalleled frequency resolution. When combined with the possibility of integrating SBS on to photonic chips, this has prompted developments translatable into applications in radar systems, communication systems, and next-generation 5G.

The significance of this publication anniversary in 2022, combined with the current research renaissance, has motivated this compilation of chapters, bringing together the world's leading scholars in the area to outline our understanding of the fundamental and theoretical aspects of SBS and explore the scientific and technological applications and implications of this knowledge. From its first conception, it has been our intention that this volume should present a broad perspective on the key thematic areas, and, thus, we have asked at least three authors from different institutions to collaborate on each chapter. Our hope is that in this way, the book will present a balanced and considered perspective on the state of the art, how the field has evolved, and where it is headed. We have been delighted by the enthusiasm of our contributors for this unusual approach to assembling a contributed book and hope that it may become more common.

Finally, with a mix of historical, theoretical, and applications-oriented chapters, we believe this book will play a significant pedagogical role in bringing new scientists and engineers into the field, whether at the beginning of their research careers or established researchers seeking to explore a new area.

With many contributions, we have organized the chapters into two volumes: Volume 1, emphasizing more fundamental aspects, and Volume 2, emphasizing applications and technological considerations.

In Chapter 1, Poulton, Steel, Wiederhecker, and Thévenaz introduce the first volume with a broad overview of Brillouin scattering, combining a historical perspective with a largely nontechnical overview of the underlying physics of SBS. With this background, we hope to give a broad context to the remainder of the volume. The historical perspective traces the history of Brillouin scattering from the second decade of the 20th century when both Brillouin and Mandelstam independently formulated theories of optical scattering by high-frequency acoustic waves. The chapter proceeds to review early efforts to observe Brillouin scattering, highlighting some of the technical challenges in the very early experiments, where Raman scattering was dominant. With the invention of the laser in the 1960s, SBS was measured and shortly thereafter was seen in optical fibers, which led to a vibrant research community that was prominent at photonics conferences for several decades. The chapter then briefly surveys the more recent period during which SBS has been extensively studied and harnessed in photonic crystal fibers and integrated waveguides. The second part of the chapter describes, at a high level and largely qualitatively, how Brillouin scattering is now understood physically, in addition to giving an overview of the categories of Brillouin scattering that may be observed in modern experiments.

In Chapter 2, Wolff, Poulton, Steel, and Wiederhecker present a comprehensive theoretical formalism for stimulated Brillouin scattering. This includes a detailed derivation of the coupled mode equations for Brillouin scattering in the popular Hamiltonian formalism, a detailed analysis of the different kinds of behaviors in the various classes of Brillouin scattering, a parallel formalism for opto-elastic scattering in resonators, and an explanation of the connection of the modern formulation to the well-known hydrodynamic treatment based on pressure fields that has dominated the fiber literature.

In Chapter 3, Schmidt, Baker, and Van Laer explore how the recent renaissance in Brillouin scattering research, which emerged in the context of sub-wavelength scale waveguides, has brought about a convergence of two separate communities studying interactions between light and sound. In the community of "cavity optomechanics," scientists are mostly concerned with interactions between localized light and sound in optically resonant and few-moded devices. In the community of "Brillouin scattering," the focus is on characterizing and harnessing interactions between traveling light and sound in optically broadband and many-moded structures. Nevertheless, there are many parallels in the physics and formalism, which this chapter explores in detail.

In Chapter 4, Dainese, Beugnot, Yang, and Pang discuss the properties of SBS in PCFs in which simultaneous guidance of both optical and mechanical waves is provided by the tight transverse confinement and relatively long propagation distances. Both solid- and hollow-core PCFs have been demonstrated as platforms for different aspects of Brillouin scattering, and numerous applications are being explored.

In Chapter 5, Pant, Madden, Eggleton, and Shin review the development of integrated planar waveguides in which SBS can be induced and harnessed on centimeter scales. They discuss the requirements for realizing and fabricating SBS in chip-scale formats and highlight emerging applications that have been demonstrated, including slow-light, frequency comb generation, and microwave photonic filters. Finally, a brief overview of hybrid integration with CMOS-compatible silicon integration is considered.

In Chapter 6, Bowen, Baker, and Harris review recent research on Brillouin scattering in quantum fluids, specifically superfluid helium, which reveals quantum behaviors that are not present in regular fluids. Most notable is the vanishing viscosity of superfluid helium at low temperatures, which, when combined with other extreme material properties, provides a regime to harness SBS for new quantum technologies, such as sensors and memories.

In Chapter 7, Zadok, Bao, Yang, and Thevenaz review the principles of harnessing SBS in waveguides for sensors, including point sensors and distributed sensors. This chapter highlights the many successful commercial applications of Brillouin scattering with examples of Brillouin fiber sensors that span hundreds of kilometers, obtain submillimeter resolution, identify sub-degree centigrade temperature changes and tens of ppm strain events, and track dynamic signals at kilohertz and even megahertz rates.

In Chapter 8, Merklein, Schneider, and Vahala review Brillouin-based radiofrequency (RF) sources. This approach relies on the use of Brillouin scattering to generate multi-tone optical signals by creating an optical Stokes wave that is separated from the optical pump by around 10 GHz and can then be used to generate a stable RF beat signal. The chapter introduces key performance metrics such as phase noise, followed by a perspective on current trends and future directions.

In Chapter 9, Marpaung and Liu discuss applications of SBS to microwave photonics. They describe how SBS has been harnessed for microwave photonic signal processing due to its distinct advantages in bridging high-resolution and coarse signal processing. These are not possible with other competing technologies; the MHz-level linewidth enabled by SBS allows for high-resolution filtering tunable over ultra-wideband optical frequencies, unmatched by most on-chip devices.

In Chapter 10, Blumenthal, Kabakova, Rakich, and Vahala review recent research on integrated and micro-Brillouin lasers to bring the robustness, light weight and small size, and low cost associated with integration to a wide range of applications that require the optical spectral purity and signal processing capabilities of Brillouin light-sound interactions.

In Chapter 11, Corcoran and Choudhary review SBS in optical communication systems, highlighting the advantages and challenges of applying SBS in this context. They discuss how SBS has been explored as a tool to not only enable novel functions in optical communication systems but also limit power transmission, which, in turn, limits achievable signal-to-noise ratio—the key metric for increasing data rates.

In Chapter 12, Stiller, Winful, Boyd, and Merklein review the principles of slow light, dynamic gratings, and light storage based on SBS. They discuss the need to delay or buffer optical signals in high-speed optical data transmission systems to permit the signal processing, switching, and rerouting that take place in the nodes of optical networks. The buffering is ideally an all-optical process as it avoids electro-optical conversion and maintains a large bandwidth.

In Chapter 13, Otterstrom, Kittlaus, Sohn, and Bahl discuss nonreciprocity in Brillouin scattering including the underlying physical and mathematical principles and demonstrations of nonreciprocal functions, which can be the basis of isolators and circulators, key components in photonic systems. Such devices, which are key components in photonic integrated circuits, are traditionally implemented using magneto-optic components, which are not always compatible with photonic integration techniques.

In Chapter 14, Li, Florez, Pan, Madiot, Sotomayor Torres, and Li discuss electromechanical Brillouin scattering (EBS), which deals with Brillouin scattering of light by electromechanically excited elastic waves. They explain that the main difference between EBS and SBS is that in EBS the elastic waves (i.e., sound or phonons) are generated by an electromechanical transducer using piezoelectric materials, which convert electromagnetic waves into propagating elastic waves, or in some cases, localized mechanical modes.

Finally, in Chapter 15, Kabakova, Scarcelli, and Yun describe applying Brillouin light scattering (BLS) to measure the mechanical properties of tissues and cells to help understand the roles of biomechanics for advancing life sciences and medical diagnosis and treatments. They show that unlike traditional mechanical testing techniques, BLS does not require application of extrinsic forces nor the use of artificially introduced labels, making this technology particularly suitable for *in situ* and *in vivo* settings.

This hugely diverse set of contributions emphasizes the ever-growing impact that Brillouin–Mandelstam scattering is delivering, a century after its first conception. We trust that readers will be excited by the range of activity and applications and inspired to drive the field forward. Finally, we want to sincerely thank all the contributors for their efforts in preparing a wide range of insightful and erudite chapters and for doing so with such enthusiasm during this most challenging couple of years. We strongly believe the result is a fitting vehicle with which to say Happy 100th Brillouin-Mandelstam scattering.

Benjamin J. Eggleton
School of Physics, University of Sydney Nano Institute (Sydney Nano), University of Sydney, NSW, Australia

Michael J. Steel
School of Mathematical and Physical Sciences, Macquarie University, NSW, Australia

Christopher G. Poulton
School of Mathematical and Physical Sciences, University of Technology Sydney, NSW, Australia

List of symbols

The following table lists a number of frequently occurring symbols in this book.

Symbol	Notes	Description	Unit
α_n, α_b	(a)	Power loss constants of optical and elastic modes along the waveguide	1/m
γ_e		Electrostrictive constant	1
Γ_n, Γ_b	(b)	Amplitude decay constants of optical and elastic modes in time	1/s
ε_{ij}	(c,d)	Relative dielectric permittivity tensor	1
η_{ijlm}		Dynamic viscosity (Kelvin–Voigt) tensor	Ns/m^2
$\boldsymbol{\pi}$		Local momentum density	Ns/m^3
ρ		(Local) mass density	kg/m^3
σ_{ij}		Electric (Ohmic) conductance tensor°	m/Ω
$\lvert\Phi\rangle$		Elastic state	–
$\chi_{ij}, \chi^{(2)}_{ijk}, \chi^{(3)}_{ijkl}$		Linear, Pockels, and Kerr susceptibilities	1, m/V, m^2/V^2
$\lvert\Psi_n\rangle$		Optical states	–
ω_n		Optical (angular) frequencies	1/s
Ω		Elastic (angular) frequency	1/s
$\hat{a}_n, \hat{a}^\dagger_n, a_n$	(e,f)	Ladder operators and classical amplitude of optical mode	Depends on argument
$\hat{b}, \hat{b}^\dagger, b$	(e,f)	Ladder operators and classical amplitude of elastic mode	Depends on argument
$\boldsymbol{B}$, $\boldsymbol{D}$, $\boldsymbol{E}$, $\boldsymbol{H}$	(g)	Real electromagnetic fields and inductions	T, C/m^2, V/m, A/m
$\boldsymbol{b}$, $\boldsymbol{d}$, $\boldsymbol{e}$, $\boldsymbol{h}$	(g)	Complex electromagnetic fields and inductions	C/m^2, V/m, A/m
c_n, c_b; c_{vac}		Phase velocity or optical and elastic modes; vacuum speed of light	m/s

Continued

—cont'd

Symbol	Notes	Description	Unit
c_{ijlm}		Mechanical stiffness tensor	N/m^2
$\mathcal{E}_n$, $\mathcal{E}_b$	(f)	Optical and elastic mode normalization energies	J
$\boldsymbol{f}$		Complex force density	N/m^3
$\boldsymbol{F}$		Lumped force	N
g	(a)	Brillouin power gain density of a material	m/W
G	(a)	Modal Brillouin power gain	1/(Wm)
i, j, k, l		Indices that can assume the Cartesian coordinates $\{x, y, z\}$	–
n	(h)	Label for numbering optical modes (band index)	–
k_n		Optical wavenumbers	1/m
L		Quantization length for QM waveguide theory	m
p_{ijkl}		Photoelastic Pockels tensor	1
$\boldsymbol{P}$	(f)	Electric polarization (mostly nonlinear)	C/m^2
$\mathcal{P}_n$, $\mathcal{P}_b$		Power flux of optical/elastic modes through x–y-plane	W
q		Elastic wavenumber	1/m
Q		Opto-mechanical coupling coefficient	J
$\boldsymbol{r}$		Position vector $\boldsymbol{r} = (x, y, z)^T$	m
$\overleftrightarrow{R}$; R_{ij}		Rotation matrix, also local rotation field	1
s_{ijkl}		Mechanical compliance tensor	m^2/N
$\overleftrightarrow{S}$; S_{ij}	(c)	Mechanical (symmetric) strain tensor	1
$\mathcal{S}_n$, $\mathcal{S}_b$		Poynting flux of optical and elastic modes	W/m^2
t		Time	s

—cont'd

Symbol	Notes	Description	Unit
$\overleftrightarrow{T}$; T_{ij}	(c)	Mechanical stress tensor	N/m^2
$\boldsymbol{u}$	(g)	Complex mechanical displacement field	m
$\boldsymbol{U}$	(g)	Real mechanical displacement field	m
v_n, v_b		Optical and elastic group velocities	m/s
$\boldsymbol{V}$		Mechanical velocity field	m/s
x, y, z	(e)	The Cartesian coordinates	m

Notes:

(a) Divide by two when used in amplitude equation.
(b) Multiply by two when used in power equation; divide by 2π for respective amplitude line width.
(c) Avoid using σ and ϵ for stress and strain, respectively.
(d) To avoid confusion: ε_{ij} is the relative dielectric tensor, explicitly include ϵ_0 when required.
(e) By convention the main direction of propagation should be along the positive z-axis.
(f) Replacing a_n, b with $\hat{a}_n$,$\hat{b}$ and setting $\mathcal{E}_n = \hbar\omega_n/L$, $\mathcal{E}_b = \hbar\Omega/L$ should yield QM formulation.
(g) We tend to use upper case for real time-domain fields, lower case for complex frequency-domain fields.
Positive frequency parts go as $e^{-i\omega t}$. So the real-valued displacement field is $\boldsymbol{U}(\boldsymbol{r}, t) = \boldsymbol{u}(\boldsymbol{r})e^{-i\omega t} + \boldsymbol{u}^*(\boldsymbol{r}) e^{i\omega t} = \boldsymbol{u}(\boldsymbol{r})e^{-i\omega t}$ + c.c.
(h) alternatively use subscripts (S) or (P) to label an optical pump and AS/S for (anti-)Stokes.

CHAPTER SEVEN

SBS-based fiber sensors

A. Zadok[a,*], X. Bao[b], Z. Yang[c], and L. Thevenaz[d]
[a]Faculty of Engineering and Institute for Nano-Technology and Applied Materials, Bar-Ilan University, Ramat-Gan, Israel
[b]Department of Physics, University of Ottawa, Ottawa, ON, Canada
[c]State Key Laboratory of Information Photonics & Optical Communications, Beijing University of Posts and Telecommunications, Beijing, China
[d]Swiss Federal Institute of Technology (EPFL), Institute of Electrical Engineering, SCI STI LT, Lausanne, Switzerland
*Corresponding author: e-mail address: avinoam.zadok@biu.ac.il

Contents

Semiconductors and Semimetals, Volume 110
ISSN 0080-8784
https://doi.org/10.1016/bs.semsem.2022.05.001

1. Background and historical perspective

1.1 Optical fiber sensors

Standard single-mode optical fibers were invented to support broadband data communication. Optical fibers also create an exceptional sensing platform (Dakin and Culshaw, 1988). The propagation of light in fibers, alongside specially designed demodulation systems, can be made sensitive to a host of environmental parameters and quantities of interest, such as temperature, static and dynamic strain and stress, sound and vibrations, electrical and magnetic fields, presence and concentration of chemical species and biological reagents, ionizing radiation and more (Dakin and Culshaw, 1988). The research and development of optical fiber sensors have been pursued for 50 years (De Waele et al., 1970), and the field supports a thriving research community and a growing industrial market.

Optical fibers bring several potential advantages with respect to other sensor technologies. They can be installed over hundreds of kilometers and support the acquisition of data from a long standoff distance. Thin fibers can be embedded within many structures with little impact on their functionality. Compared with electrical cables or wireless radio-frequency connectivity, optical fibers are less susceptible to electromagnetic interference. Fibers may be installed in harsh and hazardous environments, where the application of electrical fields and currents might be prohibited. Most importantly, optical fibers support the spatially distributed measurements of quantities of interest on a large scale. This possibility is seldom provided by other sensing platforms.

1.2 Point, position-integrated and distributed sensors

Sensors are classified in terms of the extent of spatial information they provide. The most basic and widely employed category is that of point sensors, in which parameters of interest are measured in a single, pre-defined location. The most well-known examples of optical fiber point sensors are fiber Bragg gratings: cm-scale elements whose wavelengths of peak reflectivity change with local temperature and axial strain (Kashyap, 2009; Wang et al., 2019. Point sensors may be cascaded in arrays of up to thousands of discrete elements, forming quasi-distributed networks (Wang et al., 2010). Fiber sections connecting between fiber Bragg gratings can also serve as interferometric sensor cavities (Wang et al., 2019).

A second class is that of position-integrated sensors, where a quantity of interest is aggregated over a comparatively long stretch of fiber and effectively averaged with respect to position. One example is the accumulation of very small optical phase perturbations over long arms of fiber Mach-Zehnder interferometers (Dandridge and Kersey, 1989). Position-integrated measurements may be extremely sensitive, however they wash away all local information. Both point and position-integrated sensors do not take full advantage of long optical fibers as sensor platforms.

The most sophisticated optical fiber sensor systems are those providing spatially distributed measurements (Hartog, 2017). In these systems every fiber section serves as an independent sensor node, and the quantity of interest is mapped as a function of position. A single sensing fiber may replace many point-measurement elements and their associated connectivity. Commercial distributed fiber sensing systems routinely address tens of thousands of resolution points, numbers that are extremely challenging to reach with any point-sensing technology, and provide a conceptual leap beyond the performance of other platforms. Nearly all distributed sensors rely on back-scattering effects on light that is propagating down the fiber. For example, sensors based on backwards Rayleigh and Raman scattering are widely studied and employed for decades (Barnoski and Jensen, 1976; Bolognini and Hartog, 2013). In this chapter we focus on sensors that employ Brillouin scattering.

1.3 Measurement parameters and metrics

The frequency shift between optical pump and signal waves in backward stimulated Brillouin scattering (SBS) scales with the velocity of dilatational acoustic waves in the core of the fiber (Boyd, 2020). The acoustic velocity, in turn, varies with both temperature and axial strain. The mapping of the Brillouin frequency shift as a function of position along standard single-mode fibers can therefore provide distributed measurements of both quantities. The analyses of temperature and strain dominate the research and applications of Brillouin sensing. The concept was also introduced to specialty fibers and even photonic integrated circuits. In these platforms, Brillouin sensors also map the effective index and waveguide geometry as functions of position.

Distributed fiber sensors do not benefit from direct access to optical signals in each measurement position. Instead, local scattering information must be retrieved from signals collected at the ends of the fiber, and sometimes using single-ended access only. The unambiguous assignment of observed scattering contributions to specific spatial positions, and their accurate quantitative analysis, represent the foremost challenges of distributed sensors. Much of this chapter is dedicated to localization principles, capabilities, and limitations. Performance metrics include the measurement range and spatial resolution, the number of resolution points, the accuracy of Brillouin frequency shift estimates, and the time required to obtain and process the data. The state-of-the-art in Brillouin fiber sensors research advances continuously. Current record demonstrations span hundreds of km (Gyger et al., 2014), resolve sub-millimeter long sections (Zarifi et al., 2018), address over a million independent resolution points (Denisov et al., 2016), acquire dynamic strain maps within hundreds of micro-seconds (Peled et al., 2011), and retrieve the local Brillouin frequency shifts with sub-MHz accuracy (Soto and Thévenaz, 2013). As might be expected, these results are not all obtained in a single system. Tight trade-offs prevail among performance metrics, as addressed in detail in this chapter.

Over the last 5 years, forward SBS processes have also found a role as the basis of novel fiber sensing concepts (Antman et al., 2016). Unlike backwards SBS in standard fibers, the acoustic waves stimulated in the forward process span the entire cladding cross-section and reach its outer boundary (Shelby et al., 1985). The magnitude and spectrum of forward SBS are therefore affected by the properties of media outside the cladding, and even beyond the coating, where light does not reach (Antman et al., 2016). Forward SBS fiber sensors represent a comparatively recent proposition, and their performance significantly lags behind those of the more established backwards SBS sensors. However, progress in distributed analysis of forward SBS is rapid and exciting.

1.4 Historical overview of Brillouin fiber sensors

Distributed analysis of backward Brillouin scattering was first proposed and demonstrated by Horiguchi and Tateda as a method for measuring the attenuation of optical fibers in 1989 (Horiguchi and Tateda, 1989). In the following year, Kurashima, Horiguchi and Tateda leveraged their new technique toward the distributed analysis of strain (Horiguchi et al., 1990), and temperature (Kurashima et al., 1990). The measurements were based on interactions between intense Brillouin pump pulses and a weaker, continuous Brillouin signal wave. The authors named their new protocol Brillouin optical time domain analysis, or B-OTDA for short. More than 30 years later, B-OTDA remains the workhorse of distributed Brillouin sensors, both in the research laboratory and in the field. Early performance was modest: only 10 resolved sections, each 100–200 m long (Horiguchi et al., 1990; Kurashima et al., 1990). The number of resolution points and spatial resolution of B-OTDA improved since by five orders of magnitude and four orders of magnitude, respectively. B-OTDA requires access to both ends of a fiber under test. In many scenarios such access is difficult to obtain. To work around this challenge, Kurashima, Horiguchi and co-workers proposed and demonstrated Brillouin optical time-domain reflectometry (B-OTDR) in 1993 (Kurashima et al., 1993). The technique requires the transmission of pulses and collection of back scatter from a single end of the fiber only.

In the above protocols, transient Brillouin scattering is induced throughout the entire fiber under test and the contributions of different positions are resolved in the time domain. As an alternative, in 1999 and 2000 Hotate and Hasegawa proposed a scheme in which stationary Brillouin scattering is confined to discrete and narrow locations of interest which may be scanned along the fiber (Hotate and Hasegawa, 2000). Their Brillouin optical correlation domain analysis (B-OCDA) system relied on careful frequency modulation of continuous Brillouin pump and signal waves (Hotate and Hasegawa, 2000). The spatial resolution of B-OCDA is superior to that of time-domain protocols, reaching 1.6 mm in fiber (Song et al., 2006), and hundreds of microns in photonic-integrated waveguides (Zarifi et al., 2018). Single-ended Brillouin optical correlation-domain reflectometry (B-OCDR) was introduced in 2008 by Mizuno et al. (Mizuno et al., 2008).

Landmark demonstrations of SBS sensing over the last 30 years are too numerous to be discussed here in full. The use of a single laser diode source to generate both pump and signal waves was introduced by Nikles et al. (1997). That inherently stable architecture has made B-OTDA setups far more practical and accurate. Distributed Raman amplification along the fiber under test was introduced by Cho et al. in 2003, to increase the measurement range of B-OTDR (Cho et al., 2003). Brillouin analysis over polarization maintaining fibers allows for simultaneous measurements of temperature and strain, as first shown by Zou et al. (2009). Pulse sequence coding was applied to improve the signal-to-noise ratio of B-OTDA by Soto et al. (2010). Fiber-sensing of liquids outside the cladding based on forward SBS was reported by Antman et al. (2016). Many important achievements are described throughout the following sections. Several comprehensive review papers are dedicated to Brillouin fiber sensors (Bao and Chen, 2011; Dong, 2021; Motil et al., 2016).

2. Fundamentals

As mentioned in the basics, the Brillouin gain spectrum follows very faithfully a Lorentzian distribution that can be fully characterized by 3 parameters: its central frequency ν_B designated as Brillouin frequency shift (BFS), its FWHM linewidth $\Delta\nu$ and its peak gain value at ν_B. Sensors are taking advantage of the dependence of these 3 parameters on environmental quantities—mainly temperature and strain experienced by the fiber—to deliver measured values of these quantities, which are therefore coded in the Brillouin signal.

Among these parameters the preferred avenue is to exploit the Brillouin shift changes with environmental quantities: this parameter turns out to be experimentally evaluated with a higher accuracy than the others and offers the key advantage to robustly encode the information in the frequency domain and to thus offer an intrinsic immunity to any power fluctuations. Though several reports demonstrate the use of the other parameters, their practical interest remains limited and keeps confined to restricted niche applications. So logically this section will place the focus on the dependence of the Brillouin shift ν_B on external quantities, but will also mention the impact on the other parameters, being implicitly understood that they are of lesser practical importance.

Let us recall the foundational relation giving the peak acoustic resonance frequency, or the Brillouin shift:

$$\nu_B = 2\frac{n_{eff} V_a}{c}\nu = 2\frac{n_{eff} V_a}{\lambda}. \quad (1)$$

Here V_a is the velocity of longitudinal acoustic waves in the fiber material and n_{eff} is the effective refractive index experienced by the incident light at frequency ν or wavelength λ. This shows that the acoustic resonance frequency is simply equal to the incident optical frequency scaled down by the ratio between the acoustic and optical velocities in the medium, with a factor 2 to satisfy the phase matching for the interaction between counter-propagating optical waves. Any dependence of these velocities on external quantities will be therefore reflected as a change in the Brillouin shift ν_B.

In the vast majority of cases the relative impact of external quantities on the acoustic velocity V_a is widely dominating over the observed change in n_{eff}, as described hereafter.

2.1 Effect of material composition

In an isotropic material like most glasses the longitudinal acoustic velocity is given by the following expression (Timoshenko and Goodier, 1970):

$$V_a = \sqrt{\frac{E}{\rho}}\sqrt{\frac{(1-\sigma)}{(1+\sigma)(1-2\sigma)}}, \tag{2}$$

where E is the Young modulus, σ denotes the Poisson's ratio and ρ the material density. Any small-scale change in the material properties will immediately impact on the acoustic velocity and this property can be used to identify materials and changes in their chemical composition by simply measuring the Brillouin shift. Incidentally, Brillouin scattering has been intensively used to study the glass properties through the change of their acoustic velocity (Pelous and Vacher, 1976a,b).

As a practical illustration, the Brillouin shift turns out to be fairly sensitive to the dopant concentration in the core of a silica optical fiber. This may be used as a fingerprint to associate an unknown fiber to a particular type (standard, dispersion-shifted, etc.) or production batch from a given manufacturer. A change in ν_B of -80 MHz/mol% of GeO_2 dopant concentration is extrapolated from measured data in Nikles et al. (1997), the negative slope being logically related to the higher density of GeO_2 (the refractive index change is 1.7×10^{-3} per mol% of GeO_2). The Brillouin gain linewidth is approximately increased by $+1$ MHz/mol% of GeO_2 dopant concentration.

2.2 Effect of temperature

By inspecting the expressions for V_a and ν_B given in the introductive part of this section, it can be easily deduced that nearly all quantities entering into these equations are temperature-dependent and only a thorough study may determine which are dominant. Moreover, glass is a complex medium, characterized by a disordered molecular pattern like in a fluid while maintaining its solid-state nature, with no real periodicity in the lattice structure. The notion of phonons is often related to a periodic order of the molecular lattice. Since a quasi-periodicity can still be observed over a short range, the density of states of amorphous silica shows some similarity with its crystalline counterpart. This justifies the applicability of the concept of phonons and a descriptive framework to predict the behavior of specific heat, essential to deduce the thermal dependence of the material parameters (Elliott, 1983). The most widely employed model is a two-level system characterized by a double well potential and thus tunneling states. It gives a relatively correct prediction of the behavior at low temperatures.

While in crystalline SiO_2 a phase transition at 573 °C between α- and β-quartz gives rise to a monotonic decrease of the acoustic velocity for increasing temperature up to the phase transition temperature—and a corresponding continuous increase of acoustic loss—the behavior of amorphous silica is radically different (Nikles et al., 1997; Pelous and Vacher, 1976b). Up to a cryogenic temperature of 70 K, the acoustic velocity decreases like in quartz to reach a minimum at that temperature, and then continuously increases up to the melting temperature. The acoustic loss shows a similar behavior, increasing up to a peak at 110 K and then slowly but steadily decreasing toward higher temperatures.

The Brillouin shift ν_B has been measured in optical fibers from 1 to 1000 K (Fellay, 2003; Thévenaz et al., 2003), fully confirming the results obtained in bulk silica samples. These results are presented in Fig. 1 and bring up the following comments:

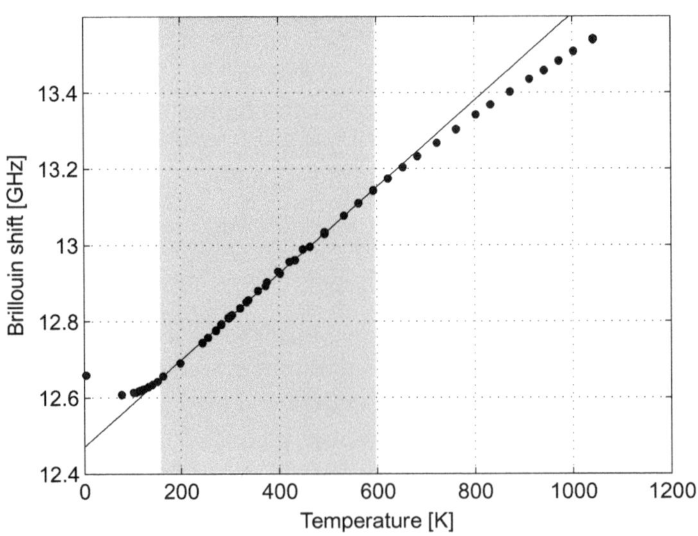

Fig. 1 Brillouin shift of a single mode fiber at a pump wavelength of 1319 nm as a function of temperatures from 4 to 1050 K (Fellay, 2003). The shaded area represents the linear response range centered around the ambient temperature.

- At low cryogenic temperatures (<70 K) the decrease of the acoustic velocity is compliant with the tunneling states model. The acoustic loss is massively reduced by one order of magnitude at very low temperatures close to 1 K.
- Around ambient temperatures a linear response with a positive slope of ~1.26 MHz/K is observed at a pump wavelength of 1550 nm. A linear range can therefore be defined, represented by the shaded area in Fig. 1 and approximately spanning from 150 to 600 K (about −120 to +330 °C). This generously covers the temperature range of most applications, but paradoxically this linear response is still widely unexplained and is resistant to any modeling.
- At higher temperature a change in the slope is observed, but the response remains linear. This response change is assumed to be due to the coating combustion that stops the coating contribution to the global fiber thermal response. The acoustic loss remains roughly constant at temperatures above 300 K.

It must be mentioned that this well-known response of 1.26 MHz/K does not only include the contribution due to the acoustic velocity change, but also a relatively minor contribution of +0.074 MHz/K from the thermal dependence of the refractive index, for which: $\frac{1}{n}\frac{dn}{dT} = 6.9 \times 10^{-6}$ per K. More importantly, the thermal effect of the jacket is estimated to result in a contribution of about +0.3 MHz/K in the case of standard acrylate coating.

2.3 Effect of strain

The effect of strain is more straightforward than what is observed for the thermal dependence, since it essentially depends on the elastic properties of silica considered as a bulk solid material, which are fairly well known and modeled, at least for moderate deformations. Like for temperature, the effect of strain on the Brillouin shift ν_B can be decomposed into two contributions: one from the dependence of the refractive index n on strain, and one from the impact of the acoustic velocity V_a. A normalized strain coefficient C_ε can be defined with these split contributions:

$$C_\varepsilon = \frac{1}{\nu_B}\frac{\partial \nu_B}{\partial \varepsilon} = \frac{1}{n}\frac{\partial n}{\partial \varepsilon} + \frac{1}{V_a}\frac{\partial V_a}{\partial \varepsilon} \tag{3}$$

The refractive index contribution may be simply calculated using the tensor representation of elasticity and photo-elasticity (Nye, 2009), and the standard agreed values for the Poisson's ratio $\sigma = 0.173$ (Fukuhara et al., 1997; Sánchez et al., 2022), and the photo-elastic coefficients $p_{11} = 0.12$ and $p_{12} = 0.27$:

$$\frac{1}{n}\frac{\partial n}{\partial \varepsilon} = -\frac{n^2}{2}\left[p_{12} - \sigma(p_{11} + p_{12})\right] = -0.215 \tag{4}$$

The contribution from the acoustic velocity contains three terms, since all quantities involved (Young's modulus E, Poisson's ratio σ and density ρ) turn out to be stress-dependent:

$$\begin{aligned}\frac{1}{V_a}\frac{\partial V_a}{\partial \varepsilon} &= \frac{1}{2E}\frac{\partial E}{\partial \varepsilon} + \frac{\sigma(2-\sigma)}{(1-\sigma^2)(1-2\sigma)}\frac{\partial \sigma}{\partial \varepsilon} - \frac{1}{2\rho}\frac{\partial \rho}{\partial \varepsilon} \\ &= 2.875 + 1.364 - 0.34 = 4.58\end{aligned} \tag{5}$$

It is important to notice that the effect of refractive index is also minor with respect to the contribution due to the acoustic velocity, and this latter is—a bit counter-intuitively—more dominated by the nonlinear elastic response of the material than by the change in the material density. The above suggests a theoretical normalized strain coefficient C_ε of 4.52 calculated from material quantities in bulk silica, which is very reasonably close to the practical values in the range 4.55–4.75 considering the potential effects of doping and coating. This gives a standard value for the change in the Brillouin shift of 500 MHz per percent of fiber elongation, which translates into a potential sensitivity in the range of 10 micro-strain (με) for a Brillouin sensor.

It should be noted that, unlike in the study of the temperature effect, the acoustic loss—and thus the Brillouin gain linewidth—does not show a significant dependence on strain and may be practically considered as strain-independent (Nikles et al., 1997; Zhang et al., 2007). This major difference has been considered for discriminating between the effects of temperature and strain, though it has remained in practice poorly applicable regarding the numerous other reasons for a spectral broadening of the gain spectrum and the large experimental uncertainties in the determination of the gain linewidth.

Standard silica fibers can only withstand axial strains up to 4%. In addition, the Brillouin gain of standard fibers drops significantly at the high end of that range. The reduction in Brillouin gain can be mitigated with the use of carbon and polyimide coatings. The BFS could be measured in the presence of up to 4.2% and 4.4% of axial strain in two specialty single-mode fibers with this type of coating, with only 20% degradation in Brillouin gain (Zhang et al., 2007). Certain applications, such as structural health monitoring, require the measurements of even larger strains. Fibers made of polymers, such as PMMA, can withstand strain levels up to 20% (Hayashi et al., 2012; Wang et al., 2020b). Polymer fibers are also compatible with biomedical applications. The Brillouin gain spectrum (BGS) of PMMA fibers becomes asymmetric above 2.6% strain, and their Brillouin gain drops considerably beyond 10%.

2.4 Effect of hydrostatic pressure

The effect of hydrostatic pressure can be derived using the same theoretical framework as for the effect of strain. In this case the applied stress is isotropic and essentially leads to a radial deformation. This turns out to be less efficient and, associated to a relatively high stiffness of the silica material, makes the sensitivity to hydrostatic pressure fairly poor, which can be improved by a proper choice of coating.

This sensitivity ranges from a change in the Brillouin shift of −0.74 MHz per MPa of pressure for a bare silica fiber (atmospheric pressure is ~0.1 MPa), to −0.91 MHz/MPa for a simply coated fiber and − 1.91 MHz/MPa for a specially-designed coating (Gu et al., 2012; Le Floch and Cambon, 2003; Zhang et al., 2013). This sensitivity can be significantly improved by using alternative materials such as polymers (up to +4.3 MHz/MPa), with the drawback of a hysteresis and a poor reversibility (Mizuno et al., 2017). The preferred avenue to measure pressure remains to exploit the birefringence change through a structural deformation by pressure of a specially-designed fiber, usually containing air holes. A technique to measure birefringence using stimulated Brillouin scattering is described in Section 5.

2.5 Specificities of forward Brillouin scattering

The resonance frequency of forward stimulated Brillouin scattering (FSBS) is also scaled by the acoustic velocity, so that the relative dependence on the environmental quantities is expected to be fairly similar to backward Brillouin scattering. Since the resonances of FSBS lie in a frequency range 10–100 times lower than classical SBS, the frequency change will be smaller in the same proportion, accordingly.

The main interest of FSBS is its dependence on the geometry of the optical fiber since it resonates transversally in the section of the fiber. In particular the resonance frequency directly depends on the fiber physical radius and determining the spectral position of this resonance precisely informs about the fiber size and its possible variations (Shelby et al., 1985).

Another important effect is the acoustic reflective properties of the interface between the fiber and the outer medium. This impacts directly on the quality factor of the resonance and it has been demonstrated that it may be a powerful tool to probe the acoustic impedance of the outer medium, giving important information about its nature (Antman et al., 2016), by deducing the acoustic impedance mismatch from the resonance linewidth.

3. Time-domain analysis and reflectometry

The BFS can be obtained by measuring the BGS or Brillouin loss spectrum (Bao et al., 1993). Two counter-propagating optical signals, a pump of optical frequency ν_0 and a probe wave at a lower frequency $\nu_s = \nu_0 - \nu$ stimulate Brillouin scattering along a fiber under test (FUT). The frequency difference ν is chosen in the vicinity of the anticipated BFS ν_B. In BGS experiments the amplification of the probe wave is measured, whereas in the Brillouin loss spectrum case the depletion of the probe wave is detected instead. In both cases, spectra are obtained by scanning the pump and probe wave frequency difference ν over tens to hundreds of MHz around an estimated BFS, depending on the foreseeable range of temperature and strain variations. The BFS is identified through curve-fitting of the raw gain or loss spectra, which helps mitigate measurement noise (Ravet et al., 2006), such as the dependence of the BGS on local residual birefringence.

As noted earlier, two classes of time-domain Brillouin sensing protocols are prevalent. B-OTDA is based on stimulated amplification or attenuation of optical waves and requires access to both ends of the fiber, whereas B-ODTR relies on spontaneous scattering and requires single-end access only. B-OTDR is simpler to employ. On the other hand, B-OTDA provides superior accuracy, higher resolution and faster acquisition over long measurement ranges (>150 km).

3.1 Brillouin optical time-domain analysis (B-OTDA)

A typical B-OTDA system is illustrated in Fig. 2. In the Brillouin gain configuration case, the continuous probe wave frequency is lower than the pump pulse and is spectrally positioned in the vicinity of the Stokes frequency. The roles are reversed in measurements of Brillouin losses and the probe frequency is higher in the vicinity of the anti-Stokes frequency. In both settings the continuous optical wave is detected at the far end of the fiber and the detector voltage is sampled and observed as a function of time. Due to the use of a pulsed pump wave, the Brillouin gain (loss)

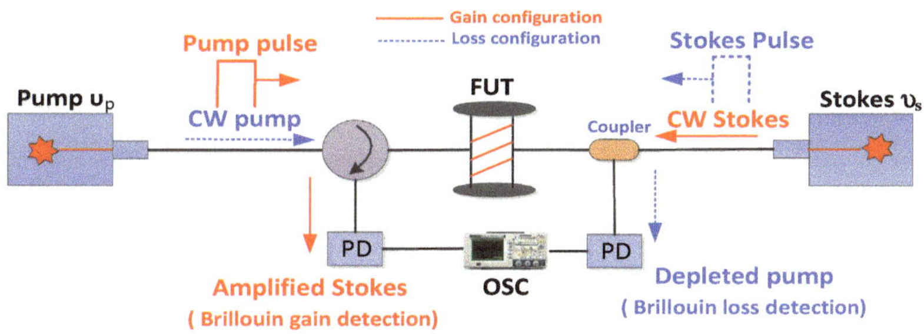

Fig. 2 Schematic illustration of a B-OTDA setup for the measurement of Brillouin gain or loss spectra, using two separate lasers for the pump and probe (referred here as Stokes) waves.

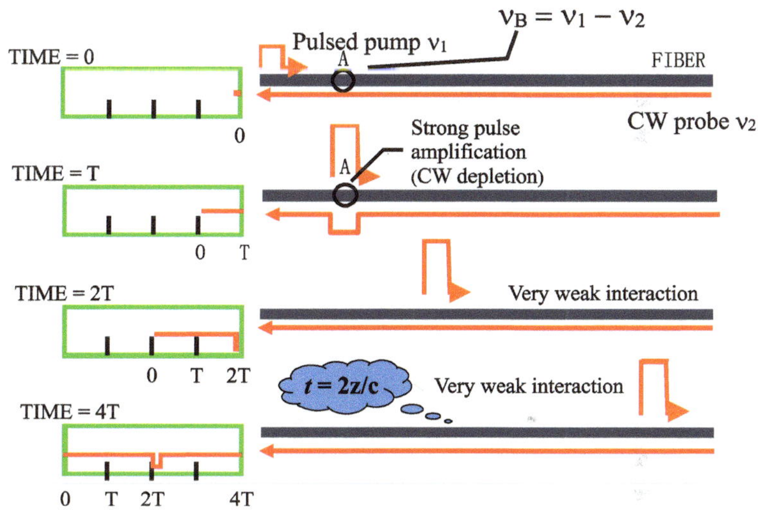

Fig. 3 The localization of Brillouin scattering contributions in B-OTDA.

imposed on the observed continuous wave at any given location can be resolved and identified based on time of arrival of detected signals (Fig. 3). The *B-OTDA trace* of collected power as a function of time is readily mapped to distance scale via $z = \frac{V_g t}{2}$. The pulse duration τ defines the spatial resolution of the sensor as: $\Delta z = \frac{1}{2} V_g \tau$, where V_g is the group velocity of light in the fiber and the factor 2 represents the two-way propagation. The strength of the Brillouin interaction depends on the pump-probe frequency offset from the local BFS of the fiber section, the product of their power levels, and the local Brillouin interaction length Δz, which is the spatial resolution for the distributed sensor.

The acquisition protocol is repeated for a number of offsets ν between the optical frequencies of pump and probe waves. Repeating acquisitions must be separated by the round-trip propagation time along the fiber and back. Data acquisition may take several minutes, depending on the number of frequency offsets ν being used. The entire set of traces then serves to plot a two-dimensional map of the detected optical power, as a function of both position and frequency offset ν (see Fig. 4). In particular, the Brillouin gain or loss spectrum may be drawn for each position z, and the local BFS can be identified.

The stimulation of the acoustic wave is maximal when the polarizations of pump and probe waves are aligned, and it vanishes entirely for orthogonal states (Zadok et al., 2008). Both states of polarizations vary arbitrarily along the fiber under test due to residual birefringence. To alleviate the impact of possible polarization fading, either the pump or probe wave requires a polarization scrambler to quickly alter their polarization states over all possibilities, to avoid zero signals in Brillouin gain/loss spectrum. The strength of the Brillouin interaction then converges to an average value throughout the

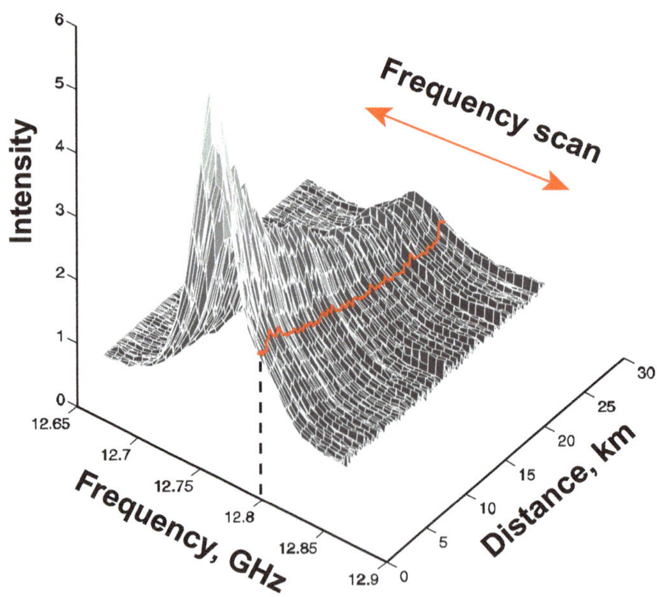

Fig. 4 An example of Brillouin optical time-domain analysis. Measured power of the probe wave as a function of position and frequency detuning between pump and probe (Thévenaz, 2010).

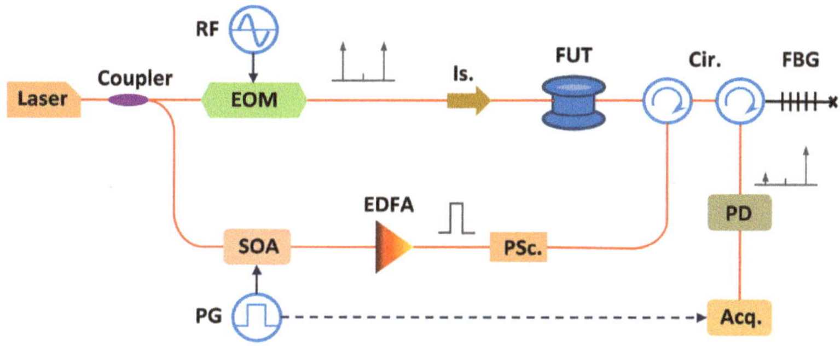

Fig. 5 Schematic illustration of a B-OTDA setup where the pump and probe waves are drawn from a single laser source (Niklès et al., 1996). EOM: electro-optic modulator; PSc.: polarization scrambler; Is.: isolator; PG: pulse generator; EDFA: erbium-doped fiber amplifier; PD: photodetector; SOA: semiconductor optical amplifier; RF: radio frequency; FBG: fiber Bragg grating; FUT: fiber under test; Cir.: circulator; Acq.: acquisition.

sensing fiber. Polarization fading can also be mitigated by two successive measurements using orthogonal states of polarization for either pump or probe (Nikles et al., 1997).

One drawback of the basic B-OTDA configuration illustrated in Fig. 2 is possible drifting of the frequency difference between two free-running lasers. Alternatively, the pump and probe waves may be drawn from a single laser source of optical frequency ν_0 (Fig. 5, Niklès et al., 1996). An electro-optic modulator (EOM) is used to offset the frequency of one replica of the optical source by a variable difference ν. In this manner, any drifting of the absolute frequency ν_0 cancels out. Standard EOMs introduce two sidebands of frequencies $\nu_0 \pm \nu$. A fiber Bragg grating can remove one sideband, or a single sideband EOM can be used instead. Keeping the two sidebands mitigates to a wide extent the impact of non-local effects (Thévenaz et al., 2013). In Fig. 5, a semiconductor optical amplifier (SOA) is used to generate the necessary optical pulses with a high extinction ratio.

3.2 Brillouin optical time-domain reflectometry (B-OTDR)

The B-OTDA setups discussed above rely on *stimulation* of acoustic waves through optical waveforms that are launched from both ends of the fiber. When light is sent from a single end of the fiber only, acoustic waves cannot be stimulated optically. Brillouin sensing in that case can only rely on the *spontaneous* scattering of the single input waveform by acoustic waves that are already there, due to thermal background activation. Spontaneous Brillouin scattering forms the basis of B-OTDR. Local information is

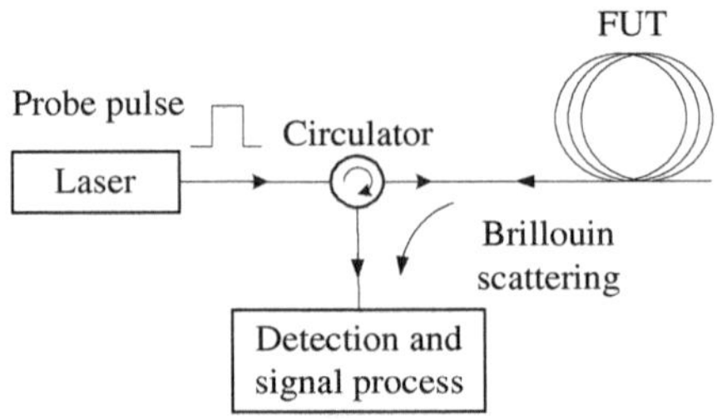

Fig. 6 Schematic illustration of a B-OTDR setup.

retrieved through the analysis of spontaneous backscatter as a function of time. Like B-OTDA, the spatial resolution of B-OTDR is also determined by the pulse duration: $\Delta z = \frac{1}{2} V_g \tau$. Due to its single-ended access, B-OTDR is very attractive for field tests, especially when the sensing fiber is broken. A B-OTDR setup is illustrated in Fig. 6.

On the downside, spontaneous Brillouin scattering is weaker than the corresponding stimulated process, hence the signal-to-noise ratios of B-OTDR are typically inferior to those of B-OTDA. This drawback degrades the spatial resolution and accuracy of the BFS measurements, as discussed later in this section. The measurement sensitivity of B-OTDR is often enhanced by coherent detection: the interference of the collected back-scatter with an intense local oscillator (Bao and Chen, 2011; Toshio et al., 1993). The electrical spectrum of the beating photo-current is proportional to the optical spectrum of the spontaneous Brillouin backscattered signal. Therefore, the Brillouin spectrum and the BFS can be identified through electrical spectrum analysis of the waveform at the detector output. No scanning of the optical frequencies of the input pulse is necessary. In order to retrieve local information, the spectral analysis is carried out within sliding temporal windows of duration τ. Rayleigh back-scattering of the input pulse and Fresnel reflections from the far end of the fiber must be effectively filtered out prior to detection, in either the optical or electrical domain.

3.3 Performance metrics of time-domain Brillouin sensing protocols

The overall performance of B-OTDA or B-OTDR is quantified primarily by four metrics: spatial resolution, BFS measurement accuracy, sensing range and acquisition duration. The BFS is estimated by finding the center frequency of measured local Brillouin gain or loss spectra. The curve fitting accuracy depends on the signal-to-noise ratio (SNR) of acquired traces, the number and step-size of the frequency offsets being scanned, the linewidth of the laser sources used, and spectral line-shape distortions. Since signal is lost with increasing fiber length, and improves with the number of repeating averages, a four-way trade-off prevails among spatial resolution, measurement accuracy, sensing range and acquisition duration. For example, short pulses provide high spatial resolution, but also smaller gain or loss signals and compromised accuracy. Averaging improves the signal-to-noise ratio but increases the experimental duration. An expression for the BFS measurement accuracy $\delta\nu_B$ in the fitting of Brillouin gain or loss spectra was provided by Soto and Thévenaz (2013):

$$\delta\nu_B(z) \approx \frac{1}{SNR(z)}\sqrt{\frac{3}{4}\delta\cdot\Delta\nu}. \tag{6}$$

Here $\delta\nu_B(z)$ is the uncertainty of the estimated BFS at position z, δ is the frequency scanning step, $\Delta\nu$ is the full width at half maximum (FWHM) of the BGS, and $SNR(z)$ denotes the local SNR. Noise sources to be considered include shot noise of the detected pump or probe wave, thermal noise of photo-receivers, relative intensity noise of laser sources, amplified spontaneous emission of optical amplifiers and quantization errors of digitizers. Position-dependent random birefringence may also add to the measurement uncertainty (Bao and Chen, 2011), and this requires mitigation through polarization diversity or scrambling. The limitations and trade-offs among performance metrics are discussed below, with Eq. (6) as a reference.

3.3.1 Spatial resolution

High spatial resolution is among the key metrics of distributed sensors. High resolution mandates short pulses: For example, 1 m resolution corresponds to a pulse duration of 10 ns. However, as the pulse duration is reduced below the acoustic lifetime $\Delta\nu^{-1}$ of 5–10 ns, the BGS broadens significantly (Bao and Chen, 2011). In addition, the extent of Brillouin gain or loss scales with pulse duration as well, and the SNRs obtained with short pulses are degraded accordingly (Bao et al., 1999). Due to the weaker Brillouin gain or loss and the broader spectra, the fitting of local BFS values becomes less accurate.

3.3.2 Measurement accuracy of the Brillouin frequency shift

The accuracy in the Brillouin sensing of strain or temperature is directly determined by $\delta\nu_B(z)$, as expressed in Eq. (6). The accuracy is evaluated experimentally as the standard deviation among repeating estimates of the BFS under the same experimental condition. The smaller the uncertainty, the better. As may be intuitively expected, accuracy is degraded by poor SNRs, broadened Brillouin spectra (as encountered, for example, when using short pulses in attempt to achieve high resolution), or coarse frequency scans. Due to the fiber losses, the SNR varies with fiber position. The accuracy of a distributed sensor system is defined at the position of worst performance. In standard B-OTDA and B-OTDR systems, this position is at the far end, at the largest distance from the input end where pulses are launched. The typical accuracy of standard B-OTDA is on the order of 0.5–1 MHz, which corresponds to a temperature error on the order of 0.5–1 °C or stain uncertainty of 10 με.

3.3.3 Sensing range

Long fiber lengths introduce large attenuation, which leads to lower SNR and reduced accuracy. The range of the sensor is defined as the longest distance from the input end of optical pulses, for which a desired accuracy is still maintained. To a certain extent, losses can be compensated by high optical power levels. However, the pump power is limited by competing nonlinear optics effects such as modulation instability (Alem et al., 2015), or self-phase modulation (Foaleng Mafang et al., 2011), and the probe power is restricted by the threshold of amplified spontaneous Brillouin scattering (Keaton et al., 2014).

In addition, the accumulated interaction between pump and probe over long distances might distort the spectrum observed toward the end of the fiber. When the BFS is nearly uniform along the fiber, and the frequency offset υ is set close to that value, the pump wave can become depleted and the Brillouin gain at the far end would be lower than expected. In contrast, when υ is detuned from that BFS, depletion does not occur, and the Brillouin gain observed at the end of the fiber would be larger. Consequently, the frequency dependence of the measured probe gain represents pump power levels that vary with υ rather than the true local BGS, and it might be interpreted incorrectly (Thévenaz et al., 2013). The phenomenon is often referred to as a first-order *non-local effect*. Higher-order non-local effects are encountered as well (Urricelqui et al., 2013). Mitigation schemes of non-local effects in standard B-OTDA setups reach 100–200 km range, with BFS accuracy on the order of 1 MHz and meter-scale resolution.

3.3.4 Acquisition duration

Four main factors limit the measurement speed of B-OTDA (Peled et al., 2012): (a) the two-way time of flight along the fiber $T_{round} = \frac{2nL}{c}$; (b) the number N_{avg} of repeating averages required to achieve a sufficient SNR for each trace; (c) the number N_{freq} of frequency offsets υ used in estimates of the BGS; and (d) the time T_{RF} required to switch between radio frequencies. The overall time necessary to acquire complete Brillouin spectra is given by:

$$T_{total} = N_{freq}\left(N_{avg}T_{round} + T_{RF}\right). \tag{7}$$

The acquisition durations of long-range B-OTDA systems may reach many minutes. Such slow response makes B-OTDA inapplicable to dynamic sensing. Times may be reduced by using fewer frequencies or a smaller number of averages, however both degrade the sensing accuracy.

3.4 Optimization of standard Brillouin optical time domain analysis and reflectometry

As discussed above, sensing performance is strongly affected by SNR. The optimization of B-OTDA and B-OTDR systems is closely linked with maximizing the Brillouin gain/loss signal, and with the reduction of noise sources.

3.4.1 Maximizing the signal

The measured Brillouin gain or loss signals are maximized by raising the power of levels of the optical pump and probe waves, up to the thresholds of detrimental competing nonlinear effects. The peak power of pump pulses in fibers that are at least tens of km long are limited to the order of +20 dBm, by the onset of self-phase modulation and modulation instability (Alem et al., 2015; Foaleng Mafang et al., 2011). Both phenomena stem from Kerr nonlinearity. In standard B-OTDA, the probe wave power is limited to the order of −13 dBm in similar fiber lengths, due to the non-local effects addressed above (Thévenaz et al., 2013). This limitation can be relaxed when a dual-tone probe wave is used, up to −6 dBm for each of the two tones (Dominguez-Lopez et al., 2016). Further mitigation of non-local effects may allow for probe power levels up to +5 dBm for each of the two tones (Dominguez-Lopez et al., 2016), limited eventually by the threshold of amplified spontaneous Brillouin scattering. In B-OTDR, the optical power of the local oscillator should be high enough to bring the detection of weak spontaneous Brillouin scattering to the shot noise limit.

Non-local effects can be mitigated by dividing the B-OTDA acquisition in time (Dong et al., 2011), between multiple pulses, or in frequency (Dong et al., 2012), using several pairs of pump and probe waves. Alternatively, sections of different nominal values of the BFS can be concatenated to form a broadened spectrum along the fiber under test. These methods might require longer acquisition durations, to accommodate a larger number of pulses and/or broader frequency scanning range. A B-OTDA system based on a frequency-modulated probe wave was free from amplified spontaneous Brillouin scattering limitations, and at the same time successfully mitigated detrimental non-local effects (Mompó et al., 2016). Measurements covered 100 km distance with 1 MHz accuracy in the BFS estimates. In 2007, Koyamada et al. developed a double-pulse B-OTDR with a spatial resolution of 20 cm and strain accuracy of 20 $\mu\varepsilon$ in a 1 km sensing range (Koyamada et al., 2007). In 2018, a wavelength diversity B-OTDR enabled 50 km sensing range with a 5 m spatial resolution (Lalam et al., 2018).

3.4.2 Minimizing the noise

The additive noise that accompanies the output waveform should be reduced as much as possible. Photo-receivers with a low power spectral density of additive noise should be chosen. In addition, an optical fiber pre-amplifier may be used to raise the power of the detected signal above the thermal noise of photo-receivers. The electrical bandwidth of the detector and associated circuitry in B-OTDA systems should match that of the optical pulses used. Exceedingly broad receiver bandwidths would elevate the overall additive noise levels with little benefit to the signal of interest. The sampling rate of digitizing equipment must be at least twice the bandwidth of the detected waveform and additive noise, to comply with the Nyquist criterion and avoid aliasing effects (Zaslawski et al., 2020). Finally, any curve fitting, cross-correlation, machine-learning or other algorithms for BFS estimates based on raw Brillouin spectra should be carefully optimized (Haneef et al., 2018; Zaslawski et al., 2020).

3.5 Performance enhancement techniques

Several means have been proposed and demonstrated toward better performance of B-OTDA and B-OTDR systems. A few of the main approaches are summarized below.

3.5.1 Circumventing the resolution limitations imposed by the acoustic lifetime

As discussed earlier, the 5–10 ns lifetime of stimulated acoustic waves restricts the duration of B-OTDA pulses and limits the spatial resolution of standard systems to the order of a meter. Since high spatial resolution is of primary interest, much effort has been dedicated to resolution enhancement in B-OTDA systems. The first B-OTDA experiment to reach 50 cm and 25 cm spatial resolution was demonstrated in 1998 by Brown et al. (Brown et al., 1998). Data analysis relied on the subtraction of Brillouin spectra collected from two and four partially overlapping sections, using 10 ns-long pulses.

This acoustic lifetime limitation can be alleviated using pump wave patterns that are more complex than single, isolated pulses. A first, comparatively long pump pulse may be used to stimulate the acoustic waves close to their steady-state magnitude. Following some delay interval, a second, short pump pulse is launched to interrogate the pre-stimulated waves. In separating the two pulses, the spectral bandwidths of the Brillouin spectra are decoupled from the spatial resolution. The non-zero background of pump waves while switched to the "off" state may also serve toward pre-excitation (Afshar et al., 2003; Beugnot et al., 2011; Lecœuche et al., 2000). Several systems have successfully employed these principles (Brown et al., 2005; Foaleng Mafang et al., 2010; Kishida et al., 2005; Song et al., 2010; Yang et al., 2016b). The sensing range of systems based on long pre-pumping is limited to 1 km by amplified spontaneous Brillouin scattering and by multiple echo phenomena that degrade the accuracy of BFS estimations.

A somewhat different solution is based on double pulse pairs (DPP) (Foaleng Mafang et al., 2010; Li et al., 2008). Each B-OTDA trace is acquired twice, with pulse durations τ and $\tau + \Delta\tau$. The baseline duration τ is sufficiently long for effective stimulation, whereas the difference $\Delta\tau$ is much shorter, typically below 1 ns. The two traces of the output probe wave power are subtracted through offline processing. Their difference provides the Brillouin gain with high resolution corresponding to $\Delta\tau$, free of the slow transient build-up limitations of the acoustic wave (Foaleng Mafang et al., 2010; Li et al., 2008). Using this technique, 1 million resolution points were addressed over 10 km of fiber with 1 cm resolution (Dominguez-Lopez et al., 2017). The acquisition duration is inevitably doubled. Finally, post-processing of B-OTDA data obtained with long pulses may recover sub-meter spatial resolution, with compromised BFS accuracy on the order of 3 MHz near transition regions (Wang et al., 2020a).

3.5.2 *Increasing the equivalent energy of optical signals*

As mentioned before, better performance of B-OTDA and B-OTDR systems requires better SNR, without reaching the threshold of competing, detrimental nonlinear effects. This objective can be met through several solution paths.

(1) Lumped optical amplification, using multiple EDFA repeaters along the sensing fiber. A sensing distance of 150 km has been reached (Dong et al., 2012), and then extended to 325 km (Gyger et al., 2014). An EDFA pre-amplifier at the detector input enabled sensing range of 120 km (Soto et al., 2011). This solution is sometimes impractical, as the power supply necessary for the repeaters may not be available.

(2) Distributed Raman amplification, in which additional pump waves are launched to amplify the SBS pump and/or probe waves through first-order (Rodriguez-Barrios et al., 2010), second-order (Martin-Lopez et al., 2010) or cascaded high-order Raman scattering processes (Fu et al., 2018). Raman gain compensates for fiber losses and improves the SNR at the far end of the fiber. The range of Raman-amplified B-OTDA systems has been successfully extended to 175 km in a single-fiber layout (Fu et al., 2018), and 240 km in a loop configuration (Angulo-Vinuesa et al., 2012). The relative intensity noise associated with Raman pumping may be mitigated using balanced detection schemes (Dominguez-Lopez et al., 2014).

(3) Optical pulse coding, in which a series of many pulses in close separation are launched into the fiber (Soto et al., 2008). Post-processing of the output raw trace can effectively compress the contributions of the entire sequence into a stronger signal, while maintaining the spatial resolution of a single pulse duration. Both periodic and aperiodic sequences have been employed (Le Floch et al., 2015; London et al., 2014; Soto et al., 2008, 2010; Sun et al., 2020; Taki et al., 2013; Yang et al., 2016a). The pulse codes can be optimized to account for non-uniform power levels along the sensing fiber (Sun et al., 2020). Care must be taken to avoid non-local effects (Yang et al., 2018a) and gain-dependent noises (Gao et al., 2021; Sun et al., 2020).

3.5.3 *Improving the frequency-scanning mechanism*

Several methods have been proposed toward faster Brillouin fiber sensing.

(1) Frequency sweeping in the microwave domain, to cover a span of several 100 MHz within short time frames (Dong et al., 2018; Peled et al., 2012). A single pump pulse was able to measure the Brillouin gain spectra of multiple dissimilar different fiber segments. The technique enabled MHz sampling rate for dynamic strain measurement (Dong et al., 2018).

(2) **Use of multi-tone probe waves**, launched in the fiber in parallel to reduce the scanning of frequencies (Voskoboinik et al., 2011). Frequency comb sources or orthogonal frequency division multiplexing protocols may be used for this purpose (Zhao et al., 2017).

(3) **Slope-assisted B-OTDA**, in which only a single frequency offset υ is being used. That frequency is chosen on a maximum slope of the BGS. Changes in the BFS manifest as intensity variations in the Brillouin amplification of the constant-frequency probe wave (Bernini et al., 2009; Peled et al., 2011). Currently, slope-assisted B-OTDA is the more popular technique for dynamic Brillouin sensing. Several techniques have been proposed to extend the range of BFS variation supported by slope-assisted analysis (Urricelqui et al., 2012; Yang et al., 2017; Zhou et al., 2017).

(4) **Local gain tracking**, for which the frequency of the probe wave is encoded to reach a similar normalized gain at any location along the fiber (Yang et al., 2018b). The frequency encoding is updated at each acquisition by calculating an error signal and applying a correction on the probe wave frequency like in a closed-loop servo system. The accuracy on BFS determination is maintained with respect to a classical B-OTDA and is inversely proportional to the tracking speed to first order, in contrast with the inverse square dependence of averaging.

The scanning over frequencies can also be eliminated by observing the polarization dependence of the Brillouin gain. The technique was employed by Bao et al. for distributed mapping of impact waves due to highway traffic in 2008 (Bao et al., 2008). Yet another approach relies on an optical chirp chain as the probe wave (Zhou et al., 2018).

4. Correlation-domain analysis and reflectometry

4.1 Brillouin-optical correlation-domain analysis

In time-domain Brillouin sensing protocols, transient scattering interactions are induced throughout the length of a fiber under test. Contributions from different locations are separated based on their time of arrival. Time-domain analysis and reflectometry provide a major advantage in addressing the entire length of fiber in each acquisition. On the other hand, the spatial resolution of the measurements is restricted by the Brillouin lifetime to the order of 1 m. As described at length in Section 3, several ingenious techniques have successfully pushed the spatial resolution of B-OTDA toward centimeter range. However, further resolution enhancement raises a formidable challenge.

An alternative concept of distributed Brillouin analysis is based on stationary scattering interactions. Rather than spanning the entire fiber, the stimulation of acoustic waves is confined to discrete and narrow fiber sections, and they are largely suppressed anywhere else. The principle was first proposed by Hotate and Hasegawa in 1999 and 2000 (Hotate and Hasegawa, 2000). In contrast to B-OTDA, the Brillouin pump wave is not pulsed. Instead, both pump and signal waves are continuous and jointly modulated in frequency or in phase. The approach is best understood through the transient build-up of the acoustic wave magnitude b as a function of time t, at a given fiber position z. Let us denote the complex envelope of an optical pump wave that is launched into the fiber under test from $z=0$ in the positive $\hat{z}$ direction as $a_p(t)$. Let us also denote the complex envelope of a signal wave coupled into the opposite end of the fiber $z=L$ as $a_s(t)$. The optical signal wave is counter-propagating in the negative $(-\hat{z})$ direction. The Brillouin stimulation of the acoustic wave at $0 \le z \le L$ is governed by the following differential equation (Antman et al., 2012a):

$$\frac{\partial b(z,t)}{\partial t} + j\frac{\Omega_B^2(z) - \Omega^2 - j\Omega\Gamma_B}{2\Omega} b(z,t) = jg_1 a_p\left(t - \frac{z}{V_g}\right) a_s^*\left(t - \frac{L-z}{V_g}\right). \tag{8}$$

In Eq. (8), $\Omega = 2\pi\nu$ denotes the difference between the central angular optical frequencies of the pump and signal waves, $\Omega_B(z) = 2\pi\nu_B$ is the local Brillouin frequency shift, V_g represents the group velocity of light in the fiber, $\Gamma_B = 2\pi \cdot \Delta\nu$ is the Brillouin linewidth, and g_1 is a constant electrostrictive parameter. For brevity, we define a local complex linewidth $\Gamma_A(\Omega, z) \equiv [\Omega_B^2(z) - \Omega^2 - j\Omega\Gamma_B]/2\Omega$, an offset time variable $t' \equiv t - z/V_g$, and a position dependent time lag $\theta(z) \equiv (2z - L)/V_g$, so that:

$$\frac{\partial b(z,t')}{\partial t'} + j\Gamma_A(\Omega, z) b(z,t') = jg_1 a_p(t') a_s^*[t' - \theta(z)]. \tag{9}$$

The stimulation of the acoustic wave is mathematically equivalent to the charging of a capacitor, with a lifetime parameter $\tau(\Omega, z) = 1/j\Gamma_A(\Omega, z)$. The lifetime reaches a maximum value of $2/\Gamma_B$ when $\Omega = \Omega_B(z)$. The driving force term, which is analogous to a charging voltage, is proportional to the inner product $a_p(t')a_s^*[t' - \theta(z)]$. In similarity to the response of a capacitor, we may distinguish between two regimes. If the driving force remains constant over time scales that are sufficiently longer than $\tau(\Omega, z)$, the local magnitude of the acoustic wave may come close to its potential steady state value of $b_{ss}(\Omega, z) = g_1 a_p a_s^* / \Gamma_A(\Omega, z)$. (The time dependence is omitted here, since the product $a_p a_s^*$ and the acoustic wave magnitude b_{ss} remain constant over the time scales of interest.) The magnitude of the acoustic wave depends on frequency through $\Gamma_A(\Omega, z)$, and its analysis for multiple choices of Ω may retrieve $\Omega_B(z)$. On the other hand, if the product $a_p(t')a_s^*[t' - \theta(z)]$ should oscillate on a time scale much faster than $\tau(\Omega, z)$, the stimulation of the acoustic wave at the specific location is strongly suppressed.

In most realizations of Brillouin sensors, the optical pump and signal waves are drawn from a common source that is split in two paths. We denote the complex envelope of the source field as $a_{in}(t)$. We disregard for the time being a possible difference in the path lengths of both replicas. The driving force of Brillouin stimulation at position z scales with $a_{in}(t')a_{in}^*[t' - \theta(z)]$. The stimulation of acoustic waves is therefore effective only where the time lag $\theta(z)$ is shorter than the coherence time of $a_{in}(t)$. In other words, *the stimulation of acoustic waves is confined to a short region, near $z = L/2$, whose extent equals the correlation length of the source* (see Fig. 7). Most B-OCDA protocols rely on the careful modulation of a laser diode source to manipulate its correlation length. The location $z = L/2$ is referred to as the *correlation peak position*. The process can be monitored through measurements of stimulated Brillouin amplification of the signal wave. Due to the suppression of Brillouin stimulation outside the correlation peak, the observed gain may be unambiguously related to SBS taking place at $z = L/2$.

Distributed analysis requires that the correlation peak position is scanned along the fiber under test. Scanning may be performed by varying the physical path length leading either the pump or the signal waves to their respective points of entry, $z = 0$, L. However, such scanning is not always convenient. Alternatively, the correlation peak position can be scanned through the modulating waveform itself. When repeating modulation with period t_{per} is used, multiple correlation peaks are formed at positions $z = L/2 + MZ$, where M is an integer and $Z = t_{per}V_g/2$. Unbalanced fiber delays may be used in the pump and signal wave paths, so that high-order correlation peaks are in overlap with the measurement region of interest, $M \gg 1$. In this case, small-scale changes in t_{per} effectively scan the correlation peak position.

The main advantage in B-OCDA is the potential for ultra-high resolution. In principle, the correlation length of the source can be made arbitrarily

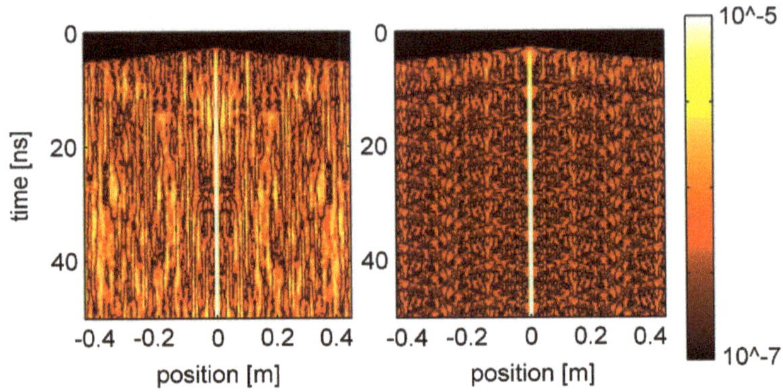

Fig. 7 Calculated magnitude of the stimulated acoustic wave density fluctuations (in units of kg $\times$ mm^{-3}, see color bar), as a function of time and position in a meter-long fiber section. The pump and signal waves were jointly modulated by a high-rate binary phase sequence. A pseudo-random bit sequence (PRBS) was used in the left panel, and a perfect Golomb code was used in the right panel. Both codes were 63 bits long with bit duration of 100 ps. Stationary stimulation of the acoustic wave may be observed in a narrow correlation peak at the center of the fiber. The magnitude of residual off-peak acoustic waves is considerably reduced with the employment of the Golomb code. *Reproduced with permission from Antman, Y., Levanon, N., Zadok, A., 2012b. Low-noise delays from dynamic Brillouin gratings based on perfect Golomb coding of pump waves. Opt. Lett. 37 (24), 5259–5261, Optical Society of America, 2012.*

short (see Section 4.3 for more details). The extent of SBS gain over very short lengths of fibers is very weak. However, unlike B-OTDA, the scattering interactions at the correlation peaks are stationary and may be acquired over arbitrarily long durations. The spatial resolution in B-OCDA is effectively de-coupled from the acoustic lifetime and from the integration time of SBS gain measurements. The main drawback of correlation-domain analysis is the need for position scanning. While B-OTDA addresses all resolution points with a single trace for every choice of Ω, the basic form of B-OCDA acquires data from one point at a time. In addition, the length of the fiber section under test must be kept shorter than the peaks separation Z to avoid ambiguity. Both limitations may be alleviated in part using hybrid analysis schemes which bring together time-domain and correlation-domain principles (see Section 4.3). However, position-scanning remains a fundamental limitation of B-OCDA.

4.2 Brillouin optical correlation-domain reflectometry

The analysis of stimulated Brillouin scattering, in either time or correlation domains, requires access to both ends of the fiber under test. In many real-world applications, however, access to the far end of the fiber might be limited. Reflectometry protocols, on the other hand, are based on single-ended access and the spontaneous Brillouin back-scatter of an incident pump wave. Brillouin reflectometry in the time domain was presented in Section 3. An equivalent correlation-domain protocol was first proposed and realized by Mizuno et al. (2008). In *Brillouin optical correlation-domain reflectometry*, or *B-OCDR*, a laser source is frequency modulated with a period t_{per} as above and split in two paths. One replica serves as a reference, and the other is launched into the fiber under test. The spontaneous Brillouin back-scatter from the fiber under test and the reference waveform are mixed in a heterodyne receiver. Let us denote the path difference between the two-way propagation distance up to point z along the fiber and the reference arm as ΔL. When $\Delta L = MZ$, the interference between the Brillouin scattering contribution from point z and the reference waveform provides a beating term centered at a stationary difference frequency of $\Omega_B(z)$. The beating current may be observed by an electrical spectrum analyzer. By contrast, the beating of back-scatter contributions from all other z locations and the reference waveform is centered at intermediate frequencies that are rapidly varying. The contributions of these terms to the observed electrical spectrum cancel out over time.

B-OCDR brings together the potential high resolution of correlation-domain analysis and the single-ended access of reflectometry. Like time-domain reflectometry, B-OCDR does not require the scanning of a frequency offset Ω: The entire Brillouin gain spectrum can be acquired in each trace. On the other hand, the method suffers from the weaker signals associated with spontaneous Brillouin scattering, as opposed to simulated interactions. Similar to B-OCDA, the technique requires point-by-point scanning of the entire fiber. B-OCDR has been successfully scaled to 134,000 resolution points (Shimizu et al., 2017), and sampling rates of 100 kHz (Mizuno et al., 2016).

4.3 Performance metrics, state-of-the-art and limitations

The first proposition of B-OCDA was based on frequency modulation of the light source at a single rate Ω_m and magnitude $\Delta\Omega$ (Hotate and Hasegawa, 2000). The spatial resolution obtained with this method equals $\Delta z = V_g \Gamma_B / (\Omega_m \Delta\Omega)$. Resolution of 1.6 mm has been achieved (Song et al., 2006), still a record for any Brillouin sensor over standard single-mode fiber. However, the original B-OCDA scheme is limited to an unambiguous sensing range $Z = \pi V_g / \Omega_m$, or $N = \pi \Delta\Omega / \Gamma_B$ resolution points. Since $\Delta\Omega$ is difficult to scale beyond tens of GHz, N is restricted to the order of 1000 positions at most. The number of resolution points can be extended using frequency modulation at multiple rates in parallel (Zou et al., 2013). The more complex modulation extends the temporal period of the waveform.

The number of points has been scaled further through binary phase coding of the common source (Zadok et al., 2012). Instead of frequency modulation, the phase of the light source is randomly chosen as either 0 or π radians during every symbol duration t_{bit}. The correlation length of the modulated source, $V_g t_{bit}$, can be reduced to centimeter-scale with multi Gbit/s modulation. The number of resolution points addressed without ambiguity equals the period of the repeating binary phase sequence, which can be made arbitrarily long. In a first demonstration of phase-coded B-OCDA, a 40 m-long fiber was monitored with 1 cm resolution (Zadok et al., 2012). Phase coding effectively removes the constraint between range and resolution.

Correlation-domain sensors are prone to noise due to residual, off-peak Brillouin interactions. Scattering contributions from off-peak locations can be described as random processes. While their expectation values vanish, the instantaneous magnitudes of the off-peak Brillouin interactions are non-zero. The standard deviation of the B-OCDA signal gain at any single off-peak location is considerably weaker than the intended, deterministic amplification taking place at the correlation peak. However, the off-peak contributions add up over many locations, and may completely mask out the signal of interest. The auto-correlation function of the optical source can be synthesized to partially suppress off-peak Brillouin interactions. For example, careful intensity modulation of a frequency-modulated B-OCDA source leads to apodization of the correlation function and reduces off-peak interactions. Binary Golomb phase codes, specifically designed to exhibit lower correlation sidelobes, can replace pseudo-random bit sequences and obtain weaker off-peak background as well (see Fig. 6, Antman et al., 2012b).

Over the last 10 years, hybrid Brillouin sensing techniques that bring together correlation-domain and time-domain principles have been proposed and demonstrated (Denisov et al., 2016; Elooz et al., 2014; Kim et al., 2015). In these methods, the common source of pump and signal waves is modulated in frequency or phase as in B-OCDA. In addition, the pump wave is also intensity-modulated by short pulses of low duty cycle in a similar way to B-OTDA. The duration of the pulses typically spans between tens and few hundreds of resolution points, and the length of fiber under test is much longer than the separation Z between adjacent correlation peaks. Due to the pulses overlay, SBS interactions in the correlation peak positions are transient, and their duration is restricted to that of the pulses. With proper choice of pulse duration with respect to the periodicity of frequency or phase modulation, SBS contributions from different peaks may be separated in time without ambiguity (Denisov et al., 2016; Elooz et al., 2014; Kim et al., 2015).

Compared with the fundamental B-OCDA approach, the hybrid analysis scheme brings about two important advantages. First, the number of position scans required to address the entire fiber is reduced to the number of resolution points within a pulse, regardless of the fiber length. Hundreds of thousands of resolution points were successfully addressed with only 200 position scans. Second, detrimental Brillouin scattering contributions from off-peak positions are restricted to the length of the pulse as well. On the other hand, the integration time of Brillouin amplification measurements is also limited to the pulse duration. The combination of time-domain and correlation-domain analyses was demonstrated over a 17 km long fiber with 8 mm resolution (Denisov et al., 2016). The number of potential resolution points, over two million, is still the largest of any Brillouin sensor.

In 2014, a B-OCDA setup has been demonstrated in which the pump and signal waves were drawn from the amplified spontaneous emission (ASE) of an EDFA (Cohen et al., 2014). The coherence length of ASE may be as short as tens of micrometers. However, the very weak Brillouin gain over such short fiber sections requires that the source spectrum is filtered, and the coherence length is extended to millimeters-scale. The stochastic intensity of ASE degrades the measurement signal to noise ratio below those of B-OCDA schemes involving the modulation of lasers (Cohen et al., 2014). In addition, the correlation peak position must be scanned through variable physical paths due to the lack of modulation periodicity. Despite these challenges, ASE-based B-OCDA reached a resolution of a few millimeters in standard single-mode fiber (Cohen et al., 2014). The method has been carried over to B-OCDA in photonic integrated waveguides, where the SBS gain per unit length is 1000 times stronger than in standard fiber. Zarifi and co-authors reported the B-OCDA of chalcogenide glass waveguides with sub-millimeter resolution (Zarifi et al., 2018, see Section 6).

5. Inter-Modal Brillouin scattering sensors

Another degree of freedom is offered by the stimulated Brillouin interaction when light has the possibility to propagate at different velocities along the exact same optical path. This is typically the case in a birefringent medium for orthogonal eigenmodes of polarization, or in a waveguide transmitting several spatial modes. In such a situation, it is possible to generate the acoustic wave at a given optical frequency and to interrogate the same acoustic wave at a different optical frequency. This extra degree of freedom offers unprecedented new avenues for sensing and for optical signal processing.

In a typical arrangement, the acoustic wave is generated by a pump wave at frequency ν_P interfering with a signal at frequency $\nu_P \pm \nu_B$, counter-propagating in the same mode (either polarization or spatial). The generated acoustic wave has a spatial period of exactly half the wavelengths of the interacting optical waves *in the medium*, its motion compensating for the small difference in frequency between the interfering waves.

The Brillouin interaction is analogous to a Bragg reflection from a moving longitudinal optical grating, which is induced by the acoustic wave through photoelasticity (hence the term *Brillouin dynamic gratings*). When the difference in frequencies between the two optical waves matches the BFS exactly, the pitch of the moving grating provides resonant reflection of the optical wave of higher frequency. The grating motion imposes a Doppler frequency shift on the reflected waveform, so that its new frequency is exactly equal to that of the interacting wave of the lower frequency. In addition, the grating also shifts the phase of the reflected wave to exactly match that of the lower frequency wave. In this way, the reflected component of the higher frequency optical wave interferes constructively with the lower frequency wave, leading to its amplification: this is the *Brillouin gain* process.

A similar effect is experienced by the lower frequency wave: the same grating associated with the acoustic wave also presents the exact optical pitch for a Bragg reflection of this wave, as a result of the grating motion in the opposite direction. The Doppler effect upshifts the frequency of the reflected wave to match that of the higher frequency wave. This time, however, the grating induces a π-phase shift, so that the reflected wave component interferes destructively with the higher frequency wave and leads to its amplitude attenuation: this is the *Brillouin loss* process.

The dynamic Bragg grating supported by the acoustic wave therefore transforms all properties of one of the interacting waves (direction, frequency, phase) to match those of the other interacting wave, except for polarization that is preserved (Zadok et al., 2008). The longitudinal, compressive nature of the acoustic wave makes it totally transversally isotropic, and of equal impact for all light polarizations.

This scheme works perfectly well if all optical waves experience the same effective refractive index and thus propagate at the same speed. Let us assume that a third optical wave is propagating along the same optical path, but at a different velocity. If ever the field of this third wave is overlapping the vibrational field of the acoustic wave, it will probe a Brillouin grating with a mechanical pitch given by the acoustic wavelength Λ, but an optical pitch that may differ since it is scaled by the refractive index experienced by this wave, as described in Fig. 8. This third wave will therefore be resonant with the Brillouin grating at a shifted frequency.

This spectral shift can be easily obtained using the Bragg condition, by equating the grating pitch given by the acoustic wavelength Λ to the half wavelength of the optical waves, considering copropagating modes experiencing different refractive indices:

$$\Lambda = \frac{1}{2}\frac{\lambda_W}{n_W} = \frac{1}{2}\frac{\lambda_R}{n_R}. \tag{10}$$

Here λ_W and n_W are the vacuum wavelength and refractive index of the "writing" wave, that is one of the interacting waves generating the acoustic wave. Since there are two counter-propagating "writing" waves, this equation considers the wavelength of the wave propagating in the same

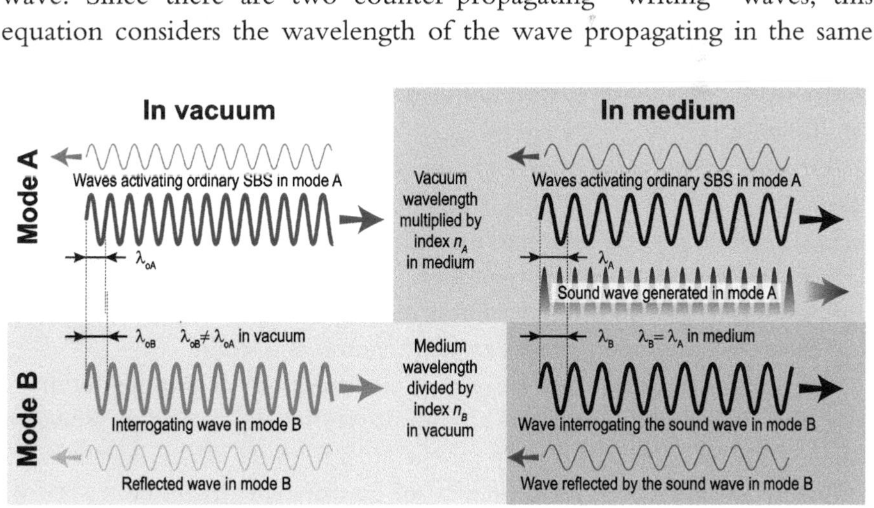

Fig. 8 Principle of the generation and interrogation of dynamic gratings using stimulated Brillouin scattering.

direction as the "reading" wave, itself characterized by λ_R and n_R, and guided in a distinct mode. This expression can be translated into the frequency domain in straightforward manner:

$$\frac{\nu_B}{V_a} = 2\frac{n_W \nu_W}{c} = 2\frac{n_R(\nu_W + \Delta\nu_{RW})}{c}. \tag{11}$$

Here ν_W denotes the optical frequency of the "writing" wave, and $\Delta\nu_{RW}$ is the difference in optical frequencies between "writing" and "reading" waves. With simple algebra and considering that the refractive indices are dispersive, the following expression is found for $\Delta\nu_{RW}$:

$$\Delta\nu_{RW} = \frac{n_W - n_R}{n_W^g}\nu_W. \tag{12}$$

Here $n_W^g = n_W + dn_W/d\nu$ is the group index for the "writing" wave. In a silica fiber at a wavelength $\lambda_W = 1550\,\text{nm}$, a refractive index difference $n_W - n_R = 10^{-3}$ results in a frequency shift of $\Delta\nu_{RW}$ of 130 GHz (or about 1 nm in wavelength units).

Brillouin dynamic gratings were first demonstrated in their most straightforward implementation using highly birefringent fibers (Song et al., 2008), and their potential for distributed sensing was soon identified and realized (Song et al., 2009). In particular, by measuring the frequency difference between the activating light wave and the reading pulse, a distributed measurement of birefringence can be achieved along the fiber. Birefringence turns out to be also dependent on temperature and strain, with an even larger sensitivity by a factor 10 to 100 with respect to the similar dependence for the Brillouin shift.

Moreover, this dependence shows a sign change for temperature in Panda fibers, as shown in Fig. 9, so that temperature and strain can be discriminated in a well-conditioned manner by simultaneously measuring the Brillouin shift ν_B and the birefringence (Zou et al., 2009). Brillouin dynamic gratings over highly birefringent fibers remain the best conditioned system to effectively discriminate between temperature and strain.

Exploiting the anisotropic cross-sectional structure of a highly birefringent fiber significantly enhances the sensitivity to hydrostatic pressure, so that a 100× increase in accuracy for the local pressure can be obtained by performing a distributed measurement of birefringence (Kim et al., 2016). Compensating for the temperature variations can be simply realized through an ordinary distributed Brillouin shift measurement.

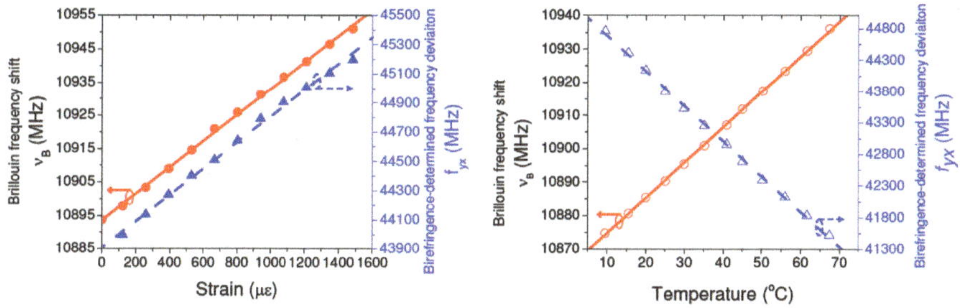

Fig. 9 Dependence of the Brillouin frequency shift and the birefringence as a function of strain and temperature, measured using a standard BOTDA and dynamic Brillouin gratings in a highly birefringent PANDA fiber. The inverted slope for strain in the birefringence measurement enables a precise discrimination between the two quantities (Zou et al., 2009).

Instead of using a highly birefringent fiber, similar performance can be obtained by using a few-mode fiber. In this case the activating optical waves are selectively launched into the same spatial mode (e.g., LP_{01}) and the reading pulse is coupled into a distinct mode (e.g., LP_{11}). Many combinations are possible using different pairs of modes and the effective refractive index difference between these modes can be obtained at all positions along the fiber. This also makes possible a discriminated measurement of temperature and strain, though with a reduced sensitivity with respect to distributed birefringence measurement (Kim and Song, 2014; Song et al., 2017). Brillouin dynamic gratings were also used for random-access coupling of light between the core and cladding modes of standard single-mode fibers (Bashan et al., 2020).

A strong point of Brillouin dynamic gratings is the possibility to activate the acoustic wave and interrogate it using distinct optical waves. This enables to use very short interrogating pulses with no impact on the spectral distribution of the Brillouin gain, so that a very sharp spatial resolution can be achieved with no spectral broadening when probing the local Brillouin gain spectrum, in total contrast with a classical configuration (Song and Yoon, 2010). A 1 cm spatial resolution and even less could be achieved using a time-domain reflectometry system (Chin et al., 2012; Song et al., 2010) with no penalty on the temperature/strain response.

A major issue experienced by this technique results from the poor uniformity of the birefringence in real-world highly birefringent fibers. The birefringence may significantly fluctuate over distances shorter than the spatial resolution, and this lowers the contrast and the accuracy of the measurements by an excess broadening of the spectral response. In addition, a reference measurement of the static birefringence distribution under constant environmental conditions is required.

The development of distributed sensing using Brillouin dynamic gratings has been largely cast aside, since it was demonstrated that similar performance and information can be obtained using phase-sensitive Rayleigh reflectometry (Chen et al., 2012; Lu et al., 2017; Soto et al., 2015). This latter approach offers the advantage of using only one pulsed optical wave, launched and collected at only one end of the fiber, whereas the generation and interrogation of Brillouin dynamic gratings mandates three interacting optical waves and access to both fiber ends. For this reason, the application of inter-modal Brillouin interactions has mostly moved to optical signal processing, where Brillouin dynamic gratings offer unique possibilities for optical storage, dynamic time delay generation, and even all-optical analogue operations on the signal (Chin and Thévenaz, 2012; Santagiustina et al., 2013).

6. Sensing with specialty fiber and waveguide platforms

SBS sensors over speciality fiber and waveguide platforms can simultaneously measure temperature and strain without ambiguity and with high resolution and address additional parameters that are inaccessible to Brillouin sensors using standard fibers (ITU-T G.652). Since both temperature T and strain ε modify the BFS, the two cannot be distinguished without additional information. Bao et al. (1994) demonstrated the separation of the two variables using two fibers: one that is isolated from strain and the other that is affected by both parameters. The two quantities may be evaluated simultaneously using fibers with two peaks in their BGS, at frequencies υ_{B1} and υ_{B2}:

$$\begin{bmatrix} \Delta\upsilon_{B1} \\ \Delta\upsilon_{B2} \end{bmatrix} = \begin{pmatrix} C_{T1} & C_{\varepsilon 1} \\ C_{T2} & C_{\varepsilon 2} \end{pmatrix} \begin{bmatrix} \Delta T \\ \Delta\varepsilon \end{bmatrix} \tag{13}$$

Here $\Delta\upsilon_{B1,\ B2}$ denote variations in $\upsilon_{B1,\ B2}$ with respect to some reference values, ΔT and $\Delta\varepsilon$ represent changes in temperature and strain, respectively, and $C_{T1,\ T2}$ and $C_{\varepsilon 1,\ \varepsilon 2}$ are coefficients that connect between changes in temperature and strain and modifications to the two BFS values. Note that the two rows of the coefficient matrix must be sufficiently different, or else the matrix becomes ill-conditioned, and its inversion is highly susceptible to inaccuracies in the assessment of $\Delta\upsilon_{B1,\ B2}$. A similar formulation would apply to the simultaneous monitoring of the BFS and the Brillouin linewidth.

The two BFS peaks may correspond to principal axes of birefringence, two guided optical modes of a few-order-mode optical fiber (Li et al., 2015), optical modes of different orbital angular momenta (Xu et al., 2016), or two acoustic modes that are guided in the core of the fiber (Afshar et al., 2005). The acoustic modes are obtained through solutions to the elastic wave equations in the core geometry, and their contributions to the BGS are determined by spatial overlap considerations (Pradhan et al., 2017). The stimulation of multiple acoustic modes typically requires higher pump power levels (Afshar et al., 2005). The fiber cross-section can be designed to enhance Brillouin interactions with multiple modes. For example, large effective area fibers (LEAFs) support four guided acoustic modes, although they only guide a single optical mode, as shown in Fig. 10 (Liu and Bao, 2012). Monitoring of the beat frequency between two BFS values in a LEAF resulted in temperature compensated strain measurements (Lu et al., 2013). Lee et al. measured temperature and strain through multiple acoustic modes in a B-OTDA over dispersion-shifted fiber (Lee et al., 2001). The spatial resolution was several meters. Wu et al. described a hybrid Raman-Brillouin sensor system in a few-order mode fiber (Wu et al., 2017).

The complex structures of photonic crystal fibers (PCFs) support many high-order acoustic modes, which give rise to multiple Brillouin scattering peaks (Dainese et al., 2006; Zou et al., 2003). The dependence of the BFS values on temperature and strain can be markedly different among distinct modes (Zou et al., 2004). Measurements of temperature and strain in

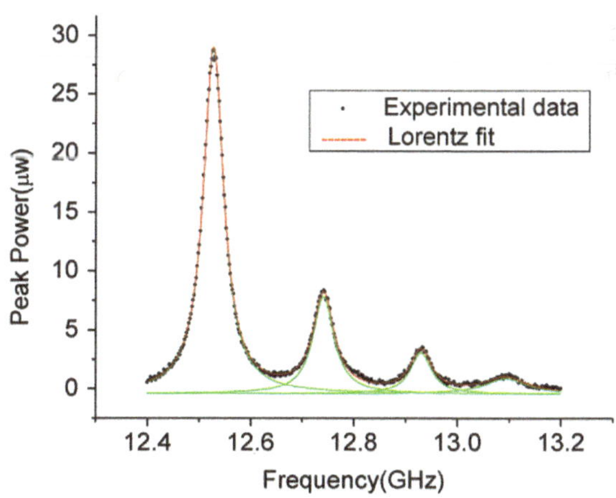

Fig. 10 The presence of multiple guided acoustic modes in a specialty, large effective area fiber results in multiple Brillouin resonances showing a distinct response to temperature and strain (Liu and Bao, 2012).

B-OTDA over a PCF, with 15 cm resolution, was reported in 2004 (Zou et al., 2004). A temperature-compensated distributed sensor of hydrostatic pressure was demonstrated in a polarization maintaining PCF based on Brillouin scattering in the two principal axes (Teng et al., 2016). Hollow-core fibers can also be exploited since a strong Brillouin gain is also observed in gases pressurized at some tens of bars. A strain-insensitive temperature distributed fiber sensor was demonstrated in such fibers with similar sensitivity to that obtained in solid silica fibers (Yang et al., 2020).

Multi-core fibers provide yet another solution for simultaneous sensing of temperature and strain. For example, Li et al. distinguished between the two quantities using a dual-core fiber (Zaghloul et al., 2018). The two cores exhibited distinct coefficients of thermal and strain dependence of their BFS values. Two cores of a seven-core fiber were used by Mizuno et al. for a similar purpose (Mizuno et al., 2015). Zhao et al. (2016) showed that the Brillouin scattering response of multi-core fibers is highly sensitive to bending, potentially allowing for three-dimensional shape sensors.

Fibers made of soft As_2Se_3 chalcogenide glass also serve in Brillouin sensing measurements. The low Young's modulus of the glass enables Brillouin sensing of transverse load on tapered fibers. BFS variations of 0.08 ± 0.02 MHz/N were observed in B-OTDA over a fiber with 2 μm As_2Se_3 core and 100 μm cladding (Saxena et al., 2019). The force sensor is based on high birefringence induced by the large difference in refractive index and Young's modulus between core and cladding.

SBS-active waveguides have been demonstrated as an enabling platform for many on-chip functionalities. A first demonstration of chip-level SBS was provided by Pant et al. in an As_2S_3 chalcogenide rib waveguide in 2011 (Pant et al., 2011). Distributed Brillouin analysis along such nano-structured waveguides can identify local variations in their geometry. Changes in waveguide widths alter the effective index of the guided optical mode and modify the BFS accordingly. Distributed Brillouin measurements were realized by the same group in a hybrid silicon-chalcogenide photonic platform (Zarifi et al., 2018). B-OCDA was performed with 2 mm spatial resolution. A record resolution of 500 μm has been reached on the same platform using ASE-based B-OCDA (Zarifi et al., 2019, see also Section 4).

7. Forward SBS sensors

7.1 Forward SBS in standard fibers

All sensors described thus far are based on backward Brillouin scattering. The interactions involve optical pump and signal waves that

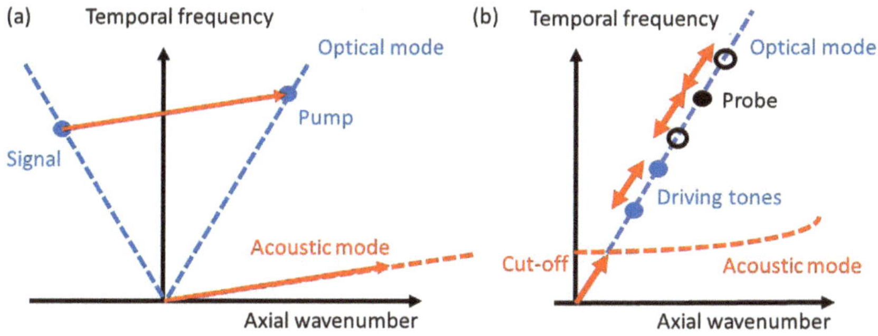

Fig. 11 SBS wavenumber matching considerations in standard single-mode fibers. (A) Backward SBS between counter-propagating optical pump and signal waves and a longitudinal acoustic mode. (B) Forward SBS between co-propagating driving optical fields (blue) and a guided acoustic mode (red) close to its cut-off frequency. The stimulated acoustic waves may also scatter an additional optical probe wave (full black circle) into upper and lower sidebands (empty black circles), giving rise to phase modulation.

are counter-propagating, and dilatational acoustic waves that also propagate in the axial direction. The acoustic waves are guided by the small elastic contrast between the fiber core and cladding. The frequencies of the acoustic waves are of the order of 11 GHz and their axial wavenumbers are very large, twice those of the optical waves (see Fig. 11). Transverse components of both the acoustic wave-vector and the material displacement are vanishingly small. Like the optical modes, the acoustic waves taking part in backward Brillouin scattering are largely confined to the core, and do not reach the outer edge of the cladding. Accordingly, backward Brillouin measurements are not directly sensitive to media outside the fiber cladding. The challenge goes beyond backward Brillouin sensing alone: most fiber sensors are unable to monitor conditions prevailing where light does not reach.

Standard optical fibers also support forward Brillouin scattering. The process was first observed by Shelby et al. (1985). Similar to the backward scattering case, the acoustic modes involved propagate in the axial direction. Unlike the backward process, however, these modes are guided by the entire cladding of the fiber. Each mode is characterized by a cut-off frequency, below which it may not propagate (see Fig. 11). The cut-off frequencies of relevant modes are hundreds of MHz. The acoustic modes may be stimulated by two optical fields components that co-propagate in the fiber. Effective stimulation requires that the axial phase velocity of the acoustic mode should match that of guided light. This condition can be met when the acoustic frequency approaches its cut-off value. Close to cut-off, the axial acoustic wavenumber becomes very small, in the range of 1 rad $\times$ m^{-1}.

The acoustic wave-vector is therefore predominantly transverse. The material displacement is almost entirely confined to the transverse plane as well.

In analogy to backward SBS, the forward Brillouin stimulation of the acoustic waves is associated with the coupling of optical power between the two driving optical fields. There is, however, another important difference between the backward and forward processes: in the latter case, the guided acoustic modes remain wavenumber-matched with co-propagating optical fields over a very broad range of wavelengths (see Fig. 11). Therefore, the acoustic waves of forward Brillouin may effectively scatter and modulate additional optical probe waves that are widely detuned from the two original stimulating fields.

Forward SBS in standard single-mode fibers takes place through two specific classes of guided acoustic modes: radially symmetric (denoted as $R_{0,\ m}$ where m is a positive integer), and torsional-radial modes of twofold azimuthal symmetry which are referred to as $TR_{2,\ m}$. Material displacement in $R_{0,\ m}$ modes is in the radial direction only and its magnitude is radially symmetric. The radial modes are purely dilatational. In $TR_{2,\ m}$ modes, the displacement vector includes both radial and azimuthal components and both depend on the radial and azimuthal coordinates. The torsional-radial modes combine dilatational and shear wave motion. Fig. 12 shows the calculated normalized displacement profiles of $R_{0,\ 8}$ and $TR_{2,\ 7}$ modes. While the fiber supports, in principle, torsional-radial guided acoustic waves of all azimuthal orders, the transverse overlap between other modes and the electrostrictive driving force cancels out.

7.2 Principles of forward SBS sensing

Forward SBS provides the basis for a new concept of fiber sensors which can successfully address media outside the cladding boundary. Let us consider

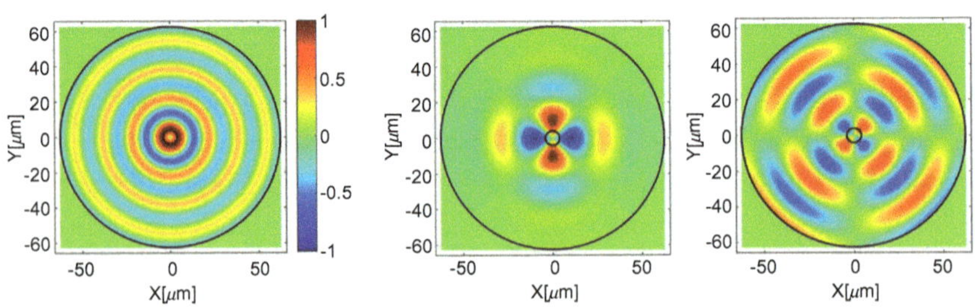

Fig. 12 Normalized transverse profiles of material displacement in guided acoustic modes of standard single-mode fibers. Left—radial displacement of radial mode $R_{0,\ 8}$ (cut-off frequency 370 MHz). Center–radial displacement of torsional-radial mode $TR_{2,7}$ (cut-off frequency 139 MHz). Right—azimuthal displacement of the same mode.

first short and isolated optical pump pulses, with nano-second scale durations, propagating in a standard fiber that was stripped of its coating layers. Beating among the constituent spectral components of the pulses stimulates a packet of guided acoustic modes which propagates outward from the core of the fiber in the radial direction. For simplicity we consider below the radial acoustic modes only. The contribution of torsional-radial modes may be suppressed with polarization scrambling of the pump pulses. The acoustic wavefront undergoes partial transmission and reflection at the outer cladding boundary, which depend on the mechanical impedance Z_{out} of the medium outside the boundary.

For example, if the bare fiber is kept in air (Z_{out} that is orders of magnitude smaller than the impedance Z_{silica} of silica), the reflectivity of the acoustic wave-packet is practically 100%. The acoustic wave-packet reconverges at the fiber axis following a two-way acoustic propagation delay to the cladding boundary and back: $\tau_{clad} = V_l d$. Here V_l is the velocity of dilatational acoustic waves in silica and d is the cladding diameter. With air outside the cladding, the magnitudes of successive acoustic impulses across the core would ideally remain the same as that of the initial one. If the fiber is embedded in a medium of non-negligible impedance Z_{out} instead, part of the acoustic wave-packet dissipates outward and the reflectivity is reduced accordingly. The acoustic impulses reforming across the core at intervals τ_{clad} would then decay with time. The decay rate Γ_{FSBS} of the acoustic wave-packet is simply related to the mechanical impedance of the outside medium:

$$\begin{aligned}\Gamma_{FSBS} &= \Gamma_{int} + \frac{1}{2\tau_{clad}} \ln\left(\left|\frac{Z_{out} - Z_{silica}}{Z_{out} + Z_{silica}}\right|\right) \\ &\approx \frac{1}{2\tau_{clad}} \ln\left(\left|\frac{Z_{out} - Z_{silica}}{Z_{out} + Z_{silica}}\right|\right)\end{aligned} \tag{14}$$

Here the first term Γ_{int} denotes the decay rate due to internal acoustic dissipation in silica. With liquid or solid media outside the cladding, the first term is typically negligible with respect to the second. The relation is equivalent to the decay of optical waves in cavities. Eq. (14) allows for the quantitative analysis of media outside the fiber cladding. The decay rates also signify the spectral linewidths of the stimulated optical modes, and they can be observed in either time or frequency domains. Measurements may rely on the stimulated Brillouin amplification and attenuation of the driving optical tones, or on forward Brillouin phase modulation of optical probe waves as discussed above. Polarization rotation of optical probe waves

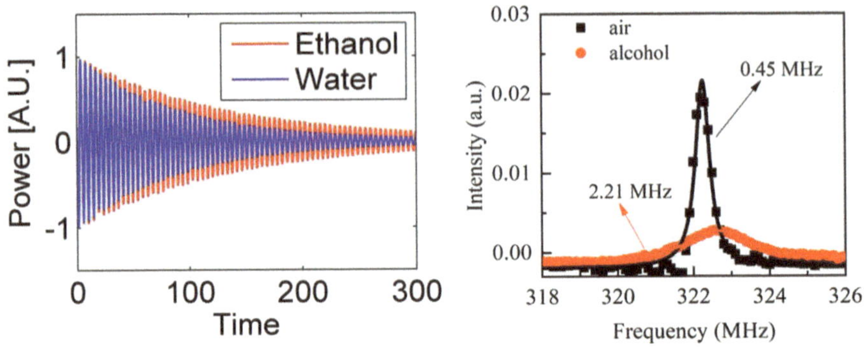

Fig. 13 (Left) Measured normalized phase modulation of an optical probe due to acoustic mode $R_{0,5}$ in a bare fiber, as a function of time. The fiber was immersed in ethanol (red) or water (blue). Decay in water is faster due to its higher mechanical impedance (Antman et al., 2016). (Right) Spectra of forward SBS through mode $R_{0,7}$ in a 2 m-long section of standard bare fiber kept in air (black) or immersed in ethanol (red). The section under test was placed at the end of a 200 m-long, coated lead-in fiber. The local contribution to the forward SBS process is successfully resolved (Pang et al., 2020).

due to $TR_{2,m}$ modes was monitored toward sensing as well (Hayashi et al., 2017). Fig. 13 shows the measured temporal decay of mode $R_{0,5}$ in a bare fiber immersed in ethanol or water (Antman et al., 2016). The forward SBS interaction was integrated over a 30 m-long section. Decay in water is faster due to the higher Z_{out}. Impedance estimates of both liquids agree with literature values to within 1%.

7.3 Distributed analysis of forward SBS

Forward scattering is inherently difficult to localize. Both time-domain and correlation-domain principles discussed earlier in this chapter with respect to backward Brillouin scattering are inapplicable to the forward effect. On the other hand, point-sensing or position-integrated measurements of media outside the cladding based on forward SBS provide only modest added value with respect to more established protocols. As in many other fiber sensing principles, the potential merit of forward SBS sensors would greatly depend on their extension toward spatially distributed measurements.

At the time of writing, distributed forward SBS sensing raises a formidable challenge. Nevertheless, several protocols have been proposed in the scientific literature starting in 2018. Most approaches proposed to-date are indirect: they rely on a complementary back-scatter process to estimate the local amplitude or phase of optical waves affected by forward SBS.

In one example, frequency-resolved optical time-domain reflectometry of Rayleigh back-scatter was used to monitor the exchange of power between the two driving tones of a forward SBS process. The spatial resolution obtained was a modest 50 m, over 3 km of fiber (Bashan et al., 2018). In few other works, multiple B-ODTA scans were employed to monitor the phase modulation of a probe wave due to forward Brillouin scattering (Chow et al., 2018). Phase modulation manifests in multiple spectral side-lobes of the probe tone. Each spectral component of the modulated probe also served as a pump wave for a B-OTDA acquisition. The B-OTDA traces recovered the magnitudes of the various spectral side-lobes and reconstructed the extent phase modulation. Pang et al. brought this technique to a spatial resolution of 2 m in 2020 (Fig. 13, Pang et al., 2020). Very recently, Zaslawski and co-workers monitored the forward SBS process through the frequency shift of probe waves (Zaslawski et al., 2021). Their protocol reached 80 cm resolution, comparable with that of the traditional B-OTDA architecture.

The distributed analysis of forward SBS is a new topic of research, and its performance is still very limited. For example, the largest number of resolution points reported at the time of writing is 100, compared with over a million in B-OTDA or B-OCDA. The indirect measurement schemes proposed thus far suffer from two major drawbacks. They require two-dimensional scanning over two degrees of freedom (in radio frequency, optical wavelength or timing), which makes the experimental durations very long. In addition, the measurements provide estimates of the cumulative effect of forward Brillouin scattering, from the point of entry up to a position z. Local information is recovered through a post-detection differentiation operation over the acquired traces, a process that is highly susceptible to noise. Despite these difficulties, forward SBS sensing carries a large promise of mapping media beyond the reach of guided light. Progress over the last 3 years has been rapid, and the current performance may well be improved upon in future research.

7.4 Coated fibers

Another difficulty facing forward SBS sensors has to do with protective coating layers. The use of bare fibers outside the research laboratory is impractical. However, the standard acrylate coating layers that are most commonly used to protect optical fibers absorb the guided acoustic waves and keep them from reaching surrounding media. Nevertheless, the analysis of outside liquids is still possible using commercially available fibers that are coated with thin layers of polyimide (Chow and Thévenaz, 2018; Diamandi et al., 2019). Compared with bare fibers, the spectra of guided acoustic modes in coated fibers are more complex. Unlike bare fibers, the decay rates may vary strongly among different acoustic modes with the same substance outside the coating. The decay rates are also sensitive to micron-scale variations in coating thickness, well within standard specifications. In spite of these difficulties, several works already reported distributed sensing based on forward SBS in polyimide coated fibers. The challenge remains a formidable one: most field applications of optical fibers require additional protection on top of thin polyimide layers. Forward SBS analysis may also evaluate the elastic properties of the coating layers themselves (Diamandi et al., 2021).

8. Applications and employment

Infrastructure decay processes represent a major problem faced by many developed countries in the world. After years of service and possibly exposure to extreme weather conditions, many structures such as bridges, tunnels, dams and power plants are showing severe deterioration. To guarantee the safety of these structures and to perform proper maintenance and rehabilitation, structural "health" conditions must be monitored over time. The monitoring of new infrastructures is also very important, as it helps identify minor degradations at an early stage.

Structural evaluations are often carried out by engineers trained in visual inspection, which sometimes can be inaccurate due to the personal experience differences. To increase the inspection efficiency and accuracy, various sensors have been developed and demonstrated in the field. Since the locations of possible failure points are usually not known in advance, distributed sensors can replace many point sensors to find potential cracks, deformations or buckling locations. Distributed fiber sensors are among the most promising candidates due to their durability, stability, small sizes and relative insensitivity to external interference. Brillouin-based systems have seen widespread application due to the long-distances they can interrogate. In addition to structural health monitoring, distributed Brillouin fiber sensors are also applicable to chemical processing plants, power lines, offshore installations and many more. The possibilities offered by distributed Brillouin fiber sensors have no equivalent in conventional sensor systems.

One major application of distributed Brillouin sensors is in the monitoring of power-transmission lines. Power distribution companies face a number of significant challenges due to harsh environmental conditions.

During cold weather conditions, for example, power transmission lines are subject to extreme temperatures, high winds, and ice. During hot weather conditions, transmission lines are subject to excessive length-dilation due to heat. In addition, power lines can be exposed to storms. These conditions can lead to large strain in transmission lines, rapid changes in the line temperature, and potentially to power disruptions, expensive repairs in remoted locations, and even catastrophic failure. B-OTDA or B-OTDR systems detect such issues at an early stage.

Specific applications of Brillouin optical fiber sensors include the oil and gas industry (monitoring of leakage and deformation in pipelines and storage tanks, temperature sensing and fault locations); power grid distribution (temperature monitoring, fault location, fire alert); construction and civil engineering (monitoring of concrete hydration, crack and fault detection in dams and bridges, preventive maintenance, damage assessment following earthquakes, landslides, debris flow and other catastrophes); the transportation sector (fire monitoring in tunnels and subways, early detection of deformation and collapse, preventive maintenance of railways and highways); and security (intrusion detection, protection of utility infrastructures, perimeter defense). As distributed fiber optic sensors become more cost-effective and the technology advances, their applications continuously expand.

9. Conclusions

Since the first demonstration of distributed Brillouin fiber sensors in 1989, they have remained the subject of extensive research efforts, which have also led to commercial products. At the time of writing, optical fiber sensors represent the most successful commercial application of Brillouin scattering. As discussed at length in the preceding sections, Brillouin fiber sensors span hundreds of kilometers, obtain sub-millimeter resolution, identify sub-degree centigrade temperature changes and tens of ppm strain events, and track dynamic signals at kilohertz and even megahertz rates. The principles of Brillouin sensing are applied to standard single-mode, polarization-maintaining, multi-core, few-order mode, tapered, photonic-crystal, polymer, and soft glass fibers, as well as photonic integrated circuits. The various platforms enable unambiguous monitoring of temperature and strain, and they extend the range of accessible quantities of interest to include waveguide geometry, birefringence, pressure, three-dimensional shapes, and more. While Brillouin sensing has been traditionally based on the backward scattering effect, forward Brillouin fiber sensors emerge as a new and promising concept for the analysis of media outside the fiber, where light does not reach.

Brillouin sensing technology is the most appealing platform for the static analysis of temperature and strain over large and expensive infrastructure, where long range is a foremost consideration. The B-OTDA architecture remains the workhorse of these applications. Brillouin sensors with high-scanning rates have been reported, reaching MHz rates with micro-strain level detection, using chirped pulses in B-OTDA or slope-assisted protocols. Yet, most dynamic fiber sensor systems (such as distributed acoustic sensors) rely on coherent Rayleigh back-scatter instead for their superior sensitivity. Correlation domain Brillouin sensors find certain niche applications in structural health monitoring and in the aerospace and automotive sectors, where high resolution is important, however their use remains restricted compared with time domain protocols. As the numbers of aging infrastructures in many countries continue to grow, and the consequences of catastrophic failures become increasingly intolerable, larger-scale field deployment of Brillouin optical fiber sensors is to be expected.

Academic research on Brillouin fiber sensors continues to flourish in many research groups worldwide. Research objectives include the traditional goals of longer range, higher resolution, better accuracy and faster acquisition. Avenues taken toward these goals begin in the optical domain, such as the mitigation of competing nonlinear propagation effects, but increasingly involve inter-disciplinary synergy with signal processing, coding, algorithmic and machine learning concepts which are combined with optics in state of the art distributed Brillouin sensors. In addition to the long-term challenges above, Brillouin fiber sensors seek and find new applications such as distributed analysis of forward scattering, gas spectroscopy, shape sensing, and more. One may safely assume that a corresponding chapter, written 20 years in the future, would be considerably different from the one ending here.

References

Afshar, S., Ferrier, G., Bao, X., Chen, L., 2003. Effect of the finite extinction ratio of EOM on the performance of the pump-probe Brillouin sensor system. Opt. Lett. 28, 1418–1420.

Afshar, S., Kalosha, V.P., Bao, X., Chen, L., 2005. Enhancement of stimulated Brillouin scattering of higher-order acoustic modes in single-mode optical fiber. Opt. Lett. 30 (20), 2685–2687.

Alem, M., Soto, M.A., Thévenaz, L., 2015. Analytical model and experimental verification of the critical power for modulation instability in optical fibers. Opt. Express 23 (23), 29514–29532.

Angulo-Vinuesa, X., Soto, M.A., Martin-Lopez, S., Chin, S., Ania-Castañon, J.D., Corredera, P., Rochat, E., Gonzalez Herraez, M., Thévenaz, L., 2012. Brillouin optical time-domain analysis over a 240 km-long fiber loop with no repeater. In: Optical Fiber Sensors. Proc. SPIE, vol. 8421. International Society for Optics and Photonics (SPIE), p. 8421C9.

Antman, Y., Primerov, N., Sancho, J., Thévenaz, L., Zadok, A., 2012a. Localized and stationary dynamic gratings via stimulated Brillouin scattering with phase modulated pumps. Opt. Express 20 (7), 7807–7821.

Antman, Y., Levanon, N., Zadok, A., 2012b. Low-noise delays from dynamic Brillouin gratings based on perfect Golomb coding of pump waves. Opt. Lett. 37 (24), 5259–5261.

Antman, Y., Clain, A., London, Y., Zadok, A., 2016. Optomechanical sensing of liquids outside standard fibers using forward stimulated Brillouin scattering. Optica 3 (5), 510–516.

Bao, X., Chen, L., 2011. Recent progress in Brillouin scattering based fiber sensors. Sensors 11 (4), 4152–4187.

Bao, X., Webb, D.J., Jackson, D.A., 1993. 32-km distributed temperature sensor based on Brillouin loss in an optical fiber. Opt. Lett. 18 (18), 1561–1563.

Bao, X., Webb, D.J., Jackson, D.A., 1994. Combined distributed temperature and strain sensor based on Brillouin loss in an optical fiber. Opt. Lett. 19, 141–143.

Bao, X., Brown, A., DeMerchant, M., Smith, J., 1999. Characterization of the Brillouin gain/loss linewidth for single mode fibers using very short pulses. Opt. Lett. 24 (8), 510–512.

Bao, X., Zhang, C., Li, W., Eisa, M., Ei-Gamal, S., Benmokrane, B., 2008. Monitoring the distributed impact wave on a concrete slab due to the traffic based on polarization dependence on stimulated Brillouin scattering. Smart Mater. Struct. 17, 015003.

Barnoski, M.K., Jensen, S.M., 1976. Fiber waveguides: a novel technique for investigating attenuation characteristics. Appl. Opt. 15 (9), 2112–2115.

Bashan, G., Diamandi, H.H., London, Y., Preter, E., Zadok, A., 2018. Optomechanical time-domain reflectometry. Nat. Commun. 9 (1), 2991.

Bashan, G., London, Y., Diamandi, H.H., Zadok, A., 2020. Distributed cladding mode fiber-optic sensor. Optica 7 (1), 85–92.

Bernini, R., Minardo, A., Zeni, L., 2009. Dynamic strain measurement in optical fibers by stimulated Brillouin scattering. Opt. Lett. 34 (17), 2613–2615.

Beugnot, J.-C., Tur, M., Foaleng Mafang, S., Thévenaz, L., 2011. Distributed Brillouin sensing with sub-meter spatial resolution: modeling and processing. Opt. Express 19, 7381–7397.

Bolognini, G., Hartog, A., 2013. Raman-based fibre sensors: trends and applications. Opt. Fiber Technol. 19 (6), 678–688.

Boyd, R.W., 2020. Nonlinear Optics. Academic press.

Brown, A., DeMerchant, M., Bao, X., Bremner, T., 1998. Advances in distributed sensing using Brillouin scattering. Proc. SPIE 3330, 294–300.

Brown, A.W., Colpitts, B.G., Brown, K., 2005. Distributed sensor based on dark-pulse Brillouin scattering. IEEE Photon. Technol. Lett. 17, 1501–1503.

Chen, T., Wang, Q., Chen, R., Zhang, B., Jewart, C., Chen, K.P., Maklad, M., Swinehart, P.R., 2012. Distributed high-temperature pressure sensing using air-hole microstructural fibers. Opt. Lett. 37, 1064–1066.

Chin, S., Thévenaz, L., 2012. Tunable photonic delay lines in optical fibers. Laser Photonics Rev. 6, 724–738.

Chin, S., Primerov, N., Thevenaz, L., 2012. Sub-centimeter spatial resolution in distributed Fiber sensing based on dynamic Brillouin grating in optical fibers. IEEE Sensors J. 12, 189–194.

Cho, Y.T., Alahbabi, M., Gunning, M.J., Newson, T.P., 2003. 50-km single-ended spontaneous-Brillouin-based distributed-temperature sensor exploiting pulsed Raman amplification. Opt. Lett. 28 (18), 1651–1653.

Chow, D.M., Thévenaz, L., 2018. Forward Brillouin scattering acoustic impedance sensor using thin polyimide-coated fiber. Opt. Lett. 43 (21), 5467–5470.

Chow, D.M., Yang, Z., Soto, M.A., Thévenaz, L., 2018. Distributed forward Brillouin sensor based on local light phase recovery. Nat. Commun. 9 (1), 2990.

Cohen, R., London, Y., Antman, Y., Zadok, A., 2014. Brillouin optical correlation domain analysis with 4 millimeter resolution based on amplified spontaneous emission. Opt. Express 22 (10), 12070–12078.

Dainese, P., St. J. Russell, P., Joly, N., Knight, J.C., Wiederhecker, G.S., Fragnito, H.L., Laude, V., Khelif, A., 2006. Stimulated Brillouin scattering from multi-GHz-guided acoustic phonons in nanostructured photonic crystal fibres. Nat. Phys. 2 (6), 388–392.

Dakin, J., Culshaw, B., 1988. Optical Fiber Sensors: Principles and Components. vol. 1 Artech House.

Dandridge, A., Kersey, A.D., 1989. Overview of Mach-Zehnder sensor technology and applications. In: Fiber Optic and Laser Sensors VI. Proc. SPIE, vol. 985. International Society for Optics and Photonics (SPIE).

De Waele, W., Degrieck, J., Taerwe, W.M.L., Baets, R., 1970. Non-destructive monitoring of composite elements by means of embedded optical fibre Bragg-sensors. In: WIT Transactions on Engineering Sciences. vol. 28. WIT Press.

Denisov, A., Soto, M.A., Thévenaz, L., 2016. Going beyond 1000000 resolved points in a Brillouin distributed fiber sensor: theoretical analysis and experimental demonstration. Light Sci. Appl. 5 (5), e16074.

Diamandi, H.H., London, Y., Bashan, G., Zadok, A., 2019. Distributed opto-mechanical analysis of liquids outside standard fibers coated with polyimide. APL Photonics 4 (1), 016105.

Diamandi, H.H., London, Y., Bashan, G., Shemer, K., Zadok, A., 2021. Forward stimulated Brillouin scattering analysis of optical fibers coatings. J. Lightwave Technol. 39 (6), 1800–1807.

Dominguez-Lopez, A., Lopez-Gil, A., Martin-Lopez, S., Gonzalez-Herraez, M., 2014. Strong cancellation of RIN transfer in a Raman-assisted BOTDA using balanced detection. IEEE Photonics Technol. Lett. 26 (18), 1817–1820.

Dominguez-Lopez, A., Yang, Z., Soto, M.A., Angulo-Vinuesa, X., Martin-Lopez, S., Thévenaz, L., Gonzalez-Herraez, M., 2016. Novel scanning method for distortion-free BOTDA measurements. Opt. Express 24 (10), 10188–10204.

Dominguez-Lopez, A., Soto, M.A., Martin-Lopez, S., Thevenaz, L., Gonzalez-Herraez, M., 2017. Resolving 1 million sensing points in an optimized differential time-domain Brillouin sensor. Opt. Lett. 42, 1903–1906.

Dong, Y., 2021. High-performance distributed Brillouin optical fiber sensing. Photonic Sensors 11 (1), 69–90.

Dong, Y., Chen, L., Bao, X., 2011. Time-division multiplexing-based BOTDA over 100km sensing length. Opt. Lett. 36, 277–279.

Dong, Y., Chen, L., Bao, X., 2012. Extending the sensing range of Brillouin optical time-domain analysis combining frequency-division multiplexing and in-line EDFAs. J. Lightwave Technol. 30, 1161–1167.

Dong, Y., Wang, B., Pang, C., Zhou, D., Ba, D., Zhang, H., Bao, X., 2018. 150 km fast BOTDA based on the optical chirp chain probe wave and Brillouin loss scheme. Opt. Lett. 43 (19), 4679–4682.

Elliott, S.R., 1983. Physics of Amorphous Materials. Longman Inc., New York.
Elooz, D., Antman, Y., Levanon, N., Zadok, A., 2014. High-resolution long-reach distributed Brillouin sensing based on combined time-domain and correlation-domain analysis. Opt. Express 22 (6), 6453–6463.
Fellay, A., 2003. Extreme Temperature Sensing Using Brillouin Scattering in Optical Fibers. EPFL thesis, vol. 2728, https://doi.org/10.5075/epfl-thesis-2728.
Foaleng Mafang, S., Tur, M., Beugnot, J.C., Thevenaz, L., 2010. High spatial and spectral resolution long-range sensing using Brillouin echoes. J. Lightwave Technol. 28, 2993–3003.
Foaleng Mafang, S., Rodríguez-Barrios, F., Martin-Lopez, S., González-Herráez, M., Thévenaz, L., 2011. Detrimental effect of self-phase modulation on the performance of Brillouin distributed fiber sensors. Opt. Lett. 36, 97–99.
Fu, Y., Rao, Y., Zhu, R., Wang, Z., Han, B., Wu, H., Lu, C., 2018. 175km repeaterless BOTDA with hybrid 2nd-and 3rd-order Raman random fiber laser amplification. In: Optical Fiber Sensors. Proc. Optical Society of America, Optical Society of America (OSA) (p. TuD2).
Fukuhara, M., Sanpei, A., Shibuki, K., 1997. Low temperature-elastic moduli, Debye temperature and internal Dilational and shear frictions of fused quartz. J. Mater. Sci. 32, 1207.
Gao, X., Yang, Z., Wang, S., Hong, X., Sun, X., Soto, M.A., Wu, J., Thévenaz, L., 2021. Impact of optical noises on unipolar-coded Brillouin optical time-domain analyzers. Opt. Express 29 (14), 22146–22158.
Gu, H., Dong, H., Zhang, G., He, J., Xu, N., Brown, D.J., 2012. Pressure dependence of Brillouin frequency shift in bare silica optical fibers. Chin. Opt. Lett. 10 (10), 100604.
Gyger, F., Rochat, E., Chin, S., Niklès, M., Thévenaz, L., 2014. Extending the sensing range of Brillouin optical time-domain analysis up to 325 km combining four optical repeaters. In: 23rd International Conference on Optical Fibre Sensors, Proc. SPIE 9157, 91576Q.
Haneef, S.M., Yang, Z., Thévenaz, L., Venkitesh, D., Srinivasan, B., 2018. Performance analysis of frequency shift estimation techniques in Brillouin distributed fiber sensors. Opt. Express 26 (11), 14661–14677.
Hartog, A.H., 2017. An Introduction to Distributed Optical Fibre Sensors. CRC Press.
Hayashi, N., Mizuno, Y., Nakamura, K., 2012. Brillouin gain spectrum dependence on large strain in perfluorinated graded-index polymer optical fiber. Opt. Express 20 (19), 21101–21106.
Hayashi, N., Mizuno, Y., Nakamura, K., Set, S.Y., Yamashita, S., 2017. Experimental study on depolarized GAWBS spectrum for optomechanical sensing of liquids outside standard fibers. Opt. Express 25 (3), 2239–2244.
Horiguchi, T., Tateda, M., 1989. Optical-fiber-attenuation investigation using stimulated Brillouin scattering between a pulse and a continuous wave. Opt. Lett. 14 (8), 408–410.
Horiguchi, T., Kurashima, T., Tateda, M., 1990. A technique to measure distributed strain in optical fibers. IEEE Photon. Technol. Lett. 2 (5), 352–354.
Hotate, K., Hasegawa, T., 2000. Measurement of Brillouin gain spectrum distribution along an optical fiber using a correlation-based technique—proposal, experiment and simulation. IEICE Trans. Electron. 83 (3), 405–412.
Kashyap, R., 2009. Fiber Bragg Gratings. Academic Press.
Keaton, G.L., Leonardo, M.J., Byer, M.W., Richard, D.J., 2014. Stimulated Brillouin scattering of pulses in optical fibers. Opt. Express 22, 13351–13365.
Kim, Y.H., Song, K.Y., 2014. Mapping of intermodal beat length distribution in an elliptical-core two-mode fiber based on Brillouin dynamic grating. Opt. Express 22 (14), 17292–17302.
Kim, Y.H., Lee, K., Song, K.Y., 2015. Brillouin optical correlation domain analysis with more than 1 million effective sensing points based on differential measurement. Opt. Express 23 (26), 33241–33248.

Kim, Y.H., Kwon, H., Kim, J., Song, K.Y., 2016. Distributed measurement of hydrostatic pressure based on Brillouin dynamic grating in polarization maintaining fibers. Opt. Express 24 (19), 21399–21406.

Kishida, K., Li, C., Nishiguchi, K., 2005. Pulse pre-pump method for cm-order spatial resolution of BOTDA. Proc. SPIE 5855, 559–562.

Koyamada, Y., Sakairi, Y., Takeuchi, N., Adachi, S., 2007. Novel technique to improve spatial resolution in Brillouin optical time-domain reflectometry. IEEE Photon. Technol. Lett. 19 (23), 1910–1912.

Kurashima, T., Horiguchi, T., Tateda, M., 1990. Distributed-temperature sensing using stimulated Brillouin scattering in optical silica fibers. Opt. Lett. 15 (18), 1038–1040.

Kurashima, T., Horiguchi, T., Izumita, H., Furukawa, S.I., Koyamada, Y., 1993. Brillouin optical-fiber time domain reflectometry. IEICE Trans. Commun. 76 (4), 382–390.

Lalam, N., Ng, W.P., Dai, X., Wu, Q., Fu, Y.Q., 2018. Performance improvement of Brillouin ring laser based BOTDR system employing a wavelength diversity technique. J. Lightwave Technol. 36 (4), 1084–1090.

Le Floch, S., Cambon, P., 2003. Study of Brillouin gain spectrum in standard single-mode optical fiber at low temperatures (1.4-370 K) and high hydrostatic pressures (1-250 bars). Opt. Commun. 219, 395–410.

Le Floch, S., Sauser, F., Llera, M., Rochat, E., 2015. Novel Brillouin optical time-domain analyzer for extreme sensing range using high-power flat frequency-coded pump pulses. J. Lightwave Technol. 33 (12), 2623–2627.

Lecœuche, V., Webb, D.J., Pannell, C.N., Jackson, D.A., 2000. Transient response in high-resolution Brillouin-based distributed sensing using probe pulses shorter than the acoustic relaxation time. Opt. Lett. 25, 156–158.

Lee, C.C., Chiang, P.W., Chi, S., 2001. Utilization of a dispersion-shifted fiber for simultaneous measurement of distributed strain and temperature through Brillouin frequency shift. IEEE Photon. Technol, Lett. 13 (10), 1094–1096.

Li, W., Bao, X., Li, Y., Chen, L., 2008. Differential pulse-width pair BOTDA for high spatial resolution sensing. Opt. Express 16, 21616–21625.

Li, A., Wang, Y., Fang, J., Li, M.J., Kim, B.Y., Shieh, W., 2015. Few-mode fiber multi-parameter sensor with distributed temperature and strain discrimination. Opt. Lett. 40 (7), 1488–1491.

Liu, X., Bao, X., 2012. Brillouin Spectrum in LEAF and Simultaneous temperature and strain measurement. J. Lightwave Technol. 30, 1053–1059.

London, Y., Antman, Y., Cohen, R., Kimelfeld, N., Zadok, A., 2014. High-resolution long-range distributed Brillouin analysis using dual-layer phase and amplitude coding. Opt. Express 22 (22), 27144–27158.

Lu, Y., Qin, Z., Lu, P., Zhou, D., Chen, L., Bao, X., 2013. Distributed strain and temperature measurement by Brillouin beat Spectrum. IEEE Photon. Technol. Lett. 25 (11), 1050–1053.

Lu, X., Soto, M.A., Thévenaz, L., 2017. Temperature-strain discrimination in distributed optical fiber sensing using phase-sensitive optical time-domain reflectometry. Opt. Express 25 (14), 16059–16071.

Martin-Lopez, S., Alcon-Camas, M., Rodriguez, F., Corredera, P., Gonzalez-Herraez, M., 2010. Brillouin optical time-domain analysis assisted by second-order Raman amplification. Opt. Express 18 (18), 18769–18778.

Mizuno, Y., Zou, W., He, Z., Hotate, K., 2008. Proposal of Brillouin optical correlation-domain reflectometry (BOCDR). Opt. Express 16 (16), 12148–12153.

Mizuno, Y., Hayashi, N., Tanaka, H., Wada, Y., Nakamura, K., 2015. Brillouin scattering in multi-core optical fibers for sensing applications. Sci. Rep. 5 (1), 1–9.

Mizuno, Y., Hayashi, N., Fukuda, H., Song, K.Y., Nakamura, K., 2016. Ultrahigh-speed distributed Brillouin reflectometry. Light Sci. Appl. 5 (12), e16184.

Mizuno, Y., Lee, H., Hayashi, N., Nakamura, K., 2017. Hydrostatic pressure dependence of Brillouin frequency shift in polymer optical fibers. Appl. Phys. Exp. 11 (1), 012502.

Mompó, J.J., Urricelqui, J., Loayssa, A., 2016. Brillouin optical time-domain analysis sensor with pump pulse amplification. Opt. Express 24, 12672–12681.

Motil, A., Bergman, A., Tur, M., 2016. State of the art of Brillouin fiber-optic distributed sensing. Optics Laser Technol. 78, 81–103.

Niklès, M., Thévenaz, L., Robert, P.A., 1996. Simple distributed fiber sensor based on Brillouin gain spectrum analysis. Opt. Lett. 21, 758–760.

Nikles, M., Thevenaz, L., Robert, P.A., 1997. Brillouin gain spectrum characterization in single-mode optical fibers. J. Lightwave Technol. 15 (10), 1842–1851.

Nye, J.F., 2009. Physical Properties of Crystals: Their Representation by Tensors and Matrices. Clarendon Press, Oxford.

Pang, C., Hua, Z., Zhou, D., Zhang, H., Chen, L., Bao, X., Dong, Y., 2020. Opto-mechanical time-domain analysis based on coherent forward stimulated Brillouin scattering probing. Optica 7 (2), 176–184.

Pant, R., Poulton, C.G., Choi, D.Y., Mcfarlane, H., Hile, S., Li, E., Thevenaz, L., Luther-Davies, B., Madden, S.J., Eggleton, B.J., 2011. On-chip stimulated Brillouin scattering. Opt. Express 19, 8285–8290.

Peled, Y., Motil, A., Yaron, L., Tur, M., 2011. Slope-assisted fast distributed sensing in optical fibers with arbitrary Brillouin profile. Opt. Express 19 (21), 19845–19854.

Peled, Y., Motil, A., Tur, M., 2012. Fast Brillouin optical time domain analysis for dynamic sensing. Opt. Express 20 (8), 8584–8591.

Pelous, J., Vacher, R., 1976a. Thermal Brillouin crystalline and fused quartz from 20 to 1000 degrees C. Solid State Commun. 18 (5), 657–661.

Pelous, J., Vacher, R., 1976b. Behavior of thermal phonons in amorphous media from 4 to 300K. Phys. Rev. B 14 (2), 823–828.

Pradhan, P., Sengupta, D., Wang, L., Tremblay, C., LaRochelle, S., Ung, B., 2017. The Brillouin gain of vector modes in a few-mode fiber. Sci. Rep. 7 (1), 1–7.

Ravet, F., Bao, X., Zou, L., Yu, Q., Li, Y., Kalosha, V., Chen, L., 2006. Accurate strain detection and localisation with the distributed Brillouin sensor based on phenomenological signal processing approach. In: Nondestructive Evaluation and Health Monitoring of Aerospace Materials, Composites, and Civil Infrastructure V. Proc. SPIE 6176, 61761C, International Society for Optics and Photonics (SPIE).

Rodriguez-Barrios, F., Martin-Lopez, S., Carrasco-Sanz, A., Corredera, P., Ania-Castanon, J.D., Thévenaz, L., Gonzalez-Herraez, M., 2010. Distributed Brillouin Fiber sensor assisted by first-order Raman amplification. J. Lightwave Technol. 28 (15), 2162–2172.

Sánchez, L.A., Díez, A., Cruz, J.L., Andrés, M.V., 2022. High accuracy measurement of Poisson's ratio of optical fibers and its temperature dependence using forward-stimulated Brillouin scattering. Opt. Express 30, 42.

Santagiustina, M., Chin, S., Primerov, N., Ursini, L., Thévenaz, L., 2013. All-optical signal processing using dynamic Brillouin gratings. Sci. Rep. 3 (1), 1–6.

Saxena, B., Baker, C., Bao, X., Chen, L., 2019. High birefringent Brillouin frequency shifts in a single-mode As2Se3-PMMA microtaper induced by a transverse load. Opt. Lett. 44, 4789–4792.

Shelby, R.M., Levenson, M.D., Bayer, P.W., 1985. Guided acoustic-wave Brillouin scattering. Phys. Rev. B 31 (8), 5244.

Shimizu, R., Kishi, M., Hotate, K., 2017. Enlargement of measurement function in Brillouin optical correlation domain reflectometry with combining four performance improvement schemes. In: 25th International Conference on Optical Fiber Sensors, Proc. SPIE 10323, 103239O.

Song, K.Y., Yoon, H.J., 2010. Observation of narrowband intrinsic spectra of Brillouin dynamic gratings. Opt. Lett. 35 (17), 2958–2960.

Song, K.Y., He, Z., Hotate, K., 2006. Distributed strain measurement with millimeter-order spatial resolution based on Brillouin optical correlation domain analysis. Opt. Lett. 31 (17), 2526–2528.
Song, K.Y., Zou, W., He, Z., Hotate, K., 2008. All-optical dynamic grating generation based on Brillouin scattering in polarization-maintaining fiber. Opt. Lett. 33 (9), 926–928.
Song, K.Y., Zou, W., He, Z., Hotate, K., 2009. Optical time-domain measurement of Brillouin dynamic grating spectrum in a polarization-maintaining fiber. Opt. Lett. 34 (9), 1381–1383.
Song, K.Y., Chin, S., Primerov, N., Thévenaz, L., 2010. Time-domain distributed fiber sensor with 1 cm spatial resolution based on Brillouin dynamic grating. J. Lightwave Technol. 28 (14), 2062–2067.
Song, K.Y., Kazuo, H., Weiwen, Z., He, Z., 2017. Applications of Brillouin dynamic grating to distributed Fiber sensors. J. Lightwave Technol. 35 (16), 3268–3280.
Soto, M.A., Thévenaz, L., 2013. Modeling and evaluating the performance of Brillouin distributed optical fiber sensors. Opt. Express 21 (25), 31347–31366.
Soto, M.A., Sahu, P.K., Bolognini, G., Di Pasquale, F., 2008. Brillouin-based distributed temperature sensor employing pulse coding. IEEE Sensors J. 8 (3), 225–226.
Soto, M.A., Bolognini, G., Di Pasquale, F., Thévenaz, L., 2010. Simplex-coded BOTDA fiber sensor with 1 m spatial resolution over a 50 km range. Opt. Lett. 35 (2), 259–261.
Soto, M.A., Bolognini, G., Pasquale, F.D., 2011. Long-range simplex-coded BOTDA sensor over 120km distance employing optical pre amplification. Opt. Lett. 36, 232–234.
Soto, M.A., Lu, X., Martins, H.F., Gonzalez-Herraez, M., Thévenaz, L., 2015. Distributed phase birefringence measurements based on polarization correlation in phase-sensitive optical time-domain reflectometers. Opt. Express 23 (19), 24923–24936.
Sun, X., Yang, Z., Hong, X., Zaslawski, S., Wang, S., Soto, M.A., Gao, X., Wu, J., Thévenaz, L., 2020. Genetic-optimised aperiodic code for distributed optical fibre sensors. Nat. Commun. 11 (1), 5774.
Taki, M., Muanenda, Y., Oton, C.J., Nannipieri, T., Signorini, A., Di Pasquale, F., 2013. Cyclic pulse coding for fast BOTDA fiber sensors. Opt. Lett. 38 (15), 2877–2880.
Teng, L., Zhang, H., Dong, Y., Zhou, D., Jiang, T., Gao, W., Lu, Z., Chen, L., Bao, X., 2016. Temperature-compensated distributed hydrostatic pressure sensor with a thin-diameter polarization-maintaining photonic crystal fiber based on Brillouin dynamic gratings. Opt. Lett. 41, 4413–4416.
Thévenaz, L., 2010. Brillouin distributed time-domain sensing in optical fibers: state of the art and perspectives. Front. Optoelectron. China 3, 13–21.
Thévenaz, L., Fellay, A., Scandale, W., 2003. Brillouin gain spectrum characterization in optical fibres from 1 to 1000 K. In: 16th International Conference on Optical Fiber Sensors, Nara Japan, Technical Digest, Paper Tu2–2, pp. 38–41.
Thévenaz, L., Foaleng Mafang, S., Lin, J., 2013. Effect of pulse depletion in a Brillouin optical time-domain analysis system. Opt. Express 21, 14017–14035.
Timoshenko, S.P., Goodier, J.N., 1970. Theory of Elasticity (Chapter 14). MacGraw-Hill, New York.
Toshio, K., Tsuneo, H., Hisashi, I., Furukawa, S.I., Yahei, K., 1993. Brillouin optical-fiber time domain reflectometry. IEICE Trans. Commun. E76-B(4), 382–390.
Urricelqui, J., Zornoza, A., Sagues, M., Loayssa, A., 2012. Dynamic BOTDA measurements based on Brillouin phase-shift and RF demodulation. Opt. Express 20 (24), 26942–26949.
Urricelqui, J., Sagues, M., Loayssa, A., 2013. BOTDA measurements tolerant to non-local effects by using a phase-modulated probe wave and RF demodulation. Opt. Express 21 (14), 17186–17194.

Voskoboinik, A., Wang, J., Shamee, B., Nuccio, S.R., Zhang, L., Chitgarha, M., Willner, A.E., Tur, M., 2011. SBS-based Fiber optical sensing using frequency-domain simultaneous tone interrogation. J. Light. Technol. 29 (11), 1729–1735.
Wang, Y., Gong, J., Wang, D.Y., Dong, B., Bi, W., Wang, A., 2010. A quasi-distributed sensing network with time-division-multiplexed fiber Bragg gratings. IEEE Photonics Technol. Lett. 23 (2), 70–72.
Wang, C., Li, Z., Gui, X., Fu, X., Wang, F., Wang, H., Wang, J., Bao, X., 2019. Micro-cavity array with high accuracy for fully distributed optical fiber sensing. J. Lightwave Technol. 37 (3), 927–932.
Wang, S., Yang, Z., Zaslawski, S., Thévenaz, L., 2020a. Short spatial resolution retrieval from a long pulse Brillouin optical time-domain analysis trace. Opt. Lett. 45 (15), 4152–4155.
Wang, H., Gao, S., Baker, C., Wang, Y., Chen, L., Bao, X., 2020b. Stimulated Brillouin scattering in a tapered dual-core As_2Se_3-PMMA fiber for simultaneous temperature and strain sensing. Opt. Lett. 45 (12), 3301–3304.
Wu, H., Tang, M., Wang, M., Zhao, C., Zhao, Z., Wang, R., Liao, R., Fu, S., Yang, C., Tong, W., Shum, P.P., Liu, D., 2017. Few-mode optical fiber based simultaneously distributed curvature and temperature sensing. Opt. Express 25 (11), 12722–12732.
Xu, Y., Ren, M., Lu, Y., Lu, P., Bao, X., Wang, L., Messaddeq, Y., LaRochelle, S., 2016. Multi-parameter sensor based on stimulated Brillouin scattering in inverse-parabolic graded-index fiber. Opt. Lett. 41 (6), 1138–1141.
Yang, Z., Soto, M.A., Thévenaz, L., 2016a. Increasing robustness of bipolar pulse coding in Brillouin distributed fiber sensors. Opt. Express 24 (1), 586–597.
Yang, Z., Hong, X., Lin, W., Wu, J., 2016b. Evaluating and overcoming the impact of second echo in Brillouin echoes distributed sensing. Opt. Express 24 (2), 1543–1558.
Yang, G., Fan, X., He, Z., 2017. Strain dynamic range enlargement of slope-assisted BOTDA by using Brillouin phase-gain ratio. J. Lightwave Technol. 35 (20), 4451–4458.
Yang, Z., Li, Z., Zaslawski, S., Thévenaz, L., Soto, M.A., 2018a. Design rules for optimizing unipolar coded Brillouin optical time-domain analyzers. Opt. Express 26 (13), 16505–16523.
Yang, Z., Soto, M.A., Chow, D.M., Ray, P., Thévenaz, L., 2018b. Brillouin distributed optical Fiber sensor based on a closed-loop configuration. J. Lightwave Technol. 36 (5), 1239–1248.
Yang, F., Gyger, F., Thévenaz, L., 2020. Intense Brillouin amplification in gas using hollow-core waveguides. Nat. Photonics 14, 700–708.
Zadok, A., Zilka, E., Eyal, A., Thévenaz, L., Tur, M., 2008. Vector analysis of stimulated Brillouin scattering amplification in standard single-mode fibers. Opt. Express 16 (26), 21692–21707.
Zadok, A., Antman, Y., Primerov, N., Denisov, A., Sancho, J., Thevenaz, L., 2012. Random-access distributed fiber sensing. Laser Photonics Rev. 6 (5), L1–L5.
Zaghloul, M.A., Wang, M., Milione, G., Li, M.J., Li, S., Huang, Y.K., Wang, T., Chen, K.P., 2018. Discrimination of temperature and strain in Brillouin optical time domain analysis using a multicore optical fiber. Sensors 18 (4), 1176.
Zarifi, A., Stiller, B., Merklein, M., Li, N., Vu, K., Choi, D.Y., Ma, P., Madden, S.J., Eggleton, B.J., 2018. Highly localized distributed Brillouin scattering response in a photonic integrated circuit. Appl. Phys. Lett. Photon. 3, 036101.
Zarifi, A., Stiller, B., Merklein, M., Liu, Y., Morrison, B., Casas-Bedoya, A., Ren, G., Nguyen, T.G., Vu, K., Choi, D.-Y., Mitchell, A., Madden, S.J., Eggleton, B.J., 2019. On-chip correlation-based Brillouin sensing: design, experiment, and simulation. J. Opt. Soc. Am. B 36 (1), 146–152.
Zaslawski, S., Yang, Z., Thévenaz, L., 2020. On the 2D post-processing of Brillouin optical time-domain analysis. J. Lightwave Technol. 38 (14), 3723–3736.

Zaslawski, S., Yang, Z., Thévenaz, L., 2021. Distributed optomechanical fiber sensing based on serrodyne analysis. Optica 8 (3), 388–395.
Zhang, C., Li, W., Bao, X., Chen, L., Du, M., 2007. Tensile strain dependence of the Brillouin gain spectrum in carbon/polyimide coated fibers. Opt. Lett. 32 (17), 2565–2567.
Zhang, G., Gu, H., Dong, H., Li, L., He, J., 2013. Pressure sensitization of Brillouin frequency shift in optical fibers with double-layer polymer coatings. IEEE Sensors J. 13 (6), 2437–2441.
Zhao, Z., Soto, M.A., Tang, M., Thévenaz, L., 2016. Distributed shape sensing using Brillouin scattering in multi-core fibers. Opt. Express 24 (22), 25211–25223.
Zhao, C., Tang, M., Wang, L., Wu, H., Zhao, Z., Dang, Y., Wu, J., Fu, S., Liu, D., Shum, P., 2017. BOTDA using channel estimation with direct-detection optical OFDM technique. Opt. Express 25 (11), 12698–12709.
Zhou, D., Dong, Y., Wang, B., Jiang, T., Ba, D., Xu, P., Zhang, H., Lu, Z., Li, H., 2017. Slope-assisted BOTDA based on vector SBS and frequency-agile technique for wide-strain-range dynamic measurements. Opt. Express 25 (3), 1889–1902.
Zhou, D., Dong, Y., Wang, B., Pang, C., Ba, D., Zhang, H., Lu, Z., Li, H., Bao, X., 2018. Single-shot BOTDA based on an optical chirp chain probe wave for distributed ultrafast measurement. Light Sci. Appl. 7 (1), 32.
Zou, L., Bao, X., Chen, L., 2003. Brillouin scattering spectrum in photonic crystal fiber with a partially germanium-doped core. Opt. Lett. 28 (21), 2022–2024.
Zou, L., Bao, X., Chen, L., 2004. Dependence of the Brillouin frequency shift on strain and temperature in a photonic crystal fiber. Opt. Lett. 29 (13), 1485–1487.
Zou, W., He, Z., Hotate, K., 2009. Complete discrimination of strain and temperature using Brillouin frequency shift and birefringence in a polarization-maintaining fiber. Opt. Express 17 (3), 1248–1255.
Zou, W., He, Z., Hotate, K., 2013. Range elongation of distributed discrimination of strain and temperature in Brillouin optical correlation-domain analysis based on dual frequency modulations. IEEE Sensors J. 14 (1), 244–248.

CHAPTER EIGHT

Brillouin-based radio frequency sources

Moritz Merklein[a,b,*], Thomas Schneider[c], and Kerry John Vahala[d]
[a]The University of Sydney Nano Institute (Sydney Nano), The University of Sydney, Sydney, NSW, Australia
[b]Institute of Photonics and Optical Science (IPOS), School of Physics, The University of Sydney, Sydney, NSW, Australia
[c]THz Photonics Group, Technische Universität Braunschweig, Braunschweig, Germany
[d]T.J. Watson Laboratory of Applied Physics, California Institute of Technology, Pasadena, CA, United States
*Corresponding author: e-mail address: moritz.merklein@sydney.edu.au

Contents

1. Introduction

High frequency, low-noise Radio Frequency (RF) sources are of great importance for fundamental science as well as many applications ranging from sensing to communications and RADAR to name a few. Due to the effects of ohmic losses and residual capacitances it is challenging to scale electronic sources to higher frequencies without degradation of the signal quality. To address this challenge multiple methods were developed to

Semiconductors and Semimetals, Volume 110
ISSN 0080-8784
https://doi.org/10.1016/bs.semsem.2022.05.002

generate RF signals via optical means that utilize the wide bandwidth of optical systems. On a basic level the optical method relies on beating two optical tones separated by the RF frequency of interest on a photodetector that converts the optical signals in an RF signal.

In general, these methods can be broken down into two steps: the generation of optical signals at multiple frequencies, and where there are more than two optical signals, selecting the two optical signals of interest before the photodetector. Stimulated Brillouin scattering (SBS) can be utilized in both processes. It can be used to generate multi-tone optical signals by creating an optical Stokes wave that is separated by the optical pump by around 10 GHz, which depends on the properties of the waveguide material. The generated Stokes wave can then be used to generate a stable RF beat signal.

Other methods to generate multiple optical waves that are utilized for the generation of RF signals are multiple laser sources, optical frequency combs, or resonant modes in a cavity as for the case of the opto-electronic oscillator (OEO). The latter approaches offer advantages over the multi-laser approach in terms of complexity and stability (the lasers need to be phase locked to avoid relative frequency drift), however, they rely on optical filter techniques to ensure that only the optical signals of interest beat with each other on the photodetector.

Stimulated Brillouin scattering (SBS) provides unique capabilities to selectively amplify or filter optical signals within a narrow linewidth while also offering wide frequency tunability. The narrowband Brillouin amplification can be used to select certain comb teeth in a frequency comb or provide selective gain within a resonator to ensure only one or two modes have enough gain to overcome the roundtrip losses and oscillate. Linewidth narrowing of the Stokes wave above the Brillouin lasing threshold offers further advantages in terms of noise performance of the generated RF signals.

This chapter describes the chronological development of RF sources harnessing photonics, and highlights the role Brillouin scattering played in their evolution, starting with fiber-based approaches and moving toward small footprint chip-based schemes. The chapter introduces key performance metrics, such as phase-noise, to provide the reader with enough context to follow the discussion and comparisons of the different schemes (for a more comprehensive discussion of performance metrics and noise of RF sources the reader is referred to the specialist literature). At the end of the chapter, we provide a perspective on current trends and future directions.

2. Background

Traditionally, microwave signals are generated by electronic means; with an electronic oscillator for instance. Such electronic oscillators, however, have a limited frequency tunability and—especially for high frequencies in the millimeter (10–1 mm; 3–300 GHz) and THz (1–0.1 mm; 0.3–3 THz) range—the phase noise can become quite high. For the generation of millimeter and THz waves by electronics, usually a high-quality but low-frequency signal is generated by an oscillator and then upconverted by a nonlinear element (Kallfass et al., 2011). Thus, the nonlinear element generates n harmonics of the input wave up to the desired millimeter or THz frequency. The phase noise, however, increases quadratically with the number of harmonics n, leading to a severe performance degradation.

In contrast, optical technologies can easily generate very high-quality and high-frequency waves and they can be made widely tunable from the microwave to the THz range, just limited by the bandwidth of the element used for the conversion from the optical into the radio frequency (RF) domain. Please note that from now on we will no longer distinguish between micro-, millimeter and THz waves and call everything radio frequency (RF) waves. Thus, especially for the generation of high frequency, wide tuning range, high resolution, low phase noise and high purity RF waves, required in applications like communications, radar and signal processing optical means are by far superior compared to electrical ones. One possibility to generate high-quality RF waves is stimulated Brillouin scattering (SBS).

2.1 Heterodyning of optical signals

The basic idea behind all methods to generate high-quality RF waves with SBS is the heterodyning, i.e., the mixing, of two optical waves in an element with a second order nonlinearity. A photodiode, like all optical instruments, is just able to measure the intensity resulting in a quadrature of the optical field. Thus, a photodiode can be used as a second order nonlinearity. By a second order nonlinearity, optical rectification, the second harmonic and sum and difference frequencies can be generated. However, because of the limited bandwidth of the photodiode, the output current can just follow the difference frequency between the two optical waves, with a frequency that corresponds to the desired RF. In most cases SBS is either used as a narrowband filter for the selection of one or two single optical frequencies from already existing sources or to directly generate these two frequencies; the electrical field of the single frequencies can be written as:

$$E_1(t) = E_{10}e^{j(\omega_1 t)} \tag{1a}$$

$$E_2(t) = (E_{20} + \Delta E_{20}(t))e^{j([\omega_2+\Delta\omega_2(t)]t+\Delta\varphi(t))}. \tag{1b}$$

For the sake of simplicity, it was assumed that the first field $E_1(t)$ is constant, whereas all time dependent changes of amplitude $\Delta E_{20}(t)$, frequency $\Delta\omega_2(t)$ and phase $\Delta\varphi(t)$ are due to the second field. If two photodiodes (a balanced detector) are used for the conversion, the beating between the two optical fields that results in the electrical output current of the photodiode is:

$$I_{PD} \propto 4E_{10}(E_{20} + \Delta E_{20}(t))^* \cos([\omega_{RF} + \Delta\omega_2(t)]t - \Delta\varphi(t)). \tag{2}$$

With the asterisk (*) representing the conjugate complex and $\omega_{RF} = \omega_1 - \omega_2$ as the desired frequency in the RF domain. Thus, if for instance two free running lasers are used for the generation, the RF wave has a time dependent amplitude, frequency and phase. For a high-quality and stable RF wave, however, all time dependent terms in Eq. (2) should be constant, leading to $I_{PD} = I_0 \cos(\omega_{RF} t)$. By generating multiple frequencies or a frequency comb from a single laser source by modulation or in a cavity such as a mode-locked laser, a Brillouin laser, or an oscillator, it is possible to come quite close to this goal.

3. Key performance metrics of radio frequency sources

In this section we will provide the reader with a short introduction of the key performance metrics of RF sources. As this book is dedicated to Brillouin scattering, we will mainly focus on phase noise and keep the discussion at a high level that the reader who is more familiar with Brillouin physics rather than RF sources can follow the later discussion and is able to contextualize presented results. For a more in-depth discussion the reader is referred to (Rubiola, 2008).

There are many metrics one can consider when looking at an RF source, such as output power, dynamic range, frequency, tuning range and speed, and in particular the stability, i.e., the noise of the output RF signal. Here, one can distinguish between amplitude and phase noise as illustrated in Fig. 1.

We will focus the discussion on phase noise as it correlates to frequency noise, which is critical for many applications of RF sources. There are

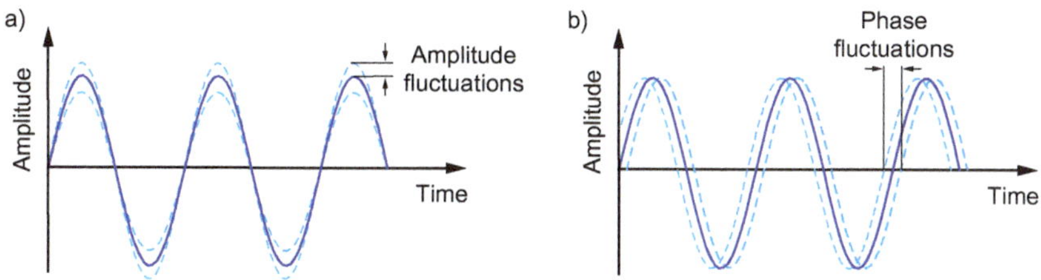

Fig. 1 Schematic illustration of (A) amplitude and (B) phase noise (Merklein, 2018).

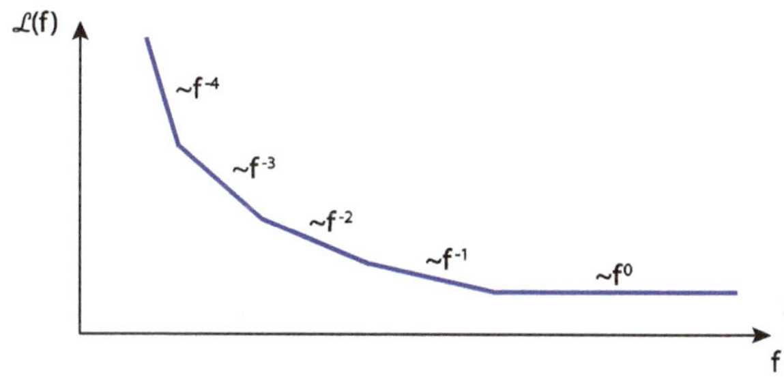

Fig. 2 Typical phase noise power spectral density plot with frequency dependent phase noise processes.

different methods and metrics used to analyze the phase noise of an RF signal. A common metric is the single sideband power spectral density $\mathcal{L}(f) = S\varphi(f)/2$ with $S_\varphi(f)$ being the spectral density of the random phase fluctuations $\varphi(t)$. The units commonly used are dBc/Hz, meaning dB relative to the carrier within a 1 Hz bandwidth.

There are, in fact, multiple sources of phase noise that are dominant in different frequency ranges and scale differently with frequency. The most common phase noise processes are random walk of frequency ($\sim f^{-4}$), flicker frequency noise ($\sim f^{-3}$), white frequency noise ($\sim f^{-2}$), flicker phase noise ($\sim f^{-1}$), and white phase noise ($\sim f^{0}$). Fig. 2 shows an illustration of the different noise components as they would appear in a typical phase noise measurement.

4. Fiber-based SBS radio frequency sources

As described in Section 1, due to the unique properties of SBS, it can be used as a narrowband optical filter and gain to select one or two frequencies, which usually come from a comb source. Historically, and due to the simplicity of the approach, this was first done in optical fibers.

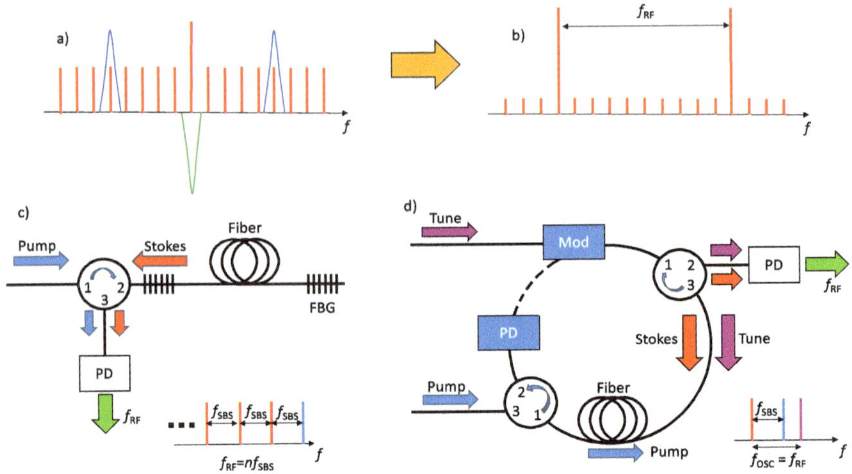

Fig. 3 Basic ideas for the generation of RF frequencies with SBS in fibers. In (A) and (B) the incorporation of SBS gains (blue) and a loss (green) to filter two optical frequencies from a comb (red) is shown. In (C) the generation of RF frequencies with a Brillouin fiber laser and in (D) with an optoelectronic oscillator (OEO) loop is depicted. The solid lines in (D) are optical and the dashed, electrical connections. Mod: modulator, PD: photodiode, FBG: fiber Bragg grating.

Here an overview of these approaches will be given. The aim of this overview is to present the basic ideas behind SBS based RF sources. Thus, it is subjective and not comprehensive. A selection of the basic ideas that have been proposed up to now can be seen in Fig. 3.

4.1 Generation of tunable RF frequencies with SBS based gains and losses

Fig. 1A and B illustrates how Brillouin gains and losses can be incorporated to filter out single frequencies from a comb. This comb can be generated by modulators or a mode-locked laser (MLL), for instance. If a phase or intensity modulator is modulated with a radio frequency Δf with sufficient power, several sidebands with a frequency spacing Δf are generated. However, usually the carrier has a much higher amplitude then the sidebands. To reduce the amplitude of the carrier, a Brillouin loss can be used, as shown with the green line in Fig. 1A. The two Brillouin gains (blue) amplify two of the sidebands, while the other sidebands are attenuated due to the propagation in the fiber, or further reduced by SBS based polarization pulling (Preußler et al., 2013; Zadok et al., 2008). These two sidebands can be heterodyned in a photodiode to generate an RF frequency of

$f_{RF} = (n+1) \times \Delta f$, with n as the number of sidebands between the two selected ones, as shown in Fig. 1B. Thus, a coarse tunability of the RF wave can be achieved by the selection of another sideband and a fine tuning by changing the driving frequency Δf.

Since both frequencies come from the same source, they can show a very high quality and they can be phase-locked to each other. The phase noise, amplitude and frequency stability is defined by the source used to drive the modulator. These sources are usually high-quality electronic oscillators with a very low phase noise and jitter in the femtosecond range even for integrated ones (140×10^{-15} s @ 8 GHz) (Keysight Technologies, 2020).

The main advantage of using this approach is the fast and simple tunability and that multiples of the input frequency can be generated. However, another very important advantage is that the RF frequencies can be brought directly to the place where they are needed. Waveguides for RF frequencies, usually coax cables or hollow waveguides, show a very high attenuation of several dB/m especially for higher frequencies. Optical fibers instead, have an attenuation of just around 0.2 dB/km for optical waves at a wavelength of 1550 nm. Thus, for wireless access points or the cell sites of a cellular system, for instance, the optical sidebands can be delivered over kilometers of fiber, with almost no loss, and just at the access point itself, the waves are transformed to the wireless domain by a photodiode directly connected to an antenna. Additionally, the two sidebands can be separated and one of the two can be modulated, directly with an optical modulator. Such optical modulators can have a much higher bandwidth than electrical ones and higher order modulation formats are quite easily achievable. If the modulated and non-modulated sideband are superimposed in a photodiode, the wireless RF wave modulated with a high bandwidth information can be generated.

In one of the first experiments (Yao, 1998a) SBS was used to amplify the first order sideband generated by an intensity modulator. This sideband was heterodyned with the carrier to achieve the desired RF wave, which of course, was the same as that used for driving the modulator. To achieve multiples of the input frequency, this idea was extended in Park et al. (2007) to amplify the third order sideband and thus, achieve a frequency tripling. A further extension was shown in Schneider et al. (2004, 2006) and Yao (1998b) where two higher order sidebands were amplified by SBS and heterodyned. In Yao (1998b) the frequency comb was generated by a phase modulator and only the even order sidebands were amplified. In Schneider et al. (2004, 2006), an amplitude modulator, driven with a high-power RF wave was used, a phase noise of −60 dBc/Hz@100 kHz has been reported for the generation of a 40 GHz signal.

Brillouin losses were incorporated in Chan and Minasian (2013, 2014), Chen and Chan (2017), and Zheng et al. (2014). In Chan and Minasian (2013) a microwave photonic mixer from 0.2 to 20 GHz based on two consecutive phase modulators was proposed. The SBS loss is used to suppress the carrier after the modulation. In Chan and Minasian (2014) SBS gain and loss were incorporated together for an RF frequency shifter. First the signal was modulated with a double-sideband suppressed carrier modulation. One of the two sidebands was suppressed by the loss, while the other was amplified by the gain. The achieved frequency shift was 20 GHz and the signal to noise ratio 25 dB. Gains and losses were as well used for a frequency shifter in Chen and Chan (2017) to amplify the third order sideband while suppressing the others. Thus, the frequency shift was three times the input electrical driving signal.

Another source for the generation of frequency combs is a mode locked laser (MLL). An MLL can produce frequency combs with such an extremely high precision, that it has revolutionized the metrology of fundamental physical constants (Ye and Cundiff, 2005). A typical fiber MLL producing optical pulses with a duration of 100 fs has a repetition rate of around 100 MHz, defined by the cavity length. Thus, the spectrum consists of 100,000 lines spanning over a bandwidth of around 10 THz with a frequency separation of 100 MHz between the lines (see Fig. 4A). The heterodyning of two of these lines can produce RF waves with a possible tuning range of up to 10 THz, a coarse tuning is possible by the line separation and a fine tuning by a change of the repetition rate. However, typical optical filters have minimum bandwidths of around 10 GHz, leading to the separation of 100 lines at once. Stimulated Brillouin scattering, instead, has a gain bandwidth of around 30 MHz for pump wavelength of 1550 nm in standard single mode fibers, which can be further reduced down to 3 MHz by several methods (Preussler et al., 2011; Preußler and Schneider, 2012; Wiatrek et al., 2012). Thus, in Preußler et al. (2013) SBS together with polarization pulling was used for the separation of two lines from the frequency comb of a fiber femtosecond laser to produce RF waves with frequencies up to 1 THz (see Fig. 4B). The linewidth measured at 24 GHz was <1 Hz and the phase noise −107 dBc/Hz at an offset of 10 kHz from the carrier. For the 110 GHz wave (see Fig. 4C) the measured phase noise was 3 dB higher, potentially due to the RF mixer that was required for the measurement.

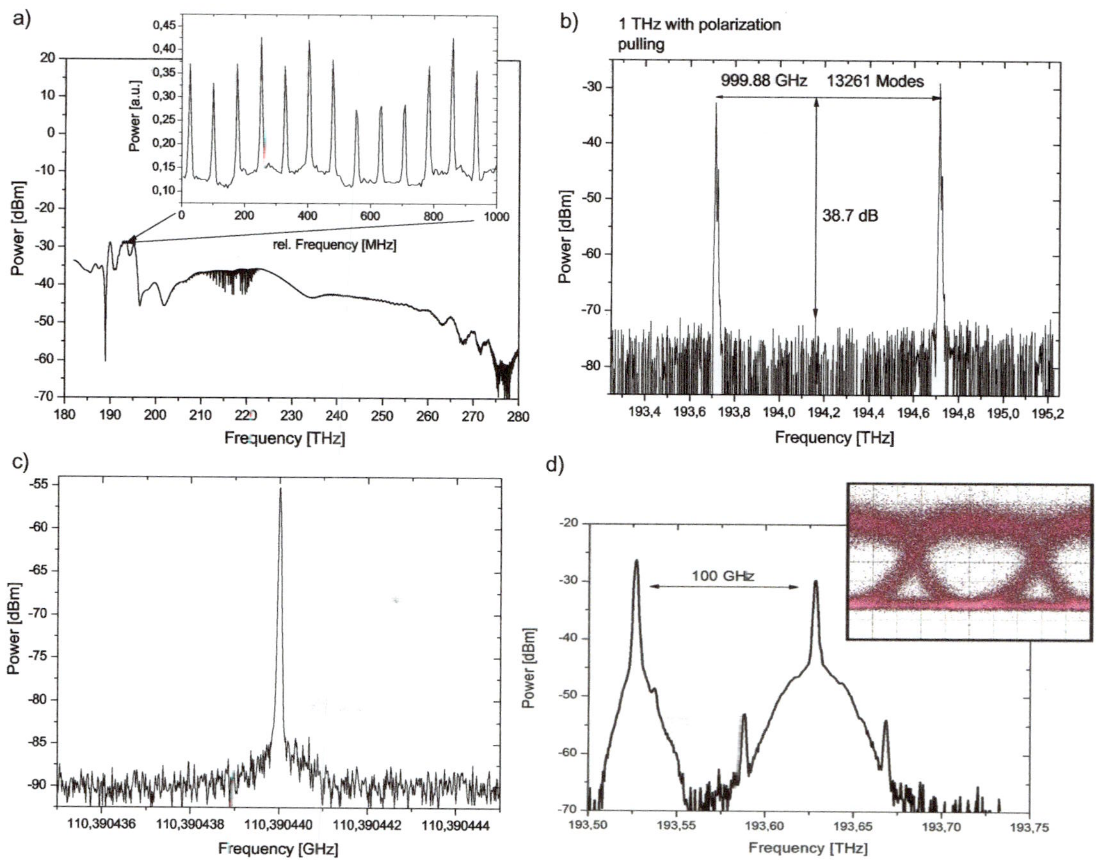

Fig. 4 See figure legend on next page.

Polarization pulling is a very interesting effect of SBS, which can be used for the suppression of unwanted spectral components in SBS based RF sources, spectrometers or filters (Schneider, 2005; Stern et al., 2014; Zadok et al., 2008). Basically, the spectral component which falls into the gain bandwidth of SBS will be amplified and—at the same time—the polarization of this amplified part will be pulled toward the polarization of the pump wave. If the comb generated by the MLL has a perpendicular polarization to the pump wave, and if in front of the photodiode a polarization filter with a polarization parallel to the pump is placed, just the two amplified sidebands can pass this filter, leading to a very high suppression (see Fig. 4B).

4.2 The generation of tunable RF frequencies with Brillouin fiber lasers

A Brillouin fiber laser (BFL) is shown in Fig. 1C. Since Brillouin scattering is a gain effect, a laser can simply be built by adding a feedback to the gain medium. Such a feedback is conventionally built with a cavity. In Fig. 1C this is achieved by two Bragg gratings, which can be directly written into the fiber used for the Brillouin interaction (Shen et al., 2005). The special advantage of BFL is their extremely narrow linewidth, which can be in the range of a few Hz or even lower.

Usually, for the generation of RF waves with a BFL, the first or higher order Stokes waves are heterodyned (mixed) with the pump wave in a photodiode. The higher order Stokes waves are produced by generating the first order Stokes wave from noise. Then this Stokes wave is used as a pump for the generation of the second order Stokes and so on. Thus, several Stokes lines can propagate in the BFL. However, only multiples of the Brillouin

Fig. 4 RF wave generation by the amplification of two frequencies from a frequency comb produced by a fiber fs laser (Preußler et al., 2013). The frequency comb generated by the MLL and measured with a conventional optical spectrum analyzer (OSA) can be seen in (A). Since the resolution of the OSA is a few GHz, the distinct lines of the comb cannot be seen. The inset shows a 1 GHz broad part of the spectrum measured with a high resolution OSA based on SBS (Preussler et al., 2012). In (B) two of these frequencies are amplified by SBS while the others are suppressed by polarization pulling. The frequency distance between the two lines is 1 THz. Two frequency lines with a separation of around 110 GHz were heterodyned in a photodiode to produce the RF wave in (C). In (D) one of the two frequency lines was optically modulated with a 40 Gbit/s pseudo random bit sequence. The inset shows the eye diagram of the modulation. *Reprinted with permission from Preußler, S., Wenzel, N., Braun, R.-P., Owschimikow, N., Vogel, C., Deninger, A., Zadok, A., Woggon, U., Schneider, T., 2013. Generation of ultra-narrow, stable and tunable millimeter- and terahertz-waves with very low phase noise. Opt. Express 21, 23950–23962. doi:10.1364/OE.21.023950 © The Optical Society.*

frequency shift can be generated by a BFL. A limited tunability is possible by a change of the temperature or the pump wavelength. In Shen et al. (2005) the BFL had a similar setup as depicted in Fig. 1C and a tuning range of 100 MHz for a generated 33 GHz signal has been achieved by temperature tuning, for instance. The 33 GHz were generated by heterodyning the 3rd order Stokes with the pump wave.

In Shee et al. (2010) a loop was used for the setup of a BFL. Here a second order Stokes wave was heterodyned with the pump, achieving an RF wave of 21 GHz and a tuning was achieved by a change of the pump wavelength. The reported phase noise was −58.67 dBc/Hz.

Stokes signals up to the 10th order were used in Liu et al. (2013) in a so-called multi-wavelength BFL. The Brillouin frequency shift between pump and Stokes was around 11 GHz, enabling the generation of RF waves in 11 GHz steps.

5. Optoelectronic-oscillators: Fiber- and chip-based approaches

The basic sketch of an optoelectronic oscillator (OEO) is shown in Fig. 1D. The feedback loop is built by an optical (solid line) and electrical (dashed line) connection. The SBS provides the gain for oscillation and at the same time it filters out just a part or even only a single spectral mode in the loop. With feedback, long cavities and due to the filtering of just one single mode, the generated RF waves can show a very high purity and low phase noise. Additionally, due to the SBS process, the tuning range of such an OEO can be quite high and since the SBS gain in the loop gets saturated, the noise can be low as well. Since the RF wave is generated in an oscillator loop, the phase noise of the OEO is defined by the fluctuation of the loop length due to temperature changes or other influences. Therefore, for the generation of very high quality RF signals, temperature stabilization of the OEO is required.

The working principle of such an OEO is depicted in Fig. 1D. The pump wave injected via a circulator into the loop produces a counter propagating Stokes wave in the fiber via SBS. This Stokes wave propagates clockwise in the loop and thus in the same direction as an additional laser wave, called the tuning laser. The pump wave, instead, cannot circulate, since its blocked by the circulators. In the photodiode the Stokes and the tune wave are heterodyned, generating an oscillation frequency $f_{OSC} = f_{tune} - f_{Stokes}$. This oscillation frequency in the RF domain is driving a modulator, closing the loop via an electrical connection (dashed line in Fig. 1D).

Since the frequency shift of the Stokes wave is defined by the SBS process $f_{\text{Stokes}} = f_{\text{pump}} - f_{\text{SBS}}$, the oscillation frequency of the OEO is:

$$f_{OSC} = f_{RF} = f_{tune} - \left(f_{pump} - f_{SBS}\right) \tag{3}$$

Thus, the oscillator can be easily and widely tuned by a change of the frequency of the tuning laser or the pump laser. If tuning and pump laser are completely independent, however, the frequency stability of the OEO is rather poor.

Such an OEO was first proposed in 1997 (Yao, 1997). In the experiment a phase modulator was used to eliminate the bias drift of an amplitude modulator. Additionally, the tuning laser was not incorporated. Thus, the oscillation frequency was fixed to the SBS frequency shift of 12.8 GHz, for the arrangement used in the experiment. Since the bandwidth of the SBS gain is broader than the frequency distance between different modes in the fiber loop, mode hopping occurred. The bandwidth of the SBS gain is usually around 30 MHz, while the mode spacing in a loop with the length l is c/nl, with c as the speed of light in vacuum and n as the refractive index. If a fiber is used as the SBS medium, usually kilometers are required and thus, lots of modes oscillate in the SBS gain bandwidth. This problem was solved in Shi et al. (2018) and Peng et al. (2015) by using a dual loop setup. The length of the two loops was made different and so the mode intervals are different as well. Within the Brillouin gain bandwidth for the two loops and the two modes, just one mode can be made coincident with the Brillouin shift. Therefore, just this one will superimpose and oscillate (Shi et al., 2018).

In Shi et al. (2018) a phase modulator was also used. However, this phase modulator was driven with a high RF power so that many sidebands have been produced. These different sidebands can be selected as pump waves for the SBS process in the loop. Thus, a coarse tuning is possible by selecting different sidebands and a fine tuning by changing the driving frequency for the phase modulator. A tuning range of 40 GHz with a resolution of 10 MHz has been achieved. The reported phase noise was independent of the selected sideband and −120 dBc/Hz@100 kHz offset. In Peng et al. (2015) a tuning range from dc to 60 GHz with a phase noise of −100 dBc/Hz@10 kHz offset was reported.

Brillouin-based OEO has also been shown in chip-based architecture (Merklein et al., 2016). A schematic of the hybrid optic and electronic cavity with the chip in the loop is shown in Fig. 5A. In this demonstration,

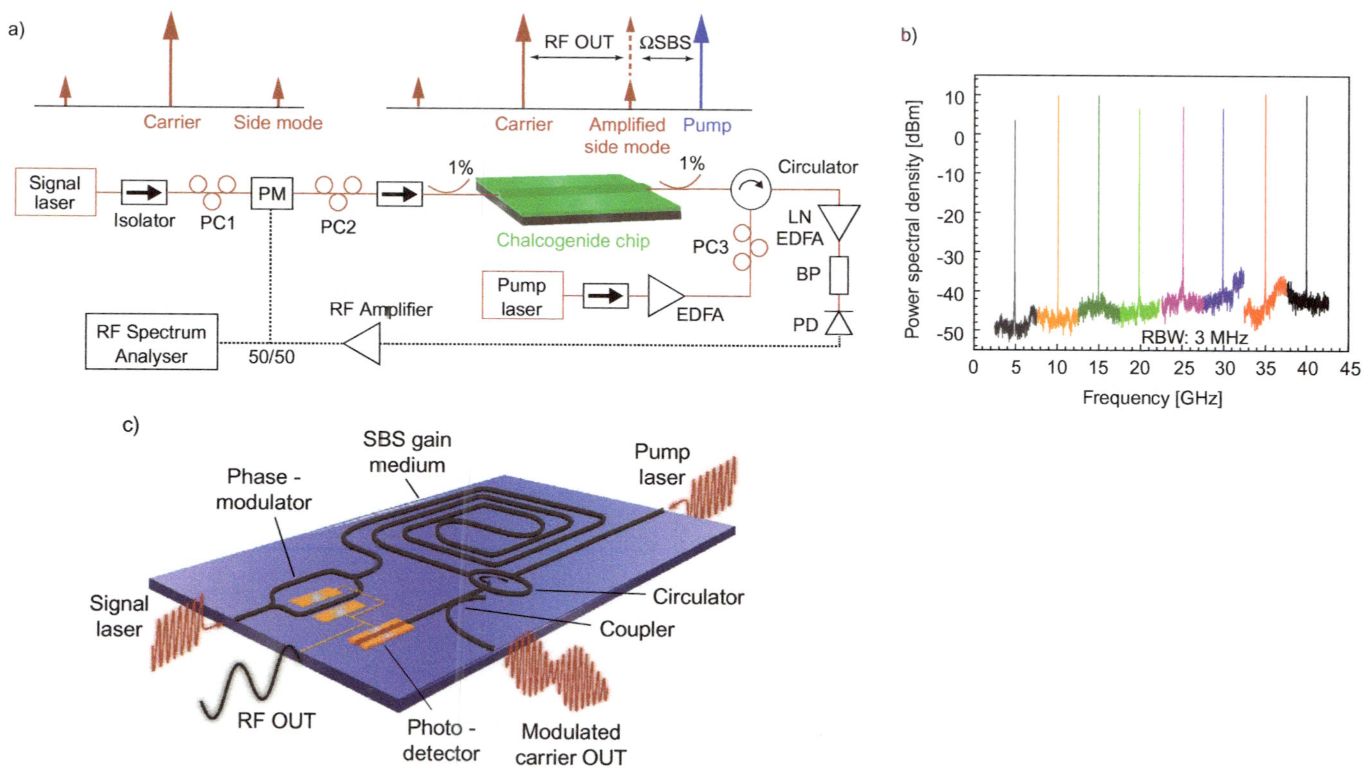

Fig. 5 (A) Experimental setup and diagram of Brillouin-amplified sideband of chip-based OEO. (B) Frequency tunability of generated RF signal. (C) Vision of a fully integrated Brillouin OEO (Merklein et al., 2016). *Reprinted with permission from Merklein, M., Stiller, B., Kabakova, I. V., Mutugala, U.S., Vu, K., Madden, S.J., Eggleton, B.J., Slavík, R., 2016. Widely tunable, low phase noise microwave source based on a photonic chip. Opt. Lett. 41, 4633. doi:10.1364/OL.41.004633 © The Optical Society.*

the Brillouin gain that selectively amplifies one of the sidebands above the threshold was provided by a chalcogenide glass waveguide (Pant et al., 2011). It had been shown previously that chalcogenide glass waveguides are able to deliver large Brillouin gain of over 50 dB in a cm length scale due to its roughly two orders of magnitude larger gain than standard single mode fiber. This large Brillouin gain provides enough amplification to overcome the losses of the OEO cavity and hence produce stable RF signals in a small footprint.

Besides the potential to reduce the overall size, weight, and power (SWaP) that is crucial for many applications, the shorter cavity also relaxes the performance requirements on the Brillouin filter. As described in the previous section, long OEO cavities impose challenges for Brillouin filtering as cavity modes are tightly spaced and hence one can observe mode hopping. In contrast the demonstrated Brillouin OEO operates in a single mode. However, the reduction in length of the cavity comes at an overall increase in the phase noise which is directly related to the fiber length of the OEO. On the other hand, the shorter cavity has a shorter roundtrip time and hence a faster build-up of the oscillating mode that can be utilized for rapid frequency tuning, taking full advantage of the frequency tunability that is one of the major benefits of Brillouin-based OEOs. Frequency tuning of the chip-based Brillouin OEO is shown in Fig. 5B.

Going forward a fully integrated Brillouin OEO is feasible, and an artist's impression is depicted in Fig. 5C. We will further discuss the integration of RF sources onto a photonic ship in the Section 7 of this chapter.

6. RF sources based on high-Q-resonators

In this section we discuss Brillouin-based RF sources that harness high-Q-resonators. We start with an overview of Brillouin lasing in microcavities, discuss the main sources of phase noise, and show how those systems can be harnessed to synthesize stable RF signals.

6.1 Microcavity Brillouin laser overview

Over the last decade another approach for generation of Brillouin lasers has emerged with implications for miniaturization and integration with other photonic devices as well as electronics. These devices generate Brillouin gain in compact optical cavities that are most often of the whispering gallery mode (WGM) type wherein optical waves execute closed circular paths (Vahala, 2003). When the round-trip optical loss in these resonators is very low (equivalently, the optical Q is very high), resonant enhancement of input coupled power is large, making possible access to the stimulated Brillouin process at remarkably low pumping power levels that can be less than 1 mW (Lee et al., 2012). Such power levels are readily attainable with III-V semiconductor lasers. Also significant is that the high optical Q reduces the fundamental laser linewidth to below 1 Hz levels. As discussed below, this fundamental noise is the dominant source of phase noise in microwave generation by SBS.

Because Brillouin scattering is ubiquitous in low loss dielectrics, laser action has been demonstrated in discrete resonators based on many materials including silica (Lee et al., 2012; Li et al., 2012; Tomes and Carmon, 2009), CaF_2 (Grudinin et al., 2009), chalcogenides (Morrison et al., 2017), silicon (Otterstrom et al., 2018), and silicon nitride (Gundavarapu et al., 2018). Moreover, the broad optical window of most dielectrics enables a wide wavelength operating range for these devices. Besides telecom operation, SBS micro-lasers have been demonstrated at 1064 nm in silica devices (Li et al., 2014a). It is also important to note that beyond resonators used for high-coherence microwave generation, there are other microwave related processes that benefit from chip-based stimulated Brillouin scattering. An excellent overview is provided in reference (Eggleton et al., 2019). In what follows, the microfabrication of chip-based Brillouin laser devices and requirements for efficient and low-noise microwave generation will be considered using results for silica-based resonators. In general, however, the same operating and design principles apply for all SBS micro-lasers.

6.2 Microcavity design for Brillouin laser action

Although forward Brillouin scattering has been studied in WGM resonators containing many transverse optical modes, the most typical Brillouin laser configuration is the one described above in optical fiber systems. There, a forward pump wave scatters into a Stokes wave within the same transverse mode family, but propagating in the opposite direction to the pump wave. With this configuration in mind, the Brillouin phase matching condition imposes a requirement on micro-resonator fabrication precision that ensures both the pumping and Stokes wave frequencies coincide with WGM resonant frequencies. In practice, this condition is satisfied when the resonator free-spectral-range (FSR) is equal to the Brillouin frequency shift ($\Omega_B = 4\pi$ n V_S/λ_P) where V_S is the sound velocity for the phonons involved in the scattering process, n is the refractive index and λ_P is the pumping wavelength. For telecom-band pumping wavelengths (near 1.55 μm), the Brillouin shift in Hertz units ($\Omega_B/2\pi$) is around 10.8 GHz in silica. By comparison, the SBS gain features a relatively narrow spectral width (typically 10s of MHz) so that frequency matching in the WGM system requires some care so as to control resonator dimension and in turn the FSR. Referring to Fig. 6A, silica resonators with diameters of about 6 mm feature FSRs that are close to the required 10.8 GHz (X-band) rate and this dimension must be controlled to at least one part in one thousand (1:1000) to ensure a reasonable degree of spectral alignment of SBS gain with the Stokes resonance. Fortunately, precise microfabrication control of the device dimensions is possible so that precision of 1 part in 10,000 is achievable.

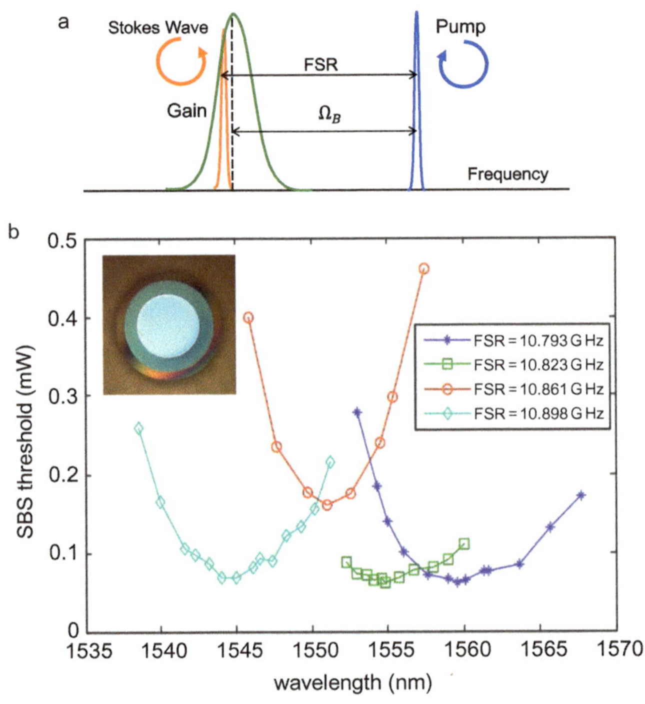

Fig. 6 (A) Illustration showing Brillouin gain spectrum generated by an optical pumping wave. Efficient laser action requires that the Stokes wave be located near the maximum of the Brillouin gain. This, in turn, requires that the Brillouin shift (Ω_B) be closely matched to the resonator free-spectral-range (FSR). (B) Measured laser threshold plotted versus pumping wavelength for four different resonators featuring slightly different FSRs (see legend). Phase matching occurs at slightly different pumping wavelengths for each FSR. When properly matched, threshold pumping power is less than 0.1 mW on account of the high optical Q factor of the resonators. Inset: photograph of typical silica wedge resonator used in this measurement. *With permission of Lee, H., Chen, T., Li, J., Yang, K.Y., Jeon, S., Painter, O., Vahala, K.J., 2012. Chemically etched ultrahigh-Q wedge-resonator on a silicon chip. Nat. Photonics 6, 369–373. doi:10.1038/nphoton.2012.109.*

To illustrate the importance of dimensional control for efficiency Brillouin laser action, the effect of intentional variation of resonator diameter on pumping threshold is illustrated in Fig. 6B. In the data plot, four resonators having slightly different FSR values are pumped at a series of telecom wavelengths. The minimum threshold levels occur for $\Omega_B = FSR$ which happens at a distinctly different pumping wavelength for each resonator. The increase in threshold away from the threshold minimum for each resonator reflects reduction of Brillouin gain caused by frequency mismatch of the Stokes frequency with the Brillouin gain maximum. Besides arrangements in which $\Omega_B = FSR$ it is also possible to satisfy the phase matching condition in larger resonators wherein $\Omega_B = m\ FSR$ (i.e., diameter is m times larger where m is an integer). Such a design featuring m = 6 was used recently to realize a laser gyroscope based upon counter-propagating Brillouin lasers (Lai et al., 2020). In these designs, the Stokes wave lasing occurs in the mth nearest neighbor mode to the pump mode.

Finally, once the proper cavity diameter is achieved, the resonant alignment of the Stokes wave with the Brillouin gain produces a useful cascade effect. Here, the initial Stokes laser field can act as a pumping wave to achieve laser action at a neighboring (lower-frequency) cavity resonance (see Fig. 7A). This cascading effect has been observed in microcavity-based Brillouin lasers out to 9 orders as shown in Fig. 7B.

6.3 Sources of phase noise in microcavity Brillouin lasers

As discussed earlier, microwave generation using SBS lasers relies upon optical mixing of the Stokes laser waves on a photodetector. In general, the phase noise of the radio-frequency signal can have contributions that are of technical origin or that result from the fundamental physics of the Brillouin process. In dielectric microcavities that are optical pumped, the two most significant sources of technical noise are thermo-refractive noise (TRN) and photo-thermal noise (PTN) (Gorodetsky and Grudinin, 2004; Matsko et al., 2007). The former results from thermodynamically driven changes in the refractive index of the dielectric, while the latter is the result of changes in temperature driven by power fluctuations of the pumping laser. The impact of these two fluctuations on the absolute phase of the laser signal is most significant at low offset frequencies (typically below 100 Hz) as set by thermal time constants. At these frequencies, TRN and PTN become the dominant source of optical phase noise providing noise contributions many orders larger than SBS fundamental noise. Importantly, however, the co-generation of lasing Stokes waves within the same transverse mode

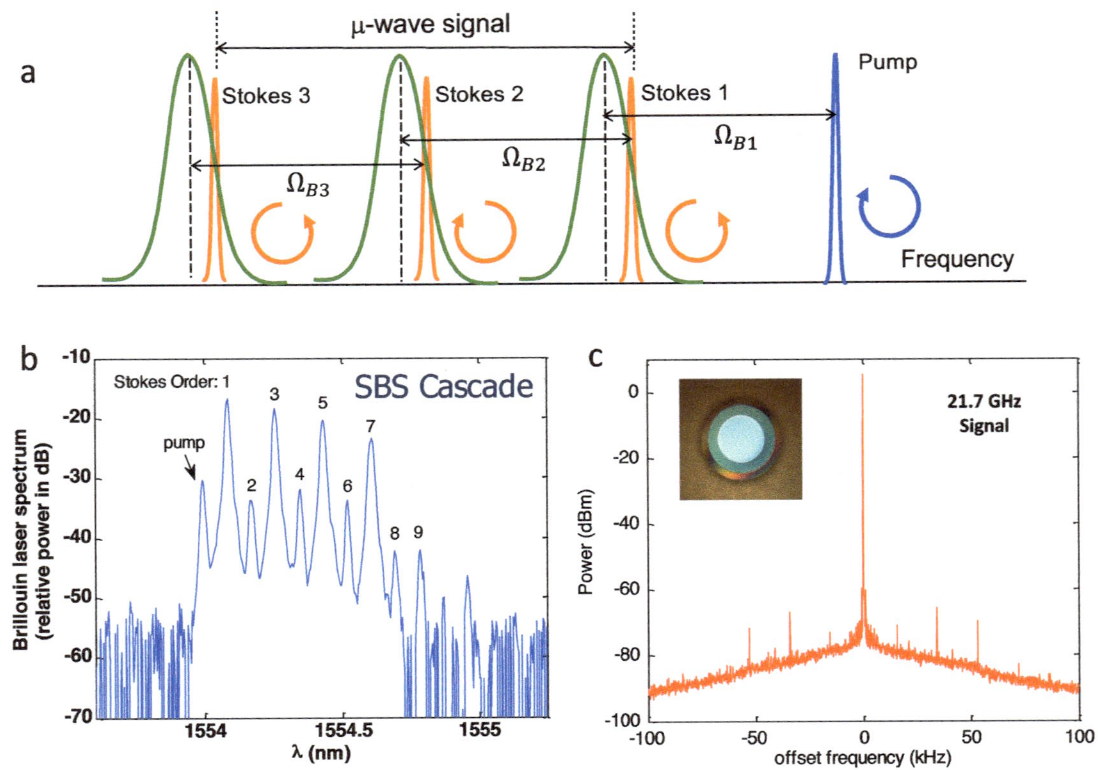

Fig. 7 See figure legend on opposite page.

family of the resonator causes the spatially dependent thermal fluctuations generated by either TRN or PTN sources to be largely common mode in nature. As a result, TRN and PTN noise contributions are not observed to contribute significantly to radio-frequency signal phase noise when Stokes co-generation is used.

Another important source of noise of technical origin is a result of the parametric nature of the Brillouin process. Here, optical phase noise contributed by the pump wave couples into the Stokes wave. This process is most easily discussed in terms of a frequency pulling induced on the Stokes wave optical frequency when the pump frequency is tuned (Debut et al., 2000). The pulling contribution scales like the cavity linewidth divided by the Brillouin gain bandwidth so in ultra-high-Q optical resonators featuring linewidths less 1 MHz, the pulling is suppressed by up to 2 orders of magnitude. Any frequency noise (or equivalent phase noise) on the pumping wave is then suppressed by the square of this pulling factor. However, even this suppression level, which can be over 30 dB, is frequently not sufficient to completely suppress pump noise. To provide extra suppression, it has been found that cascaded pumping of Stokes waves offers one solution. Here, an existing Stokes wave acts to pump a second Stokes wave and so forth. Each step in the cascade further suppresses the initial pump frequency noise. Cascaded pumping has been used to generate X and K band signals (the former in a Brillouin laser gyroscope; Li et al., 2017) and shows excellent suppression of any pump noise. Another approach that has been applied in gyroscopes is to counter-pump the clockwise and counter-clockwise

Fig. 7 (A) Spectral illustration of cascaded Brillouin laser action in which Stokes waves function as pumping waves for subsequent Stokes waves. Phase matching causes each Stokes wave to propagate in a direction opposite to its pumping wave (see orange circular arrows). Stokes frequency shifts (Ω_{B1}, Ω_{B2}, Ω_{B3}) depend upon the wavelength of their corresponding pump (see text). Two co-propagating waves (Stokes 1 and Stokes 3) are readily photo-detected to generate a K-band signal (see panel C). (B) Measurement of cascaded Brillouin laser action to 9th order. The alternating power levels of the even and odd orders is an artifact of collection of optical power in the direction backward relative to the pumping direction. This favors all of the odd Stokes orders. (C) Photodetection of Stokes orders 1 and 3 provides a K-band signal at 21.7 GHz when a silica resonator is optically pumped near 1550 nm. This specific spectrum is from the Brillouin VCO (see Fig. 3) and has also been locked to a quartz oscillator for improved long term stability. *Panel (B) With permission of Li, J., Lee, H., Chen, T., Vahala, K.J., 2012. Characterization of a high coherence, Brillouin microcavity laser on silicon. Opt. Express 20, 20170. doi:10.1364/oe.20.020170. Panel (C) With permission of Li, J., Lee, H., Vahala, K.J., 2013. Microwave synthesizer using an on-chip Brillouin oscillator. Nat. Commun. 4, 2097. doi:10.1038/ncomms3097.*

Stokes laser waves using two pump frequencies that are themselves derived from the same laser (Lai et al., 2020). Here, any pump noise transfers as common mode noise to the two Stokes waves.

In summary to this point, all sources of technical noise can be strongly suppressed in the final radio frequency signal when co-lasing operation of the Stokes waves is employed. What then remains is the fundamental phase (or frequency noise) of the Brillouin laser process itself. Interestingly, the way this noise impacts Brillouin laser phase noise was not fully appreciated until quite recently (Li et al., 2012). Specifically, prior work assumed that a Schawlow-Townes (ST) linewidth (Lax, 1967) applied in the Brillouin laser, and this assumption produced strange results such as the need for enormous spontaneous emission factors to explain measured linewidths. The recently derived correct linewidth formula is given by (Li et al., 2012; Yuan et al., 2020),

$$\Delta \upsilon_{SBS} = \Delta \upsilon_{ST} \left[n_T(\Omega_B, T) + 2 \right]$$

where for purposes of comparison with the ST formula, the factor $\Delta \upsilon_{ST}$ is the conventional ST formula. Within the term in brackets, the factor of "2" is purely quantum mechanical in origin. It can be interpreted as zero-point contributions arising from both photon and phonon contributions. In the ST formula, it is only the optical zero-point contribution that plays a role (the "2" is replaced by a "1"). Also, and of greater importance, is the second term in brackets ($n_T(\Omega_B, T)$). This term, which does not enter the ST linewidth formula, is the number of thermo-mechanical quanta in the phonon mode that participates in the Brillouin process. At non-cryogenic temperatures this is given by, $n_T(\Omega_B, T) = k_B T / \hbar \Omega_B$ and is by far the most important contribution to linewidth. For room temperature X-band phonons $n_T(\Omega_B, T) \approx 600$, a number that has been verified by direct measurement of Brillouin laser linewidth versus power (Li et al., 2012). The temperature dependence of this linewidth has also been verified separately by linewidth measurement (Suh et al., 2017). Physically, this major difference between conventional laser linewidth and Brillouin laser linewidth is readily understood when one considers that all Stokes waves are generated by scattering of the pump wave from a microwave phonon. This scattering produces a channel into which thermal noise from the phonon population is able to couple into the lasing field. Despite the very large enhancement in noise relative to the ST linewidth case, it has been possible to achieve sub Hertz fundamental linewidth in chip-based Brillouin lasers on account of the large optical Q factor of the resonators (above 100 million), which enters inverse quadratically into $\Delta\upsilon_{ST}$. Moreover, gyro measurements have shown that this fundamental source of noise is dominant even down to offset frequencies as low as 0.1 Hz (Lai et al., 2020) when employing proper pump noise suppression methods (either cascade suppression or counter-pumped suppression).

Finally, we comment on two additional effects that, while less significant in their impact on fundamental linewidth, are nonetheless observable. First, is the presence of the so-called Henry linewidth enhancement factor which is present when the Brillouin laser is not operating at perfect phase matching (Yuan et al., 2020). Second, is an added term to the linewidth that is present when cascaded operation is present. This latter term does not increase the linewidth of the last Stokes wave in a cascade, but does add noise to intermediate waves that are functioning as pump waves (Behunin et al., 2018).

6.4 A Brillouin microwave synthesizer

To illustrate the performance possible using microcavity-based Brillouin lasers for generation of microwave signals, consider the application of a silica microdisk wedge resonator to co-generation of two Stokes waves separated by a K-band frequency (22 GHz). The Stokes waves operate in cascade and represent the 1st and 3rd laser lines in the cascade as illustrated in Figs. 7A and B and 8A. Also, the 22 GHz signal can be fine-tuned by adjustment of the pumping frequency. This occurs through the pulling mechanism that was discussed above. The fine tuning adjustment in combination with electrical frequency division of 22 GHz signal enables frequency locking of the microwave signal to a quartz oscillator for long term stability (see Fig. 8B). In essence, the Brillouin generated device is functioning like a high-frequency voltage controlled oscillator (VCO) by providing short term stability that is augmented with the long term stability of the quartz oscillator (Li et al., 2013). Moreover, the fully stabilized 22 GHz oscillator can then function as the microwave source for generation of other frequencies by introduction of additional frequency division. The corresponding phase noise spectra are illustrated in Fig. 8C. In analogy to a conventional VCO used in a synthesizer, the low offset frequency noise in the spectra is determined by the quartz oscillator while the high offset frequency noise level is set by the free running SBO noise (see Fig. 8C). Moreover, the phase noise levels decrease as $1/N^2$ where N gives the frequency division. Examples of waveforms generated by the synthesizer are also provided in Fig. 8D.

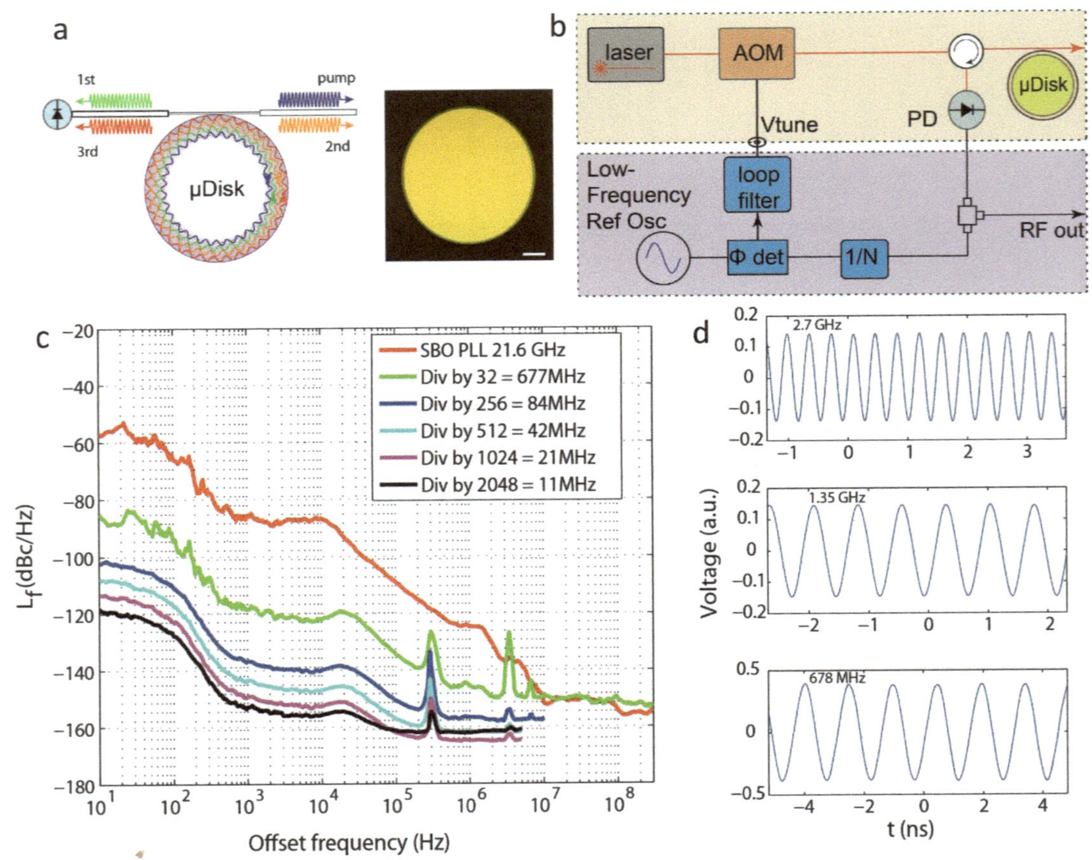

Fig. 8 See figure legend on opposite page.

6.5 Brillouin frequency reference in electro-optical frequency division

The excellent relative frequency stability of co-lasing Brillouin Stokes waves has so far been shown to produce high stability microwave signals by optical mixing on a photodetector. However, the relative frequency stability of the optical Stokes waves is actually much greater than indicated by the measurements presented so far. In particular, because the Stokes waves are optical carriers it is possible to increase the frequency separation between the carriers with negligible impact on the fundamental phase noise properties of either Stokes wave. As a result, the relative frequency stability of the system scales like $\Delta f/|f_{s2}-f_{s1}|$ where f_{s1} and f_{s2} are the Stokes frequencies and Δf is the linewidth of their beat note. For example, by setting the difference frequency to 100 GHz with Δf fixed, the inherent frequency stability is improved by 5× relative to the results presented in Fig. 8 (22 GHz frequency separation). Upon electrical frequency division this translates into a 25× improvement in phase noise relative to any of the phase noise plots presented in Fig. 8. Unfortunately, this strategy eventually encounters a practical limit imposed by detector bandwidth as well as the lack of availability of electrical frequency dividers to operate at the higher millimeter-wave frequencies. However, another recently developed approach called electro-optical frequency division (eOFD) removes these constraints so that much greater frequency separations can be conveniently divided down using Stokes waves that have frequency separations in excess of 1 THz (Li et al., 2014b). The method has been commercialized and is now the basis for the highest performance commercial X to K band oscillators. Further details on the eOFD approach are described in Li et al. (2014b).

Fig. 8 (A) Pumping and Stokes emission schematic used in the microwave synthesizer. 1st and 3rd Stokes waves are photodetected to produce a VCO signal at 21.6 GHz. (B) Schematic of microwave wave synthesizer showing Brillouin laser VCO in the upper panel and phase locked loop control in the lower panel. (C) Phase noise spectra for VCO signal and representative frequency-divided signals at frequencies provided in the legend. Phase noise levels decrease as $1/N^2$ where N is the division radion (see legend). (D) Sinusoidal signals measured on an oscilloscope for three of the synthesized frequencies. *With permission of Li, J., Lee, H., Vahala, K.J., 2013. Microwave synthesizer using an on-chip Brillouin oscillator. Nat. Commun. 4, 2097. doi:10.1038/ncomms3097.*

7. Discussion and outlook

As shown in this chapter Brillouin scattering can be harnessed in multiple ways to generate stable high-frequency RF signals that can be tuned over a wide frequency range. The generated RF signals can have a very high quality. Besides microwave signal generation for radar and signal processing the high Brillouin coherence (both optical and the beat frequency coherence) is useful in other applications. For example, the high relative stability of counter-propagating (CP) Stokes waves has been used to create on-chip ring-laser gyroscopes (Gundavarapu et al., 2018; Li et al., 2017) with sensitivity sufficient to detect the Earth's rotation (Lai et al., 2020). Here, the frequency stability of the beat signal produced by the CP Stokes waves enables detection of minute frequency shifts produced by the Sagnac effect upon resonator rotation. In another application, long term drift in absolute optical frequency stability is dramatically reduced using the beat frequency of Stokes waves lasing on orthogonal polarizations (Loh et al., 2019). Here, the beat frequency is sensitive to modal temperature change. This provides a way to servo control temperature and thereby reduce long-term optical frequency drift. This approach has been applied to stabilize a Brillouin laser for interrogation of a strontium transition as required in an optical clock (Loh et al., 2020).

The low linewidth in the sub-Hz range enabled by Brillouin lasers is accompanied with a low phase noise. Since noise is a very big problem for all quantum experiments and applications, low-linewidth Brillouin lasers might increase the reach for quantum communications, the number of qubits for quantum computing and the sensitivity in quantum sensing.

We have shown the role Brillouin scattering plays in integrated photonic-based approaches. These can enable miniaturization and ultimately reduce SWaP, which is so crucial for a wide range of applications.

Work toward further reduction of the footprint of those integrated photonic based sources is currently ongoing research and we foreshadow will be crucial for the future development of the technology. Many of the demonstrations so far, despite being based on integrated photonics, are still rather bulky since many components remain off-chip. Integrating active components, such as lasers, EO-modulators, photodetectors, together with low-loss waveguides and high-Q resonators is still a challenge, especially in a monolithic platform. Important steps toward this goal have been achieved recently with the advent of high-Q micro-resonators that combine the resonator with an on-chip waveguide for integration with other components (Gundavarapu et al., 2018; Yang et al., 2018). Even more recent work has enabled fabrication of high-Q silicon nitride resonators in a CMOS foundry (Jin et al., 2021; Puckett et al., 2021). These latter devices greatly expand the potential for on-chip systems.

Acknowledgments

KJV thanks Jiang Li, Hansuek Lee, Tong Chen, Myoung-Gyun Suh, Yu-Hung Lai, Zhiquan Yuan, and Heming Wang; and also acknowledges financial support from DARPA.

References

Behunin, R.O., Otterstrom, N.T., Rakich, P.T., Gundavarapu, S., Blumenthal, D.J., 2018. Fundamental noise dynamics in cascaded-order Brillouin lasers. Phys. Rev. A 98, 1–17. https://doi.org/10.1103/PhysRevA.98.023832.

Chan, E.H.W., Minasian, R.A., 2013. High conversion efficiency microwave photonic mixer based on stimulated Brillouin scattering carrier suppression technique. Opt. Lett. 38, 5292. https://doi.org/10.1364/ol.38.005292.

Chan, E.H.W., Minasian, R.A., 2014. All-optical frequency shifter based on stimulated Brillouin scattering in an optical Fiber. IEEE Photonics J. 6, 1–10. https://doi.org/10.1109/JPHOT.2014.2312921.

Chen, H., Chan, E.H.W., 2017. Optical frequency shifter employing a Brillouin-assisted filtering technique. IEEE Photonics J. 9, 1–11. https://doi.org/10.1109/JPHOT.2017.2688441.

Debut, A., Randoux, S., Zemmouri, J., 2000. Linewidth narrowing in Brillouin lasers: theoretical analysis. Phys. Rev. A 62, 023803. https://doi.org/10.1103/PhysRevA.62.023803.

Eggleton, B.J., Poulton, C.G., Rakich, P.T., Steel, M.J., Bahl, G., 2019. Brillouin integrated photonics. Nat. Photonics 13. https://doi.org/10.1038/s41566-019-0498-z.

Gorodetsky, M.L., Grudinin, I.S., 2004. Fundamental thermal fluctuations in microspheres. J. Opt. Soc. Am. B 21, 697. https://doi.org/10.1364/josab.21.000697.

Grudinin, I.S., Matsko, A.B., Maleki, L., 2009. Brillouin lasing with a CaF2 whispering gallery mode resonator. Phys. Rev. Lett. 102, 043902. https://doi.org/10.1103/PhysRevLett.102.043902.

Gundavarapu, S., Brodnik, G.M., Puckett, M., Huffman, T., Bose, D., Behunin, R., Wu, J., Qiu, T., Pinho, C., Chauhan, N., Nohava, J., Rakich, P.T., Nelson, K.D., Salit, M., Blumenthal, D.J., 2018. Sub-hertz fundamental linewidth photonic integrated Brillouin laser. Nat. Photonics 13. https://doi.org/10.1038/s41566-018-0313-2.

Jin, W., Yang, Q.F., Chang, L., Shen, B., Wang, H., Leal, M.A., Wu, L., Gao, M., Feshali, A., Paniccia, M., Vahala, K.J., Bowers, J.E., 2021. Hertz-linewidth semiconductor lasers using CMOS-ready ultra-high-Q microresonators. Nat. Photonics 15, 346–353. https://doi.org/10.1038/s41566-021-00761-7.

Kallfass, I., Antes, J., Schneider, T., Kurz, F., Lopez-Diaz, D., Diebold, S., Massler, H., Leuther, A., Tessmann, A., 2011. All active MMIC-based wireless communication at 220 GHz. IEEE Trans. Terahertz Sci. Technol. 1, 477–487. https://doi.org/10.1109/TTHZ.2011.2160021.

Keysight Technologies, 2020. Datasheet: E8257D PSG Microwave Analog Signal Generator. https://www.keysight.com/de/de/assets/7018-01211/data-sheets/5989-0698.pdf.

Lai, Y.H., Suh, M.G., Lu, Y.K., Shen, B., Yang, Q.F., Wang, H., Li, J., Lee, S.H., Yang, K.Y., Vahala, K., 2020. Earth rotation measured by a chip-scale ring laser gyroscope. Nat. Photonics. https://doi.org/10.1038/s41566-020-0588-y.

Lax, M., 1967. Quantum noise. X. Density-matrix treatment of field and population-difference fluctuations. Phys. Rev. 157, 213–231. https://doi.org/10.1103/PhysRev.157.213.

Lee, H., Chen, T., Li, J., Yang, K.Y., Jeon, S., Painter, O., Vahala, K.J., 2012. Chemically etched ultrahigh-Q wedge-resonator on a silicon chip. Nat. Photonics 6, 369–373. https://doi.org/10.1038/nphoton.2012.109.

Li, J., Lee, H., Chen, T., Vahala, K.J., 2012. Characterization of a high coherence, Brillouin microcavity laser on silicon. Opt. Express 20, 20170. https://doi.org/10.1364/oe.20.020170.

Li, J., Lee, H., Vahala, K.J., 2013. Microwave synthesizer using an on-chip Brillouin oscillator. Nat. Commun. 4, 2097. https://doi.org/10.1038/ncomms3097.

Li, J., Lee, H., Vahala, K.J., 2014a. Low-noise Brillouin laser on a chip at 1064 nm. Opt. Lett. 39, 287–290. https://doi.org/10.1364/OL.39.000287.

Li, J., Yi, X., Lee, H., Diddams, S.A., Vahala, K.J., 2014b. Electro-optical frequency division and stable microwave synthesis. Science (80) 345, 309–313. https://doi.org/10.1126/science.1252909.

Li, J., Suh, M.-G., Vahala, K., 2017. Microresonator Brillouin gyroscope. Optica 4, 346. https://doi.org/10.1364/OPTICA.4.000346.

Liu, J., Zhan, L., Xiao, P., Wang, G., Zhang, L., Liu, X., Peng, J., Shen, Q., 2013. Generation of step-tunable microwave signal using a multiwavelength Brillouin fiber laser. IEEE Photon. Technol. Lett. 25, 220–223. https://doi.org/10.1109/LPT.2012.2234448.

Loh, W., Yegnanarayanan, S., O'Donnell, F., Juodawlkis, P.W., 2019. Ultra-narrow linewidth Brillouin laser with nanokelvin temperature self-referencing. Optica 6, 152. https://doi.org/10.1364/optica.6.000152.

Loh, W., Stuart, J., Reens, D., Bruzewicz, C.D., Braje, D., Chiaverini, J., Juodawlkis, P.W., Sage, J.M., McConnell, R., 2020. Operation of an optical atomic clock with a Brillouin laser subsystem. Nature 588, 244–249. https://doi.org/10.1038/s41586-020-2981-6.

Matsko, A.B., Savchenkov, A.A., Yu, N., Maleki, L., 2007. Whispering-gallery-mode resonators as frequency references I fundamental limitations. J. Opt. Soc. Am. B 24, 1324. https://doi.org/10.1364/JOSAB.24.001324.

Merklein, M., 2018. Controlling, storing and manipulating light using on-chip Brillouin scattering. PhD Thesis. The University of Sydney, Sydney.

Merklein, M., Stiller, B., Kabakova, I.V., Mutugala, U.S., Vu, K., Madden, S.J., Eggleton, B.J., Slavík, R., 2016. Widely tunable, low phase noise microwave source based on a photonic chip. Opt. Lett. 41, 4633. https://doi.org/10.1364/OL.41.004633.

Morrison, B., Casas-Bedoya, A., Ren, G., Vu, K., Liu, Y., Zarifi, A., Nguyen, T.G., Choi, D.-Y., Marpaung, D., Madden, S.J., Mitchell, A., Eggleton, B.J., 2017. Compact Brillouin devices through hybrid integration on silicon. Optica 4, 847. https://doi.org/10.1364/OPTICA.4.000847.

Otterstrom, N.T., Behunin, R.O., Kittlaus, E.A., Wang, Z., Rakich, P.T., 2018. A silicon Brillouin laser. Science (80) 360, 1113–1116. https://doi.org/10.1126/science.aar6113.

Pant, R., Poulton, C.G., Choi, D.-Y., Mcfarlane, H., Hile, S., Li, E., Thevenaz, L., Luther-Davies, B., Madden, S.J., Eggleton, B.J., 2011. On-chip stimulated Brillouin scattering. Opt. Express 19, 8285–8290. https://doi.org/10.1364/OE.19.008285.

Park, C.S., Lee, C.G., Park, C.S., 2007. Photonic frequency upconversion by SBS-based frequency tripling. J. Lightwave Technol. 25, 1711–1718. https://doi.org/10.1109/JLT.2007.897749.

Peng, H., Zhang, C., Xie, X., Sun, T., Guo, P., Zhu, X., Zhu, L., Weiwei, H., Chen, Z., 2015. Tunable DC-60 GHz RF generation utilizing a dual-loop optoelectronic oscillator based on stimulated Brillouin scattering. J. Lightwave Technol. 33, 2707–2715. https://doi.org/10.1109/JLT.2015.2403073.

Preußler, S., Schneider, T., 2012. Bandwidth reduction in a multistage Brillouin system. Opt. Lett. 37, 4122. https://doi.org/10.1364/ol.37.004122.

Preussler, S., Wiatrek, A., Jamshidi, K., Schneider, T., 2011. Brillouin scattering gain bandwidth reduction down to 3.4 MHz. Opt. Express 19, 8565–8570. https://doi.org/10.1364/OE.19.008565.

Preussler, S., Zadok, A., Wiatrek, A., Tur, M., Schneider, T., 2012. Enhancement of spectral resolution and optical rejection ratio of Brillouin optical spectral analysis using polarization pulling. Opt. Express 20, 14734. https://doi.org/10.1364/oe.20.014734.

Preußler, S., Wenzel, N., Braun, R.-P., Owschimikow, N., Vogel, C., Deninger, A., Zadok, A., Woggon, U., Schneider, T., 2013. Generation of ultra-narrow, stable and tunable millimeter- and terahertz-waves with very low phase noise. Opt. Express 21, 23950–23962. https://doi.org/10.1364/OE.21.023950.

Puckett, M.W., Liu, K., Chauhan, N., Zhao, Q., Jin, N., Cheng, H., Wu, J., Behunin, R.O., Rakich, P.T., Nelson, K.D., Blumenthal, D.J., 2021. 422 Million intrinsic quality factor planar integrated all-waveguide resonator with sub-MHz linewidth. Nat. Commun. 12, 1–8. https://doi.org/10.1038/s41467-021-21205-4.

Rubiola, E., 2008. Phase Noise and Frequency Stability in Oscillators. Cambridge University Press, Cambridge, https://doi.org/10.1017/CBO9780511812798.

Schneider, T., 2005. Wavelength and line width measurement of optical sources with femtometre resolution. Electron. Lett. 41, 1234. https://doi.org/10.1049/el:20053204.

Schneider, T., Junker, M., Hannover, D., 2004. Generation of millimetre-wave signals by stimulated Brillouin scattering for radio over fibre systems. Electron. Lett. 40, 1500. https://doi.org/10.1049/el:20046461.

Schneider, T., Hannover, D., Junker, M., 2006. Investigation of Brillouin scattering in optical fibers for the generation of millimeter waves. J. Lightwave Technol. 24, 295–304. https://doi.org/10.1109/JLT.2005.859826.

Shee, Y.G., Mahdi, M.A., Al-Mansoori, M.H., Yaakob, S., Mohamed, R., Zamzuri, A.K., Man, A., Ismail, A., Hitam, S., 2010. All-optical generation of a 21 GHz microwave carrier by incorporating a double-Brillouin frequency shifter. Opt. Lett. 35, 1461. https://doi.org/10.1364/ol.35.001461.

Shen, Y., Zhang, X., Chen, K., 2005. All-optical generation of microwave and millimeter wave using a two-frequency Bragg grating-based Brillouin fiber laser. J. Lightwave Technol. 23, 1860–1865. https://doi.org/10.1109/JLT.2005.846910.

Shi, M., Yi, L., Wei, W., Hu, W., 2018. Generation and phase noise analysis of a wide optoelectronic oscillator with ultra-high resolution based on stimulated Brillouin scattering. Opt. Express 26, 16113. https://doi.org/10.1364/OE.26.016113.

Stern, Y., Zhong, K., Schneider, T., Zhang, R., Ben-Ezra, Y., Tur, M., Zadok, A., 2014. Tunable sharp and highly selective microwave-photonic band-pass filters based on stimulated Brillouin scattering. Photonics Res. 2, B18. https://doi.org/10.1364/PRJ.2.000B18.

Suh, M.G., Yang, Q.F., Vahala, K.J., 2017. Phonon-limited-linewidth of Brillouin lasers at cryogenic temperatures. Phys. Rev. Lett. 119, 1–6. https://doi.org/10.1103/PhysRevLett.119.143901.

Tomes, M., Carmon, T., 2009. Photonic micro-electromechanical systems vibrating at X-band (11-GHz) rates. Phys. Rev. Lett. 102, 113601. https://doi.org/10.1103/PhysRevLett.102.113601.

Vahala, K.J., 2003. Optical microcavities. Nature 424, 839–846. https://doi.org/10.1038/nature01939.

Wiatrek, A., Preußler, S., Jamshidi, K., Schneider, T., 2012. Frequency domain aperture for the gain bandwidth reduction of stimulated Brillouin scattering. Opt. Lett. 37, 930. https://doi.org/10.1364/ol.37.000930.

Yang, K.Y., Oh, D.Y., Lee, S.H., Yang, Q.F., Yi, X., Shen, B., Wang, H., Vahala, K., 2018. Bridging ultrahigh-Q devices and photonic circuits. Nat. Photonics 12, 297–302. https://doi.org/10.1038/s41566-018-0132-5.

Yao, X.S., 1997. High-quality microwave signal generation by use of Brillouin scattering in optical fibers. Opt. Lett. 22, 1329–1331. https://doi.org/10.1364/OL.22.001329.

Yao, X.S., 1998a. Brillouin selective sideband amplification of microwave photonic signals. IEEE Photon. Technol. Lett. 10, 138–140. https://doi.org/10.1109/68.651138.

Yao, X.S., 1998b. Phase-to-amplitude modulation conversion using Brillouin selective sideband amplification. IEEE Photon. Technol. Lett. 10, 264–266. https://doi.org/10.1109/68.655379.

Ye, J., Cundiff, S.T., 2005. Femtosecond Optical Frequency Comb: Principle, Operation, and Applications. Kluwer Academic Publishers/Springer, Boston, https://doi.org/10.1007/b102450.

Yuan, Z., Wang, H., Wu, L., Gao, M., Vahala, K., 2020. Linewidth enhancement factor in a microcavity Brillouin laser. Optica 7, 1150. https://doi.org/10.1364/OPTICA.394311.

Zadok, A., Zilka, E., Eyal, A., Thévenaz, L., Tur, M., 2008. Vector analysis of stimulated Brillouin scattering amplification in standard single-mode fibers. Opt. Express 16, 21692. https://doi.org/10.1364/OE.16.021692.

Zheng, D., Pan, W., Yan, L., Zhang, W., Zou, X., Luo, B., Liu, X., 2014. Microwave photonic down-conversion based on phase modulation and Brillouin-assisted notch-filtering. Real-Time Photonic Meas. Data Manag. Process. 9279, 927914. https://doi.org/10.1117/12.2071934.

CHAPTER NINE

Stimulated Brillouin scattering for microwave photonics

David Marpaung[a,*] and Yang Liu[b,*]
[a]Nonlinear Nanophotonics Group, MESA+ Institute, University of Twente Enschede, Enschede, The Netherlands
[b]Institute of Physics, Swiss Federal Institute of Technology Lausanne (EPFL), Lausanne, Switzerland
*Corresponding authors: e-mail address: david.marpaung@utwente.nl; yang.lau@epfl.ch

Contents

1. Challenges in microwave photonics

Microwave signals occupy frequencies ranging from 300 MHz to 300 GHz in the electromagnetic spectrum. The spectrum has been utilized for various key applications, including wireless communications (Chang, 2000), radar and positioning (Levis et al., 2006; Matthews, 2014; Pan and Zhang, 2020), and sensing (Khare, 2012; Serafino et al., 2019; Zou et al., 2016). Nowadays, there are interests in the generation and processing of radio signals with high frequencies and multi-GHz bandwidth, which can be challenging for traditional electronic systems due to large loss at higher frequencies, limited bandwidth, and sensitivity to impairments such as electromagnetic interference (EMI). In the recent past, photonic systems have been explored for microwave signal processing applications such as beam forming for astronomy, Radar and defense systems, and for realizing microwave sources. This is referred as microwave photonics (MWP) (Seeds, 2002). Microwave photonics (MWP) is a field that deals with the use of optoelectronic devices and systems for enhanced radio frequency (RF) signal

Semiconductors and Semimetals, Volume 110
ISSN 0080-8784
https://doi.org/10.1016/bs.semsem.2022.05.003

manipulation, including the generation, distribution, processing, and analysis of high-speed microwave signals (Capmany and Novak, 2007; Seeds, 2002; Seeds and Williams, 2006; Yao, 2009). MWP technologies have also enabled various functionalities which are not feasible to achieve only by traditional microwave techniques, such as the generation of ultra-wide bandwidth waveforms, or filtering with reconfigurable shape and central frequency. In MWP systems, the low loss, and large bandwidth provided by photonic components and systems are exploited for high frequency and wideband RF signal generation, processing, distribution, and measurement.

Traditional microwave photonic components are bulky, and to compete with existing RF components would require miniaturization. Integrated microwave photonics (IMWP) has gained momentum for signal processing over the last decade due to several key advances in photonic integration (Marpaung et al., 2013b, 2019). Leveraging photonic integration allowed a dramatic reduction in the footprint of MWP systems with fairly high complexity, making them more comparable to RF circuits. Optical loss in MWP systems is important because an increase in this parameter translates quadratically into RF loss in RF circuits. For these reasons, efforts at the early stages of development of integrated MWP were focused on reducing on-chip losses, on integrating as many components as possible in a single chip and to demonstrate device reconfiguration. On the other hand, major advances in chip integration using a single material (monolithic) or multimaterials (hybrid or heterogeneous) have allowed the integration of all the key components of integrated MWP systems in a single chip. The synergies between integration, advanced functionalities and high performance have been the key highlights of the field in recent times. A variety of platforms have proved promising for on-chip signal processing including silicon (Zhang and Yao, 2016), indium phosphide (Fandiño et al., 2017; Liu et al., 2016a; Smit et al., 2014), silicon nitride (Moss et al., 2013; Roeloffzen et al., 2013, 2018), and recently chalcogenides (Eggleton et al., 2011, 2013, 2019).

Integrated MWP clearly promises a number of advantages compared to conventional fiber-based approach. The potential of integrating a number of different signal processing functionalities in a single chip (see Fig. 1) will lead to significant reduction in size, weight, power consumption, and cost. Moreover, the ability to integrate various functions like modulation, passive signal processing, and detection in a single chip will lead to reduction in insertion losses, which is the key for MWP signal processors to compete with RF signal processor in terms of performance.

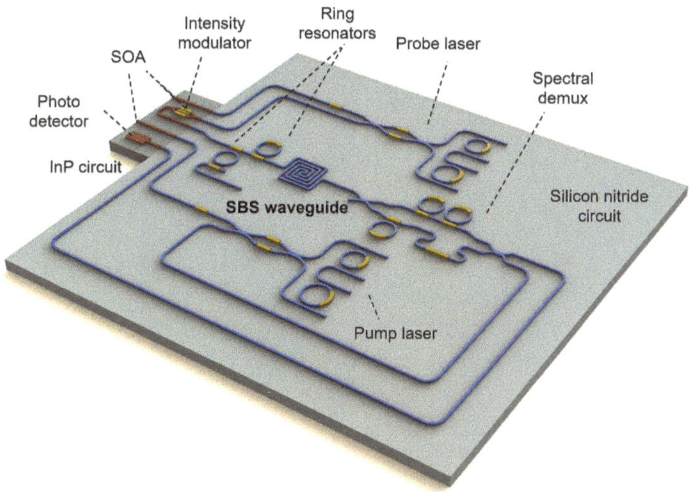

Fig. 1 A conceptual layout of a monolithic Brillouin-based integrated microwave photonic system consisting of Brillouin-active circuits, passive functional devices, light sources, modulators and photodetectors.

2. Brillouin scattering for microwave photonics

SBS has been harnessed for microwave photonic signal processing due to its distinct advantage in bridging between high-resolution and coarse signal processing, which is not possible with other competing technologies. For example, the MHz-level linewidth enabled by SBS allows for high-resolution filtering tunable over wide optical frequency range (on the level of THz) suitable for ultra-wideband microwave spectral processing, unmatched by most on-chip devices. Equally important, SBS offers spectral response reconfigurability. By pumping the SBS medium with multiple lines with shaped amplitude and frequency spacing, a broadened tailorable response can be achieved and one can generate unique filters with flat-top, sharp-edge frequency response and tunable bandwidth, as shown in Fig. 2A and C. These devices are promising for channel selection in spectrally crowded RF environments.

Apart from filtering, SBS can be used for controlling time and phase delay of microwave signals based on the induced phase responses shown in Fig. 2B and D. This is because the gain and loss responses of SBS are accompanied by sharp phase response, due to Kramers-Kronig relation

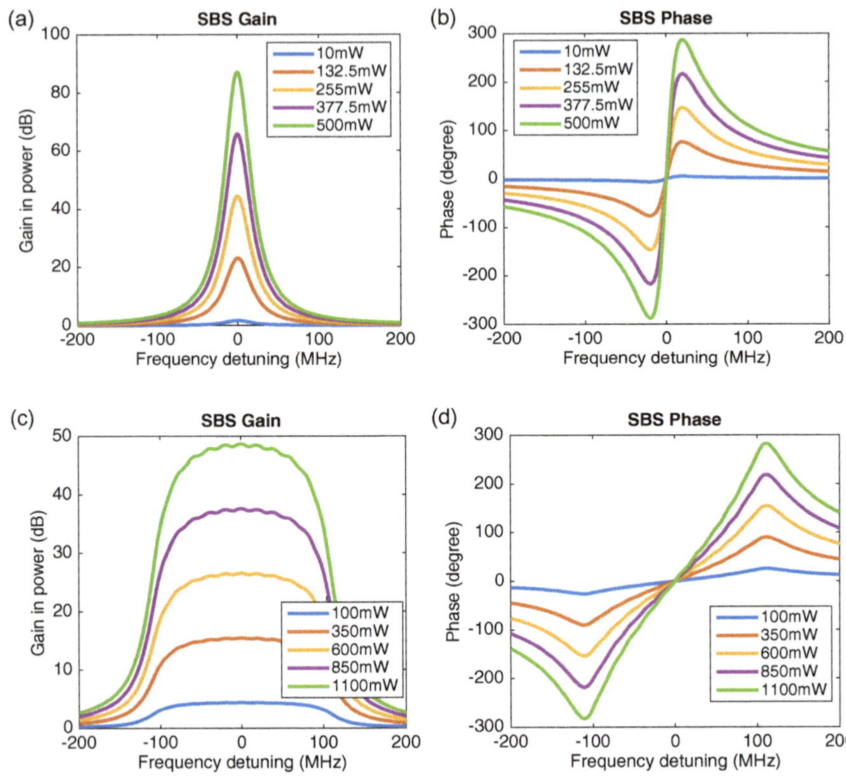

Fig. 2 Numerical calculations of Brillouin-induced intensity gain and phase responses at different pump powers in a typical integrated As_2S_3 photonic waveguide for (A) and (B) a single frequency pump, as well as (C) and (D) broadened multifrequency pumps. In simulations, the SBS gain coefficient is set to 500 m^{-1} W^{-1} and the waveguide length is set as 24 cm. For pump broadening, 10 equally-spaced pump lines with a frequency interval of 20 MHz are used.

(Kramer, 1971). These phases responses can be used to realize slow light induced tunable delay and also to control the phase of a signal. The flexibility of tailoring SBS process parameters such parametric gain and Brillouin frequency shift has enabled a number of applications in MWP (Pant et al., 2014), including tunable delay lines (Chin et al., 2010; Okawachi et al., 2005; Pant et al., 2012; Song et al., 2005), phase shifter (Loayssa et al., 2002), and filtering (Byrnes et al., 2012; Marpaung et al., 2013a; Morrison et al., 2014; Vidal et al., 2007; Zhang and Minasian, 2012a; Zhang et al., 2011). The stimulated Brillouin scattering process is therefore emerging as a potential platform for microwave photonics, having been utilized in long length (~km) optical fibers. Advanced integrated MWP signal processing, however, requires on-chip realization of SBS (Pant et al., 2011a).

3. Microwave photonic filtering

RF filters are fundamental signal processing components that play a critical role in determining the systems ultimate performance. The ultimate goal of a filter is to improve the signal-to-noise ratio of the signal of interest either by rejecting unwanted signals (i.e., band-stop filter) or by allowing only the desired signal to propagate to the receiver (i.e., band-pass filter). To operate in the increasingly crowded radio spectrum, RF receivers require frequency agility to operate in different frequency bands. A smart, reconfigurable filter that operates over a wide frequency range is a critical technology to achieve this agility.

Although optical-based RF photonic filters can be tuned over a wide frequency range, the spectral resolution of most optical filters, typically in the gigahertz, is too coarse for processing RF signals where the separation between adjacent information channels can be down to only a few tens of megahertz. SBS is suited to break this barrier, providing a basis for implementing MWP filters that can combine the strengths of photonic and electronic filters, namely, tens of gigahertz frequency tuning with megahertz spectral resolution and an ultra-high extinction. Early attempts on MWP filters with shape reconfiguration and a widely tunable pass-band and notch were demonstrated exploiting SBS in km-long optical fibers (Vidal et al., 2007; Zhang and Minasian, 2011, 2012a,b). The recent advances in photonic integration have spurred the interest in achieving chip-based Brillouin MWP filters, transferring the established fiber-based Brillouin processing capabilities (Li et al., 2011; Loayssa and Lahoz, 2006; Okawachi et al., 2005; Peng et al., 2014; Sancho et al., 2010; Schneider et al., 2010; Schneider and Preußler, 2014; Yao, 1997) to integrated platforms with significantly reduced footprint. Moreover, SBS response can overcome the filter frequency tuning range conventionally limited by periodic spectral responses in MWP filters based on multitap interferometers and ring resonators.

The recent breakthrough in As_2S_3 photonic circuits has enabled harnessing and inducing SBS in chip-scale devices with unprecedentedly high photonic–phononic interactions (Eggleton et al., 2013), in a typical waveguide cross section of As_2S_3-based Brillouin waveguide (Pant et al., 2011b) shown in Fig. 3A. It can provide a high Brillouin gain exceeding 50 dB (Aryanfar et al., 2017; Choudhary et al., 2017b), allowing for one-to-one spectral mapping of SBS responses to the RF domain for MWP filters (Morrison et al., 2014) and for high-capacity optical communications (Choudhary et al., 2017c; Giacoumidis et al., 2018). However, it becomes

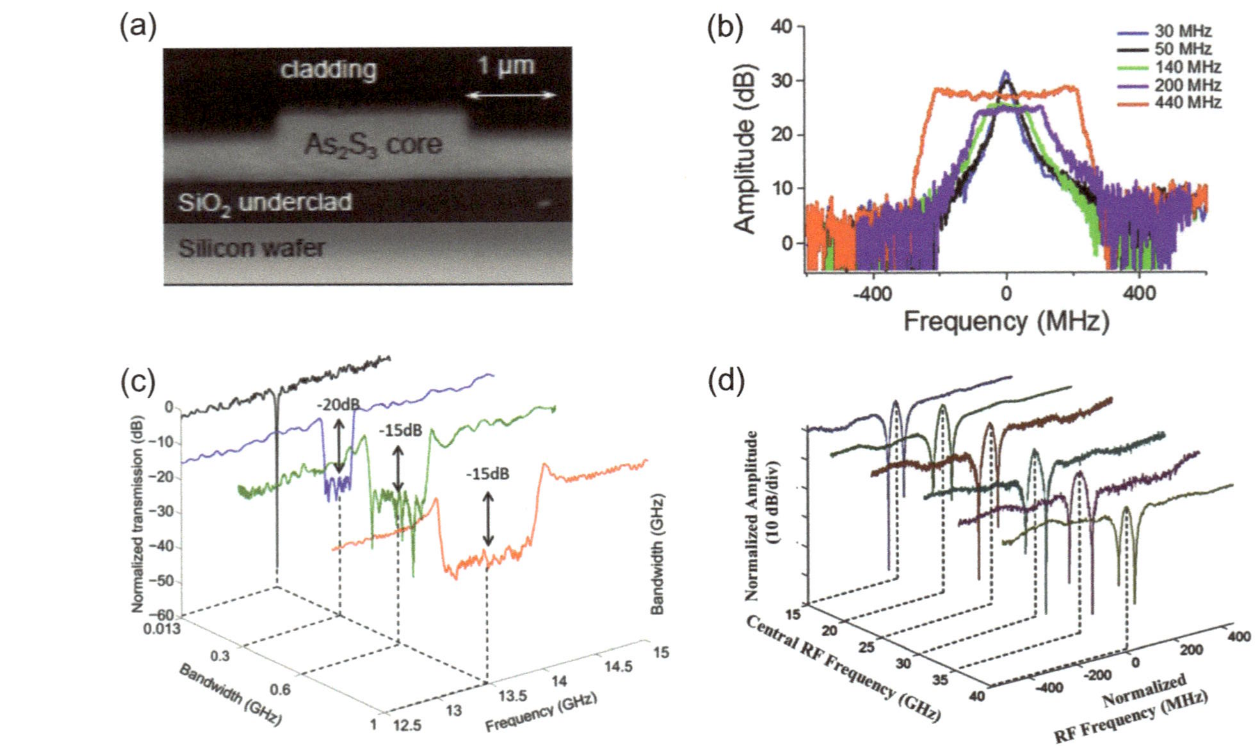

Fig. 3 See figure legend on opposite page.

extremely challenging to achieve high filter extinction, as a high filter selection requires high pump powers (hundreds of mW or even higher) which can easily exceed the power handling of photonic circuits. To circumvent this issue, a novel selective-RF-cancellation technique was proposed, achieving anomalously high filter extinction of >50 dB, while using only a few dB of SBS gain (Marpaung et al., 2015).

Attractively, a spectrally-broadened optical pump can produce broadband SBS responses. This optically programmable response makes it possible to utilize SBS as an efficient technology to implement IMWP filters with bandwidth tunability, without increasing structural complexity. Using RF photonic interference, a bandpass filter with tunable width from 30 to 440 MHz was achieved using spectrally broadened optical pump (Choudhary et al., 2016), as shown in Fig. 3B. Recently, a carrier-suppressed phase modulation scheme was applied to greatly improve the RF link performance, achieving an RF loss of only −5 dB and a high filter rejection ratio of 40 dB (Garrett et al., 2020). Conversely, the filter bandwidth can also be reduced to less than 20 MHz, through sharping the Brillouin gain resonance by two adjacent Brillouin loss resonances (Choudhary et al., 2017a; Preussler and Schneider, 2015), as

Fig. 3 As_2S_3-based Brillouin IMWP filters. (A) The cross section of an As_2S_3Brillouin-active waveguide. (B) A bandwidth tunable IMWP bandpass filter using on-chip As_2S_3 Brillouin circuit. Broadband passband can be achieved using SBS gain response tailoring. (C) A Brillouin-based bandstop filter using As_2S_3 photonic waveguide, exhibiting a maximum bandwidth of 1 GHz. (D) The tuning of the central frequency of the high-resolution filter enabled by the combination of Brillouin gain and loss resonances. *Panel (A) reprinted with permission from Pant, R., Poulton, C. G., Choi, D.-Y., McFarlane, H., Hile, S., Li, E., Thevenaz, L., Luther-Davies, B., Madden, S. J., Eggleton, B. J., 2011b. On-chip stimulated Brillouin scattering. Opt. Express 19 (9), 8285–8290. (URL http://www.opticsexpress.org/abstract.cfm?URI=oe-19-9-8285). Copyright 2011 Optical Society of America. Panel (B) reprinted with permission from Choudhary, A., Aryanfar, I., Shahnia, S., Morrison, B., Vu, K., Madden, S., Luther-Davies, B., Marpaung, D., Eggleton, B. J., 2016. Tailoring of the Brillouin gain for on-chip widely tunable and reconfigurable broadband microwave photonic filters. Opt. Lett. 41 (3), 436. (URL https://www.osapublishing.org/abstract.cfm?URI=ol-41-3-436). Copyright 2016 Optical Society of America. Panel (C) reprinted with permission from Aryanfar, I., Choudhary, A., Shahnia, S., Pagani, M., Liu, Y., Marpaung, D., Eggleton, B. J., 2016. Signal interference RF photonic bandstop filter. Opt. Express 24 (13), 14995. (URL https://www.osapublishing.org/abstract.cfm?URI=oe-24-13-14995). Copyright 2016 Optical Society of America. Panel (D) reprinted by permission from Macmillan Publishers Ltd.: Choudhary et al., Sci. Rep. 7, 5932 (2017) Choudhary, A., Liu, Y., Morrison, B., Vu, K., Choi, D.-Y., Ma, P., Madden, S., Marpaung, D., Eggleton, B. J., dec 2017a. High-resolution, on-chip RF photonic signal processor using Brillouin gain shaping and RF interference. Sci. Rep. 7 (1), 5932. URL https://doi.org/10.1038/s41598-017-06270-4 http://www.nature.com/articles/s41598-017-06270-4*

shown in Fig. 3D. A review of on-chip Brillouin-based MWP filters and applications is presented in Choudhary et al. (2018), Liu et al. (2018a)). Similarly, a bandstop filter was achieved with bandwidth tuning range up to 1 GHz, as shown in Fig. 3C. Such a filter scheme was also implemented using a small SBS gain (<1 dB) in silicon (Casas-Bedoya et al., 2015).

The RF interference technique can be also applied to the formation of notch filters with ultralow pump power consumption. The on-chip SBS process can provide optical gain responses on one optical sideband with an accompanied phase shift at the resonance frequency (Marpaung et al., 2015), which matches the amplitude of the other sideband with a π phase difference, as shown in Fig. 4A. The IMWP notch filter using a shallow SBS gain resonance showed impressive performance including >50 dB rejection, sub-30-MHz spectral resolution, and ultrawide center frequency tunability of >30 GHz (Marpaung et al., 2015), as shown in Fig. 4B.

An integrated silicon photonic-photonic emitter-receiver (PPER) architecture (Shin et al., 2015) demonstrated an alternative to produce fine-spectrum-resolution bandpass filters (Kittlaus et al., 2018), through harnessing on-chip SBS in suspended silicon membrane waveguides, as shown in Fig. 5A. The key idea of PPER-based IMWP filter is that the SBS-induced acoustic wave can transmit to an adjacent silicon waveguide and modulate the optical signal, as shown in Fig. 5B. As a result, the modulated signal will produce optical sidebands that preserve the acoustic responses, enabling a filter spectral resolution of 5 MHz and a filter extinction of 26 dB, as shown in Fig. 5C. A less obvious but very important advantage is the spatial separation of transmitter waveguide and receiver waveguide, which can potentially minimize crosstalk and noise from the optical pump signal.

To unleash the full potential of on-chip Brillouin signal processing, it is desirable to combine Brillouin-active circuits with functional linear integrated photonic devices. A high-performance IMWP reconfigurable bandstop filter based on pairing the Brillouin-active unit with passive an overcoupled ring resonator on a centimeter-scale As_2S_3 photonic circuit was recently demonstrated (Liu et al., 2018b, 2019). This on-chip implementation enables a new class of IMWP devices that are compact, and sustains all advantageous features in the earlier demonstration using SBS in a 4.6-km-long optical fiber and an integrated ring resonator (Liu et al., 2016b). Fig. 6A shows the microscope image of the fabricated As_2S_3 photonic chip consisting of an array of IMWP devices. Each device incorporates a ring resonator (Fig. 6B and C) followed by a 4-cm-long Brillouin spiral

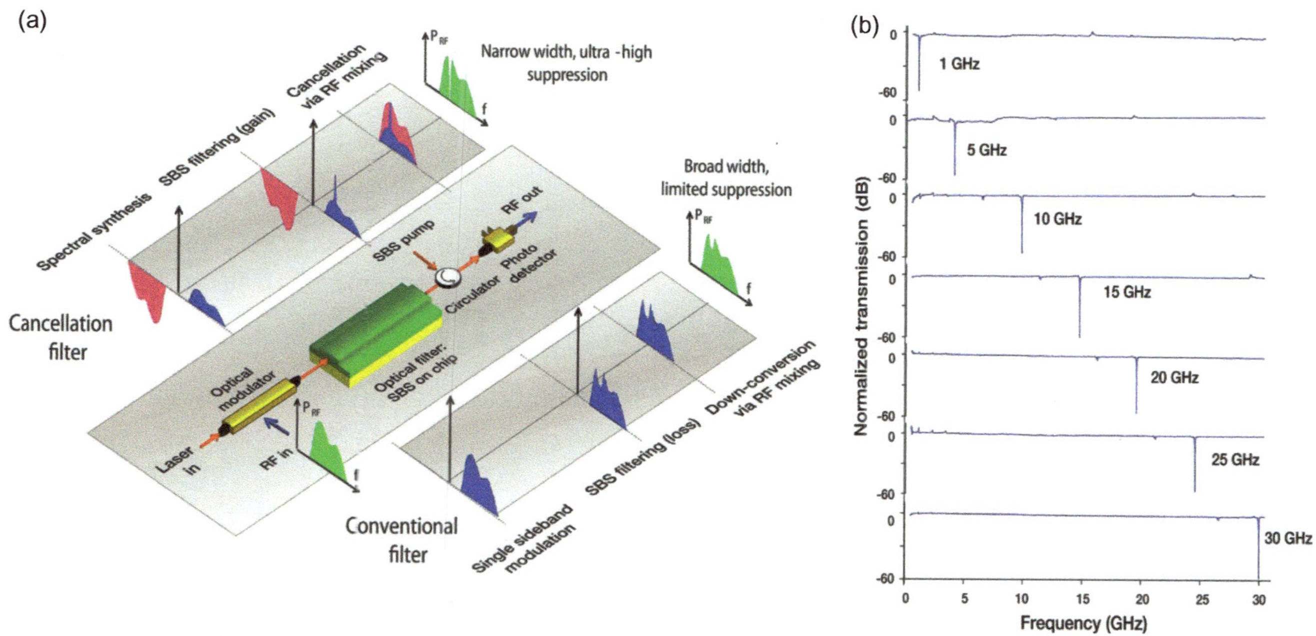

Fig. 4 (A) Schematic of SBS-chip-based MWP notch filter based on the Asym-DSB modulation scheme, compared to the filter based on single-sideband modulation scheme. (B) Measured stopband center frequency tuning over 30 GHz, maintaining a filter suppression of >50 dB and a spectral resolution of 32 MHz. *Reprinted with permission from Marpaung, D., Morrison, B., Pagani, M., Pant, R., Choi, D.-Y., Luther-Davies, B., Madden, S. J., Eggleton, B. J., 2015. Low-power, chip-based stimulated Brillouin scattering microwave photonic filter with ultra-high selectivity. Optica 2 (2), 76. (URL https://www.osapublishing.org/abstract.cfm?URI=optica-2-2-76). Copyright 2015 Optical Society of America.*

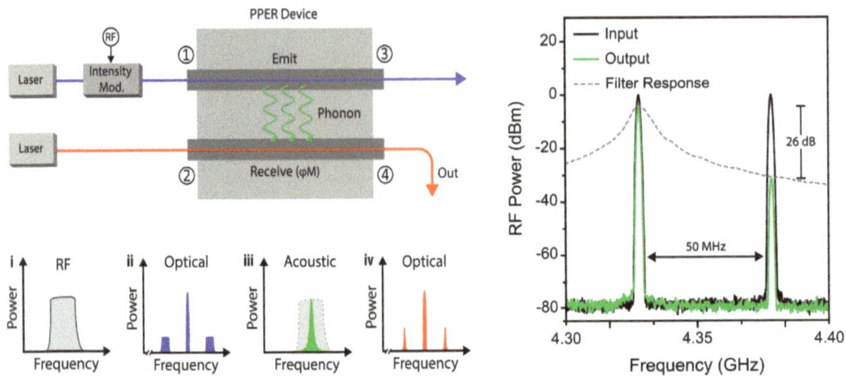

Fig. 5 A silicon photonic-photonic emitter-receiver (PPER) Brillouin-based device. (A) The schematic and SEM images of the silicon PPER device. (B) Prince of PPER-based IMWP filter. The acoustic wave emitted from the emitter waveguide modulates the signal in the receiver waveguide, resulting in (C) a bandpass filtering response. *Reprinted with permission from Kittlaus et al., J. Lightwave Technol. 36, 2803â2809 (2018).*

waveguide (Fig. 6D). The ring resonator has a total length of ~ 4.7 mm, resulting in a measured free-spectral range of around 30 GHz. The SBS spiral waveguides with a cross-section size of 1.9 μm × 0.68 μm allowed for efficient SBS gain and low optical propagation losses, as shown in Fig. 6E. Fig. 6F shows the RF notch filter responses with a filter rejection of >40 dB, when a single-frequency pump was used. By broadening the SBS gain response to compensate the amplitude suppression of the overcoupled ring resonator, a bandstop filter response can be implemented with a 20-dB bandwidth of ~100 MHz and a frequency tunability over 15 GHz, as shown in Fig. 6G. It should be noted that the 3-dB bandwidth of the synthesized RF notch responses is ~3 GHz which is limited by the OC ring resonance. Fig. 6H shows the overall RF link gain in the level of −10 dB in the filter passband with noise figure ranging from 20 to 30 dB, in the case that the notch filter response located at 9 GHz. The current performance is limited by the relatively high insertion loss of the photonic waveguide (18 dB) and optical pump reflection to the photodetector. In order to further implement more complex IMWP filter functionality and advanced performance, one promising approach is using the hybrid integration technology (Califa et al., 2015; Morrison et al., 2017; Shen et al., 2019) to integrate ultra-low-loss Si or Si_3N_4 ring resonators with Brillouin-active circuits, which can combine advantages of different materials. Following the similar principle, multinotch filters with independent tunability have recently been demonstrated by cascading three microring resonators and multi-band SBS responses (Garrett et al., 2022).

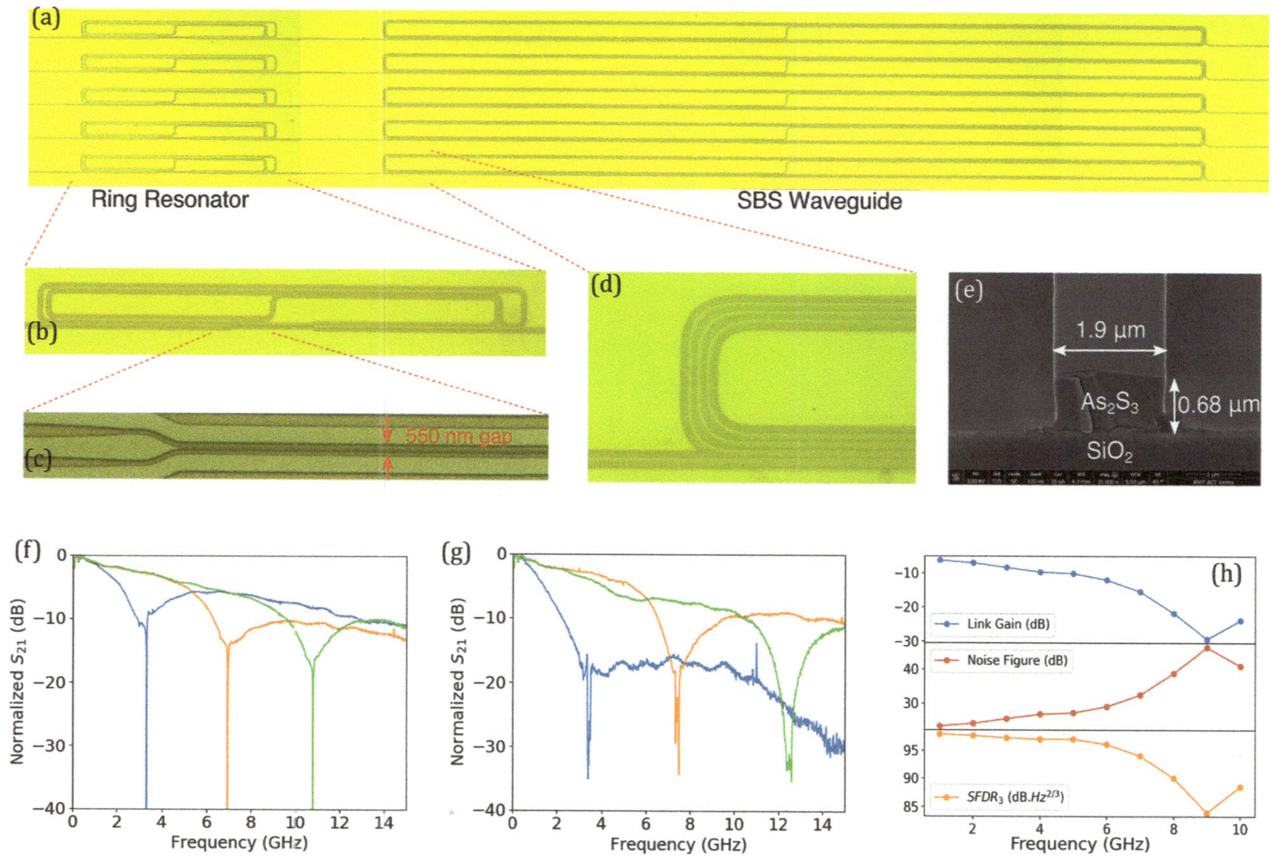

Fig. 6 Microscope images of (A) an array of the OC ring resonator cascaded with a subsequent SBS spiral waveguides on the As_2S_3. (B) An OC ring resonator, (C) the zoomed-in directional coupler with 0.85-µm-wide waveguides of the resonator for light coupling, and (D) the waveguide bends in the SBS spiral waveguide. (E) A scanning electron microscopy (SEM) image of the cross-section of the 1.9-µm-wide As_2S_3 waveguide before the silica upper cladding was sputtered. Measurement of the filter response of (F) MWP notch filters, and (G) MWP bandstop filters at different center frequencies. (H) overall link performance of a notch filter response around 9GHz. *Reproduced from Liu et al., APL Photon. 4, 106,103 (2019); licensed under a Creative Commons CCBY license.*

4. Tunable phase shifters and delay lines

Reconfigurable optical delay lines and wideband tunable phase shifters have primary importance in a number of MWP signal processing applications like optical beamforming in phased-array antenna systems. The simplest way for generating delay and phase shift in the optical domain is through physical length of optical fibers. However, this can become rather bulky. For this reason, integrated photonic solutions are used. The characteristics of an ideal phase shifter include a continuously tunable, full-360° phase shift, constant over a wide, multi-GHz frequency band. In addition, it is desirable to have low insertion loss over the whole band of operation, with a flat amplitude response across the entire phase tuning range. These specifications, however, cannot be met by conventional electrical phase shifters, which can operate over only a narrow frequency range, and cannot achieve continuous phase tuning.

The intrinsic narrowband phase response of SBS is of great interest for implementing microwave photonic phase shifters, to provide a phase shift over a broad RF frequency range. One efficient approach to achieving this functionality is to solely change the phase of the optical carrier, without affecting the signals carried by the optical sidebands. As shown in Fig. 7A, the phase shift imparted on the optical carrier can be tuned by detuning the SBS pump frequency with respect to the optical carrier frequency. Alternatively, fixing the pump frequency but varying the pump power will also provide a different phase shift (Aryanfar et al., 2017). Similarly, the phase response of the SBS loss resonance at the anti-Stokes frequency can also be used for the phase shift (Fig. 7).

However, the phase response of SBS is accompanied by the amplitude variation, which will alter the RF signal power. An improved Brillouin phase shifter scheme was introduced to eliminate the power variations, during phase tuning operations on the optical carrier (Loayssa and Lahoz, 2006). The key idea of this scheme is using the SBS loss resonance to compensate the carrier amplification induced by the SBS gain resonance, as shown in Fig. 7C. In the meantime, the SBS phase response induced by each pump will add up constructively, forming an enhanced phase shift. This optimized scheme was achieved in a As_2S_3 photonic chip (Pagani et al., 2014), reducing the footprint from kilometers-long optical fibers to centimeter-scale photonic chips. As shown in Fig. 7D and E, the chip-based Brillouin MWP phase shifter exhibits a flat amplitude response when the phase shift is

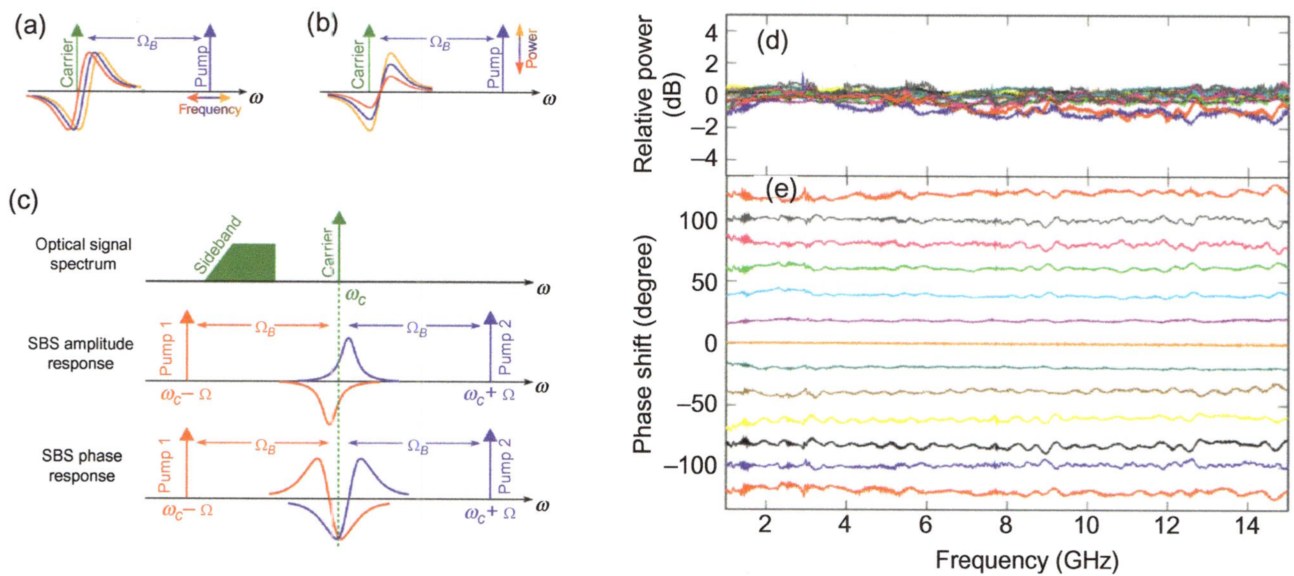

Fig. 7 The schematic of MWP photonic phase shifters based on (A) SBS gain, (B) SBS loss and (C) gain-loss balance. The amplitude (D) and phase shift (E) responses of the on-chip Brillouin-based MWP phase shifter, using gain-loss compensation. *Reprinted with permission from Pagani, M., Marpaung, D., Choi, D.-Y., Madden, S. J., Luther-Davies, B., Eggleton, B. J., 2014. Tunable wideband microwave photonic phase shifter using on-chip stimulated Brillouin scattering. Opt. Express 22 (23), 28810. (URL https://www.osapublishing.org/oe/abstract.cfm?uri=oe-22-23-28810). Copyright 2014 Optical Society of America.*

varied. A maximum phase shift of 240° was achieved, which is constrained by the power handling of the photonic chip. This scheme can achieve a lower amplitude fluctuation of 1.5 dB, when RF phase is varied (Pagani et al., 2014).

The on-chip Brillouin phase shifter only requires a small frequency tuning range of ~30 MHz to achieve a phase shift tuning to the maximum, in contrast to the frequency tuning range of several GHz when using silicon resonators (Pu et al., 2010), Si_3N_4 ring resonators (Burla et al., 2011), Bragg gratings (Burla et al., 2013) and silicon waveguides (Urick et al., 2016). The Brillouin-based phase shifter can provide signal gain, instead of introducing losses which occurs in schemes based on ring resonators and gratings. However, the maximal phase shift is limited by the available on-chip SBS gain. This issue can be potentially addressed using the phase enhancement technique via RF interference (McKay et al., 2019).

The dispersive phase response of SBS can be applied to the signal spectrum that consists of a range of frequency components to generate the relative time delay. SBS is preferably used to implement on-chip delay lines, as SBS enables flexible tuning of operation bandwidth and wavelength, less structure complexity and signal amplification (Lenz et al., 2001; Okawachi et al., 2005). For the SBS process, the delay is given by:

$$T_d = \frac{g_B P_P L}{A_{eff} \Gamma_B}, \tag{1}$$

where P_p is the pump power, A_{eff} is effective mode area, Γ_B is the full-width at half-maximum (FWHM) linewidth of the Brillouin gain, L is the medium length, and $g_B(\mathrm{m}^{-1}\ \mathrm{W}^{-1})$ is the Brillouin gain coefficient. This suggests that the time delay can be tuned by varying the pump power. Further, for a device that provides large g_B and small A_{eff}, large delay can be obtained using small power. Many of the earlier SBS delay line demonstrations (Chin et al. (2010)) used silica-based optical fibers where small g_B and large mode area A_{eff}, necessitated the use of long length (âĹijkm) to achieve large delay using small pump power (âĹij100 mW).

The first demonstration of on-chip Brillouin-induced time delay was achieved on an integrated As_2S_3 photonic waveguide with a length of 7 cm (Pant et al., 2012), using a single-frequency optical pump. A 25-ns-long optical Gaussian pulse was generated and inserted at the center frequency of the SBS gain resonance. The on-chip SBS enabled a maximum relative time delay of~ 22 ns, under an optical pump power of ~300 mW, as shown in Fig. 8A. An optical pulse with a longer period of 100 ns shows less pulse broadening and distortion, due to the narrow signal spectrum, as shown in Fig. 8B.

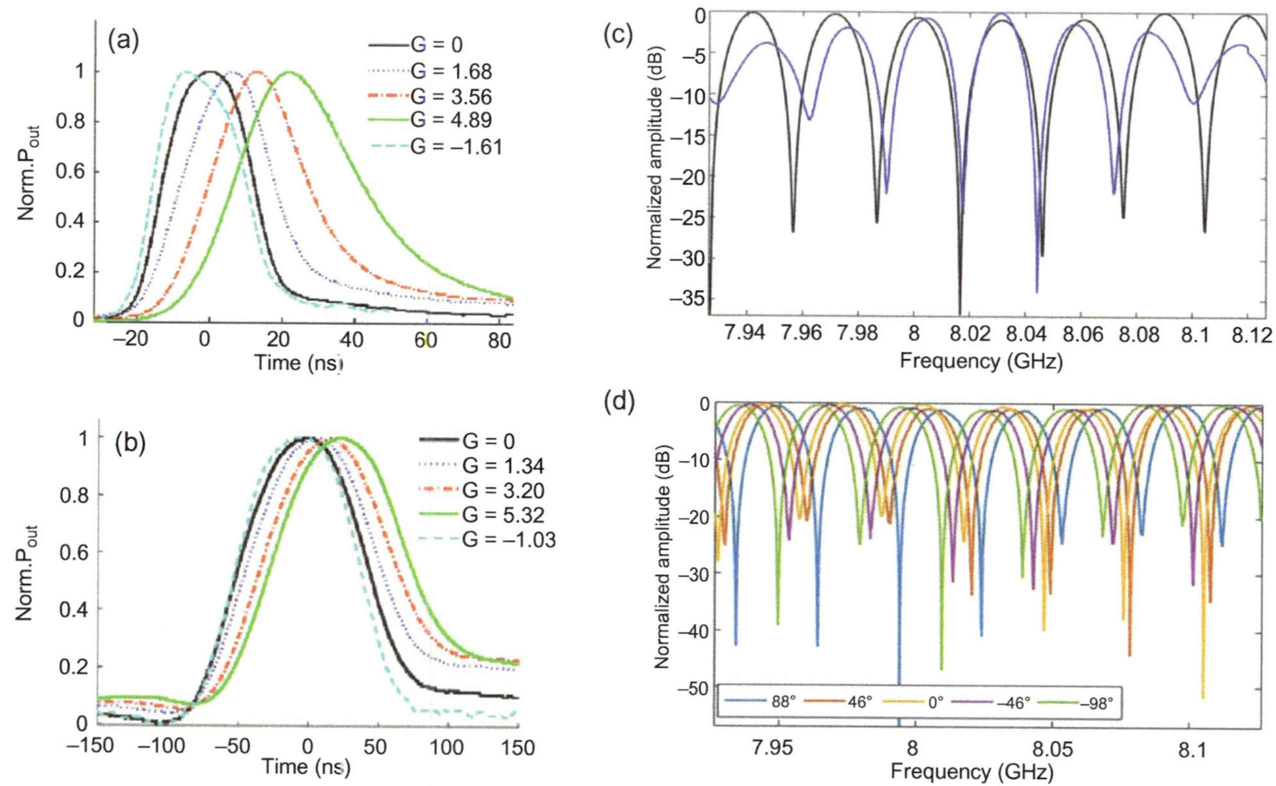

Fig. 8 Measured results of on-chip Brillouin-based MWP photonic delay lines based on (A) SBS gain resonance, (B) SBS loss resonance. (C) Measured MWP filter responses based on a two-tap topology. The time delay in one of tap is controlled by SBS-based on slow light effect. (D) Measured responses of a true-time delay line using SBS-based phase shifts and time delay. *Panel (A and B) reprinted with permission from Pant, R., Byrnes, A., Poulton, C. G., Li, E., Choi, D.-Y., Madden, S., Luther-Davies, B., Eggleton, B. J., 2012. Photonic-chip-based tunable slow and fast light via stimulated Brillouin scattering. Opt. Lett. 37 (5), 969. (URL http://ol.osa.org/abstract.cfm?URI=ol-37-5-969 https://www.osapublishing.org/abstract.cfm?URI=ol-37-5-969). Copyright 2012 Optical Society of America. Panel (D) reprinted with permission from Aryanfar, I., Marpaung, D., Choudhary, A., Liu, Y., Vu, K., Choi, D.-Y., Ma, P., Madden, S., Eggleton, B. J., 2017. Chip-based Brillouin radio frequency photonic phase shifter and wide-band time delay. Opt. Lett. 42 (7), 1313. (URL https://www.osapublishing.org/abstract.cfm?URI=ol-42-7-1313). Copyright 2017 Optical Society of America*

The RF photonic implementation of the on-chip Brillouin-based delay line can also be achieved, using the SBS resonance to process the optical sideband of a modulated optical signal. The on-chip implementation of RF photonic delay based on SBS was recently demonstrated in an integrated 24-cm-long As_2S_3 waveguide with on/off gain of up to 52 dB gain Aryanfar et al. (2017). Such a high level of the SBS gain provides the capability for broadband time delay, by distributing the SBS gain over a wide frequency range. By embedding the delay line in a two-tap topology, the FSR change indicates a time delay of ~4 ns over a frequency range of 100 MHz, as shown in Fig. 8C. Fig. 8D depicts the independent control of the carrier phase, showing the potential to achieve an on-chip true-time delay (TTD) sub-system based on the separate carrier tuning technique (Morton and Khurgin, 2009). This compact true time delay device shows a comparable performance to the previous demonstration in optical fibers (Chin et al., 2010). Compared to on-chip delay lines based on ring resonators (Cardenas et al., 2010), subwavelength gratings (Wang et al., 2016), and photonic crystals (Sancho et al., 2012), Brillouin-based delay lines enable signal gain rather than loss. More strikingly, SBS provides a flexible approach to processing both narrowband and broadband signals, while other schemes can only process broadband signals.

5. Frequency measurements

Another functionality that can be enabled by SBS is instantaneous frequency measurement (IFM). Spectrum analysis is a key functionality in modern RF systems. In particular, fast and accurate estimation of multiple unknown RF signal frequencies over a wide measurement range is crucial in defense applications. For these purposes, it is favorable to sense the frequency information of unknown intercepted signals before separately performing signal analysis and countermeasure (Burla et al., 2016; Jiang et al., 2016; Zou et al., 2016). Traditional electrical instantaneous frequency measurement (IFM) systems suffer from a limited measurement range and they are vulnerable to electromagnetic interference (EMI). Photonic measurement systems can overcome these limitations and extend the IFM receiver operation to a higher frequency range. Photonic IFM systems map the frequency information into other easily measured parameters (e.g., RF or optical power) to estimate the frequency of an unknown RF signal.

The high spectral resolution of SBS gain or loss spectra is beneficial for microwave spectrum analysis and frequency measurements. To further increase the resolution, researchers have explored bandwidth-narrowing effect through superimposed SBS gain spectrum with two loss spectra (Preussler and Schneider, 2015), yielding a reduced bandwidth from 20 to 3.4 MHz. A particular example of SBS-based IFM system was reported by Jiang et al. (2016). The principle is based on tunable bandstop filter formed by RF cancellation of asymmetric-double sideband signal processed by high-resolution SBS gain and loss spectra. The SBS bandstop filter allows for RF-frequency discrimination with simultaneous high extinction and high resolution.

In this implementation, the SBS response was provided by As_2S_3 rib waveguide, allowing partial integration of the SBS-based IFM system. The principle of operation is illustrated in the schematic diagram shown in Fig. 9. In one arm of the optical system, an RF reference signal with its frequency scanning at a fixed step is fed into a dual parallel Mach Zehnder modulator (DPMZM). A phase-modulated signal is generated, which is used as a probe signal and sent to the waveguide where the SBS will be induced in the presence of the optical pump. The SBS gain and loss spectra generated from the pump alters the phase and amplitude of the probe signal forming a bandstop filter. Scanning this filter will allow monitoring of the incoming microwave frequency based on the frequency-to-power mapping with a large-slope amplitude comparison function.

In the experiment, a frequency measurement range from 9 to 38 GHz was demonstrated with a measurement error <1 MHz (Jiang et al., 2016) due to the combination of the frequency agility of a scanning receiver and the large slope of the amplitude comparison function. A 1-MHz resolution for a frequency scanning step of 25 MHz was obtained in tens of nanoseconds due to the large slope of the amplitude comparison function.

Subsequently, several SBS-based IFM and spectrum analysis systems have been reported. In Xu et al. (2021), SBS gain from a doped-silica optical waveguide was used for IFM through single-shot frequency-power mapping scheme, without the need of averaging over time. While the frequency precision of the scheme was still limited by the width of the SBS gain response in this doped-silica waveguides (120 MHz), the scheme benefits from the CMOS-compatibility of the platform, opening the pathway to larger-scale integration of the IFM system. In Shi and Chen (2021), a photonic-assisted multiple RF measurement approach based on SBS and frequency-to-time mapping was reported. The key significance of the work is that the reduced SBS gain spectrum is used to increase the RF frequency measurement resolution. Multiple RF frequencies from 0.3 to 7.6 GHz were measured with a resolution of <10 MHz and measurement error of <1 MHz.

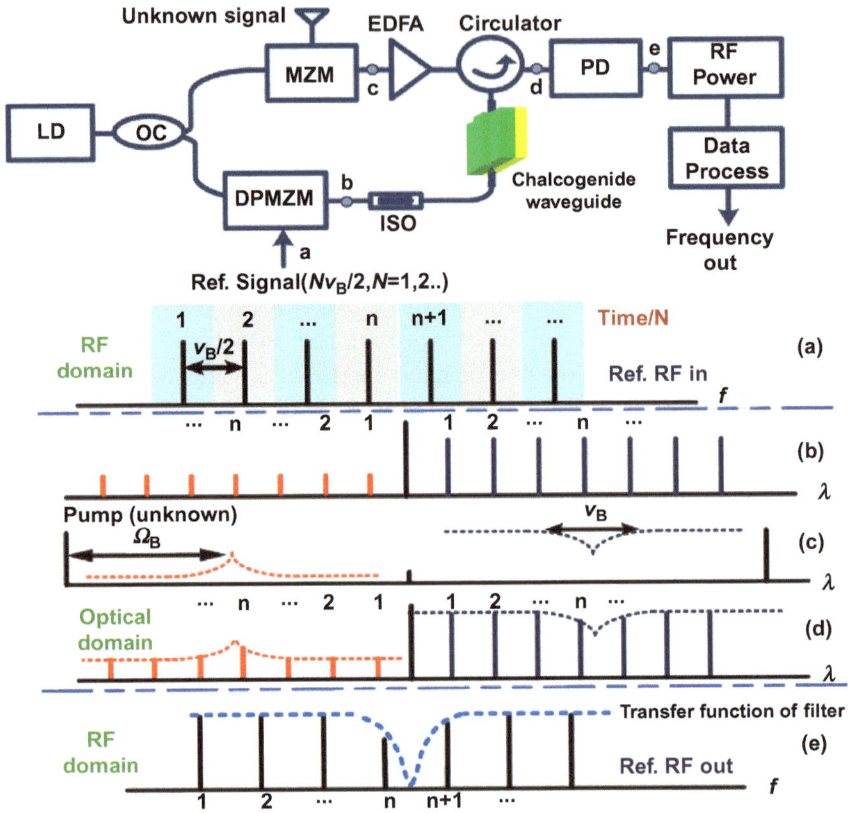

Fig. 9 Experimental setup. LD, laser; OC, optical coupler; MZM, Mach Zehnder modulator; DPMZM, dual-parallel MZM; EDFA, Er-doped fiber amplifier; ISO, inline optical isolator; PD, photodetector. (A–E) The optical/RF spectra at different points. *Reprinted with permission from Jiang, H., Marpaung, D., Pagani, M., Vu, K., Choi, D.-Y., Madden, S. J., Yan, L., Eggleton, B. J., jan 2016. Wide-range, high-precision multiple microwave frequency measurement using a chip-based photonic Brillouin filter. Optica 3 (1), 30. (URL https://www.osapublishing.org/abstract.cfm?URI=optica-3-1-30). Copyright 2016 Optical Society of America.*

6. Perspectives and outlook

On-chip stimulated Brillouin scattering has shown great advantages to offer fine spectral resolution, parametric gain and tunable phase shift or delay for microwave signal processing and manipulation, which can meet the stringent requirements for RF signal processing unattainable for conventional passive photonic integrated devices. However, on-chip Brillouin-based signal processing has thus far been realized in standalone photonic integrated devices with limited functionalities; most of the other functions such as light sources, signal modulation and photodetection still rely on off-chip components. The ultimate realization of fully-integrated Brillouin-based processors necessitates the cointegration of Brillouin devices, passive photonic circuits and active components on a single chip, in particular in the contemporary photonic integrated circuit platforms of choice such as silicon, silicon nitride and their hybrid/heterogeneous integration. Past attempts have successfully demonstrated Brillouin scattering in silicon and silicon nitride photonic circuits. However, apart from the complexity in structural engineering to prevent acoustic leakage, the realization of intense Brillouin processes are inhibited by either the two photon absorption in silicon (Kittlaus et al., 2016; Otterstrom et al., 2019; Van Laer et al., 2015) or the minuscule photon–phonon interactions in silicon nitride (Gundavarapu et al., 2019; Gyger et al., 2020). Despite these challenges, using novel waveguide designs to enhance photon–phonon interactions can possibly establish complex Brillouin-active circuits in these mainstream photonic integrated photonic circuits and motivate new research opportunities. An alternative way is to integrate established Brillouin materials such as As_2S_3 onto silicon or silicon nitride photonic platforms that have a rich library of function components. This approach has been recently proven feasible, showing appreciable Brillouin gain and RF photonic filtering function in an As_2S_3-Silicon photonic chip (Liu et al., 2019; Morrison et al., 2017), which is implemented by back-end-of-line processing of silicon photonic circuits fabricated by a commercial foundry. With advanced signal processing functionalities and enhanced performance being demonstrated, the photonic integration is of great importance for the future development of integrated Brillouin-based microwave photonic processors.

References

Aryanfar, I., Marpaung, D., Choudhary, A., Liu, Y., Vu, K., Choi, D.-Y., Ma, P., Madden, S., Eggleton, B.J., 2017. Chip-based Brillouin radio frequency photonic phase shifter and wide-band time delay. Opt. Lett. 42 (7), 1313. https://www.osapublishing.org/abstract.cfm?URI=ol-42-7-1313.

Burla, M., Marpaung, D., Zhuang, L., Roeloffzen, C., Khan, M.R., Leinse, A., Hoekman, M., Heideman, R., 2011. On-chip CMOS compatible reconfigurable optical delay line with separate carrier tuning for microwave photonic signal processing. Opt. Express 19 (22), 21475.

Burla, M., Cortés, L.R., Li, M., Wang, X., Chrostowski, L., Azaña, J., 2013. Integrated waveguide Bragg gratings for microwave photonics signal processing. Opt. Express 21 (21), 25120. https://www.osapublishing.org/oe/abstract.cfm?uri=oe-21-21-25120.

Burla, M., Wang, X., Li, M., Chrostowski, L., Azaña, J., 2016. Wideband dynamic microwave frequency identification system using a low-power ultracompact silicon photonic chip. Nat. Commun. 7 (1), 13004. http://www.nature.com/articles/ncomms13004.

Byrnes, A., Pant, R., Li, E., Choi, D.-Y., Poulton, C.G., Fan, S., Madden, S., Luther-Davies, B., Eggleton, B.J., 2012. Photonic chip based tunable and reconfigurable narrowband microwave photonic filter using stimulated Brillouin scattering. Opt. Express 20 (17), 18836. (URL http://www.opticsexpress.org/abstract.cfm?URI=oe-20-17-18836 https://www.osapublishing.org/oe/abstract.cfm?uri=oe-20-17-18836 http://www.osapublishing.org/optica/abstract.cfm?URI=optica-2-2-76 https://www.osapublishing.org/abstract.cfm?URI=optica-2-2-76 h).

Califa, R., Munk, D., Genish, H., Kaganovskii, Y., Bakish, I., Rosenbluh, M., Zadok, A., 2015. Large one-time photo-induced tuning of directional couplers in chalcogenide-on-silicon platform. Opt. Express 23 (22), 28234.

Capmany, J., Novak, D., 2007. Microwave photonics combines two worlds. Nat. Photonics 1 (6), 319–330. https://doi.org/10.1038/nphoton.2007.89. http://www.nature.com/articles/nphoton.2007.89.

Cardenas, J., Foster, M.A., Sherwood-Droz, N., Poitras, C.B., Lira, H.L.R., Zhang, B., Gaeta, A.L., Khurgin, J.B., Morton, P., Lipson, M., 2010. Wide-bandwidth continuously tunable optical delay line using silicon microring resonators. Opt. Express 18 (25), 26525. http://www.ncbi.nlm.nih.gov/pubmed/21165004. https://www.osapublishing.org/oe/abstract.cfm?uri=oe-18-25-26525.

Casas-Bedoya, A., Morrison, B., Pagani, M., Marpaung, D., Eggleton, B.J., 2015. Tunable narrowband microwave photonic filter created by stimulated Brillouin scattering from a silicon nanowire. Opt. Lett. 40 (17), 4154. https://www.osapublishing.org/abstract.cfm?URI=ol-40-17-4154.

Chang, K., 2000. RF and Microwave Wireless Systems. Vol. 7 of Wiley Series in Microwave and Optical Engineering. John Wiley & Sons, Inc., New York, USA. URL. http://doi.wiley.com/10.1002/0471224324.

Chin, S., Thévenaz, L., Sancho, J., Sales, S., Capmany, J., Berger, P., Bourderionnet, J., Dolfi, D., 2010. Broadband true time delay for microwave signal processing, using slow light based on stimulated Brillouin scattering in optical fibers. Opt. Express 18 (21), 22599. https://opg.optica.org/oe/abstract.cfm?uri=oe-18-21-22599.

Choudhary, A., Aryanfar, I., Shahnia, S., Morrison, B., Vu, K., Madden, S., Luther-Davies, B., Marpaung, D., Eggleton, B.J., 2016. Tailoring of the Brillouin gain for on-chip widely tunable and reconfigurable broadband microwave photonic filters. Opt. Lett. 41 (3), 436. https://www.osapublishing.org/abstract.cfm?URI=ol-41-3-436.

Choudhary, A., Liu, Y., Morrison, B., Vu, K., Choi, D.-Y., Ma, P., Madden, S., Marpaung, D., Eggleton, B.J., 2017a. High-resolution, on-chip RF photonic signal processor using Brillouin gain shaping and RF interference. Sci. Rep. 7 (1), 5932. https://doi.org/10.1038/s41598-017-06270-4. http://www.nature.com/articles/s41598-017-06270-4.

Choudhary, A., Morrison, B., Aryanfar, I., Shahnia, S., Pagani, M., Liu, Y., Vu, K., Madden, S., Marpaung, D., Eggleton, B.J., 2017b. Advanced integrated microwave signal processing with Giant on-Chip Brillouin gain. J. Light. Technol. 35 (4), 846–854. http://ieeexplore.ieee.org/document/7588155/.

Choudhary, A., Pelusi, M., Marpaung, D., Inoue, T., Vu, K., Ma, P., Choi, D.-Y., Madden, S., Namiki, S., Eggleton, B.J., 2017c. On-chip Brillouin purification for frequency comb-based coherent optical communications. Opt. Lett. 42 (24), 5074.

Choudhary, A., Liu, Y., Marpaung, D., Eggleton, B.J., 2018. On-Chip Brillouin filtering of RF and optical signals. IEEE J. Sel. Top. Quantum Electron. 24 (6), 1–11. (URL https://ieeexplore.ieee.org/document/8353713/).

Eggleton, B.J., Luther-Davies, B., Richardson, K., 2011. Chalcogenide photonics. Nat. Photonics 5 (3), 141–148. http://www.nature.com/articles/nphoton.2011.309.

Eggleton, B.J., Poulton, C.G., Pant, R., 2013. Inducing and harnessing stimulated Brillouin scattering in photonic integrated circuits. Adv. Opt. Photonics 5 (4), 536.

Eggleton, B.J., Poulton, C.G., Rakich, P.T., Steel, M.J., Bahl, G., 2019. Brillouin integrated photonics. Nat. Photonics 13 (10), 664–677. http://www.nature.com/articles/s41566-019-0498-z.

Fandiño, J.S., Muñoz, P., Doménech, D., Capmany, J., 2017. A monolithic integrated photonic microwave filter. Nat. Photonics 11 (2), 124–129. https://arxiv.org/abs/1612.06999%0Ahttp://doi.org/10.1038/nphoton.2016.233 http://www.nature.com/articles/nphoton.2016.233.

Garrett, M., Liu, Y., Ma, P., Choi, D.-Y., Madden, S.J., Eggleton, B.J., 2020. Low-RF-loss and large-rejection reconfigurable Brillouin-based RF photonic bandpass filter. Opt. Lett. 45 (13), 3705. https://opg.optica.org/abstract.cfm?URI=ol-45-13-3705.

Garrett, M., Liu, Y., Merklein, M., Choi, D.Y., Yan, K., Madden, S., Eggleton, B., 2022. Multi-band and frequency-agile Chip-based RF photonic filter for ultra-deep interference rejection. J. Light. Technol. 40 (6), 1672–1680. https://doi.org/10.1109/JLT.2021.3129509.

Giacoumidis, E., Choudhary, A., Magi, E., Marpaung, D., Vu, K., Ma, P., Choi, D.-Y., Madden, S., Corcoran, B., Pelusi, M., Eggleton, B.J., 2018. Chip-based Brillouin processing for carrier recovery in self-coherent optical communications. Optica 5 (10), 1191.

Gundavarapu, S., Brodnik, G.M., Puckett, M., Huffman, T., Bose, D., Behunin, R., Wu, J., Qiu, T., Pinho, C., Chauhan, N., Nohava, J., Rakich, P.T., Nelson, K.D., Salit, M., Blumenthal, D.J., 2019. Sub-hertz fundamental linewidth photonic integrated Brillouin laser. Nat. Photonics 13 (1), 60–67. https://doi.org/10.1038/s41566-018-0313-2.

Gyger, F., Liu, J., Yang, F., He, J., Raja, A.S., Wang, R.N., Bhave, S.A., Kippenberg, T.J., Thévenaz, L., 2020. Observation of stimulated Brillouin scattering in silicon nitride integrated waveguides. Phys. Rev. Lett. 124 (1), 013902. https://link.aps.org/doi/10.1103/PhysRevLett.124.013902.

Jiang, H., Marpaung, D., Pagani, M., Vu, K., Choi, D.-Y., Madden, S.J., Yan, L., Eggleton, B.J., 2016. Wide-range, high-precision multiple microwave frequency measurement using a chip-based photonic Brillouin filter. Optica 3 (1), 30. https://www.osapublishing.org/abstract.cfm?URI=optica-3-1-30.

Kittlaus, E.A., Shin, H., Rakich, P.T., 2016. Large Brillouin amplification in silicon. Nat. Photonics 10 (7), 463–467. (URL http://arxiv.org/abs/1510.08495 http://www.nature.com/articles/nphoton.2016.112).

Khare, A.S., 2012. A survey of advanced microwave frequency measurement techniques. Int. J. Electr. Comput. Eng. 1 (3), 1140–1146.

Kittlaus, E.A., Kharel, P., Otterstrom, N.T., Wang, Z., Rakich, P.T., 2018. RF-photonic filters via on-chip photonic–phononic emit-receive operations. J. Light. Technol. 36 (13), 2803–2809. https://ieeexplore.ieee.org/document/8302452/.

Kramer, L., 1971. Theory of light scattering from fluctuations of membranes and monolayers. J. Chem. Phys. 55 (1971), 2097. (URL http://link.aip.org/link/?JCPSA6/55/2097/1%5Cnpapers2://publication/uuid/8CC01EF3-3931-4732-A76D-07DF1E279C07).

Lenz, G., Eggleton, B., Madsen, C., Slusher, R., 2001. Optical delay lines based on optical filters. IEEE J. Quantum Electron. 37 (4), 525–532.

Levis, J., Sutterfield, B., Stevens, R., 2006. Fiber optic communication within the F-35 Mission systems. In: IEEE Conference Avionics Fiber-Optics and Photonics, 2006. Vol. 3. IEEE, pp. 12–13, URL. http://ieeexplore.ieee.org/document/1707476/.

Li, W., Zhu, N.H., Wang, L.X., Wang, J.S., Liu, J.G., Liu, Y., Qi, X.Q., Xie, L., Chen, W., Wang, X., Han, W., 2011. True-time delay line with separate carrier tuning using dual-parallel MZM and stimulated Brillouin scattering-induced slow light. Opt. Express 19 (13), 12312–12324.

Liu, W., Li, M., Guzzon, R.S., Norberg, E.J., Parker, J.S., Lu, M., Coldren, L.A., Yao, J., 2016a. A fully reconfigurable photonic integrated signal processor. Nat. Photonics 10 (3), 190–195.

Liu, Y., Marpaung, D., Choudhary, A., Eggleton, B.J., 2016b. Lossless and high-resolution RF photonic notch filter. Opt. Lett. 41 (22), 5306. (URL https://www.osapublishing.org/abstract.cfm?URI=ol-41-22-5306).

Liu, Y., Choudhary, A., Marpaung, D., Eggleton, B.J., 2018a. Chip-based Brillouin processing for phase control of RF signals. IEEE J. Quantum Electron. 54 (3), 1–13. (URL https://ieeexplore.ieee.org/document/8331967/).

Liu, Y., Choudhary, A., Ren, G., Vu, K., Morrison, B., Casas-Bedoya, A., Nguyen, T.G., Choi, D.-Y., Mitchell, A., Madden, S.J., Marpaung, D., Eggleton, B.J., 2018b. Integrating Brillouin processing with functional circuits for enhanced RF photonic processing. In: 2018 International Topical Meeting on Microwave Photonics (MWP). IEEE, pp. 1–4. URL. https://ieeexplore.ieee.org/document/8552914/.

Liu, Y., Choudhary, A., Ren, G., Vu, K., Morrison, B., Casas-Bedoya, A., Nguyen, T.G., Choi, D.-Y., Ma, P., Mitchell, A., Madden, S.J., Marpaung, D., Eggleton, B.J., 2019. Integration of Brillouin and passive circuits for enhanced radio-frequency photonic filtering. APL Photonics 4 (10), 106103. URL. https://doi.org/10.1063/1.5113569 http://aip.scitation.org/doi/10.1063/1.5113569.

Loayssa, A., Lahoz, F.J., 2006. Broad-band RF photonic phase shifter based on stimulated brillouin scattering and single-sideband modulation. IEEE Photon. Technol. Lett. 18 (1), 208–210.

Loayssa, A., Benito, D., Garde, M.J., 2002. Applications of optical carrier Brillouin processing to microwave photonics. Opt. Fiber Technol. 8 (1), 24–42.

Marpaung, D., Morrison, B., Pant, R., Eggleton, B.J., 2013a. Frequency agile microwave photonic notch filter with anomalously high stopband rejection. Opt. Lett. 38 (21), 4300–4303. (URL http://www.opticsinfobase.org/viewmedia.cfm?uri=ol-38-21-4300&seq=0&html=true).

Marpaung, D., Roeloffzen, C., Heideman, R., Leinse, A., Sales, S., Capmany, J., 2013b. Integrated microwave photonics. Laser Photonics Rev. 7 (4), 506–538. URL. https://doi.org/10.1002/lpor.201200032 http://doi.wiley.com/10.1002/lpor.201200032.

Marpaung, D., Morrison, B., Pagani, M., Pant, R., Choi, D.-Y., Luther-Davies, B., Madden, S.J., Eggleton, B.J., 2015. Low-power, chip-based stimulated Brillouin scattering microwave photonic filter with ultrahigh selectivity. Optica 2 (2), 76. (URL https://www.osapublishing.org/abstract.cfm?URI=optica-2-2-76).

Marpaung, D., Yao, J., Capmany, J., 2019. Integrated microwave photonics. Nat. Photonics 13 (2), 80–90. URL https://doi.org/10.1038/s41566-018-0310-5 http://www.nature.com/articles/s41566-018-0310-5.

Matthews, P.J., 2014. Analog and digital photonics for future military systems. In: Conference on Optical Fiber Communication, Technical Digest Series, pp. 3–5.

McKay, L., Merklein, M., Bedoya, A.C., Choudhary, A., Jenkins, M., Middleton, C., Cramer, A., Devenport, J., Klee, A., DeSalvo, R., Eggleton, B.J., 2019. Brillouin-based phase shifter in a silicon waveguide. Optica 6 (7), 907. (URL https://www.osapublishing.org/abstract.cfm?URI=optica-6-7-907. https://opg.optica.org/abstract.cfm?URI=optica-6-7-907).

Morrison, B., Marpaung, D., Pant, R., Li, E., Choi, D.-Y., Madden, S., Luther-Davies, B., Eggleton, B.J., 2014. Tunable microwave photonic notch filter using on-chip stimulated Brillouin scattering. Opt. Commun. 313, 85–89. (URL http://linkinghub.elsevier.com/retrieve/pii/S0030401813008869).

Morrison, B., Casas-Bedoya, A., Ren, G., Vu, K., Liu, Y., Zarifi, A., Nguyen, T.G., Choi, D.-Y., Marpaung, D., Madden, S.J., Mitchell, A., Eggleton, B.J., 2017. Compact Brillouin devices through hybrid integration on silicon. Optica 4 (8), 847 (URL http://arxiv.org/abs/1702.05233 https://www.osapublishing.org/abstract.cfm?URI=optica-4-8-847).

Morton, P.A., Khurgin, J.B., 2009. Microwave photonic delay line with separate tuning of the optical carrier. IEEE Photon. Technol. Lett. 21 (22), 1686–1688.

Moss, D.J., Morandotti, R., Gaeta, A.L., Lipson, M., 2013. New CMOS-compatible platforms based on silicon nitride and Hydex for nonlinear optics. Nat. Photonics 7 (July), 597–607.

Okawachi, Y., Bigelow, M.S., Sharping, J.E., Zhu, Z., Schweinsberg, A., Gauthier, D.J., Boyd, R.W., Gaeta, A.L., 2005. Tunable all-optical delays via brillouin slow light in an optical fiber. Phys. Rev. Lett. 94 (15), 1–4.

Otterstrom, N.T., Kittlaus, E.A., Gertler, S., Behunin, R.O., Lentine, A.L., Rakich, P.T., 2019. Resonantly enhanced nonreciprocal silicon Brillouin amplifier. Optica 6 (9). (URL http://arxiv.org/abs/1903.03907).

Pagani, M., Marpaung, D., Choi, D.-Y., Madden, S.J., Luther-Davies, B., Eggleton, B.J., 2014. Tunable wideband microwave photonic phase shifter using on-chip stimulated Brillouin scattering. Opt. Express 22 (23), 28810. (URL https://www.osapublishing.org/oe/abstract.cfm?uri=oe-22-23-28810).

Pan, S., Zhang, Y., 2020. Microwave photonic radars. J. Light. Technol. 38 (19), 5450–5484 (URL https://ieeexplore.ieee.org/document/9089228/).

Pant, R., Li, E., Choi, D.-Y., Poulton, C.G., Madden, S.J., Luther-Davies, B., Eggleton, B.J., 2011a. Cavity enhanced stimulated Brillouin scattering in an optical chip for multiorder stokes generation. Opt. Lett. 36 (18), 3687. (URL https://www.osapublishing.org/ol/abstract.cfm?uri=ol-36-18-3687).

Pant, R., Poulton, C.G., Choi, D.-Y., McFarlane, H., Hile, S., Li, E., Thevenaz, L., Luther-Davies, B., Madden, S.J., Eggleton, B.J., 2011b. On-chip stimulated Brillouin scattering. Opt. Express 19 (9), 8285–8290. (URL http://www.opticsexpress.org/abstract.cfm?URI=oe-19-9-8285).

Pant, R., Byrnes, A., Poulton, C.G., Li, E., Choi, D.-Y., Madden, S., Luther-Davies, B., Eggleton, B.J., 2012. Photonic-chip-based tunable slow and fast light via stimulated Brillouin scattering. Opt. Lett. 37 (5), 969. (URL http://ol.osa.org/abstract.cfm?URI=ol-37-5-969 https://www.osapublishing.org/abstract.cfm?URI=ol-37-5-969).

Pant, R., Marpaung, D., Kabakova, I.V., Morrison, B., Poulton, C.G., Eggleton, B.J., 2014. On-chip stimulated Brillouin scattering for microwave signal processing and generation. Laser Photonics Rev. 8 (5), 653–666. URL. http://doi.wiley.com/10.1002/lpor.201300154.

Peng, H., Sun, T., Zhang, C., Xie, X., Guo, P., Zhu, L., Hu, W., Chen, Z., 2014. Tunable DC-60 GHz RF generation based on a dual loop Brillouin optoelectronic oscillator. In: European Conference on Optical Communication, ECOC (December 2015).

Preussler, S., Schneider, T., 2015. Stimulated Brillouin scattering gain bandwidth reduction and applications in microwave photonics and optical signal processing. Optim. Eng. 55 (3), 031110. URL http://opticalengineering.spiedigitallibrary.org/article.aspx?doi=10.1117/1.OE.55.3.031110.

Pu, M., Liu, L., Xue, W., Ding, Y., Ou, H., Yvind, K., Hvam, J.M., 2010. Widely tunable microwave phase shifter based on silicon-on-insulator dual-microring resonator. Opt. Express 18 (6), 6172–6182.

Roeloffzen, C.G.H., Zhuang, L., Taddei, C., Leinse, A., Heideman, R.G., van Dijk, P.W.L., Oldenbeuving, R.M., Marpaung, D.A.I., Burla, M., Boller, K.J., 2013. Silicon nitride microwave photonic circuits. Opt. Express 21 (19), 22937. (URL http://www.opticsexpress.org/abstract.cfm?URI=oe-21-19-22937 https://www.osapublishing.org/oe/abstract.cfm?uri=oe-21-19-22937).

Roeloffzen, C.G.H., Hoekman, M., Klein, E.J., Wevers, L.S., Timens, R.B., Marchenko, D., Geskus, D., Dekker, R., Alippi, A., Grootjans, R., van Rees, A., Oldenbeuving, R.M., Epping, J.P., Heideman, R.G., Worhoff, K., Leinse, A., Geuzebroek, D., Schreuder, E., van Dijk, P.W.L., Visscher, I., Taddei, C., Fan, Y., Taballione, C., Liu, Y., Marpaung, D., Zhuang, L., Benelajla, M., Boller, K.-J., 2018. Low-loss Si3N4 TriPleX optical waveguides: technology and applications overview. IEEE J. Sel. Top. Quantum Electron. 24 (4), 1–21. (URL http://ieeexplore.ieee.org/document/8259277/).

Sancho, J., Chin, S., Sagues, M., Loayssa, A., Lloret, J., Gasulla, I., Sales, S., Thevenaz, L., Capmany, J., 2010. True time delay on tunable microwave photonic filter based on stimulated Brillouin scattering in fibers. In: 36th European Conference and Exhibition on Optical Communication. Vol. 22. IEEE, pp. 1–3. URL. http://ieeexplore.ieee.org/document/5621428/.

Sancho, J., Bourderionnet, J., Lloret, J., Combrié, S., Gasulla, I., Xavier, S., Sales, S., Colman, P., Lehoucq, G., Dolfi, D., Capmany, J., De Rossi, A., 2012. Integrable microwave filter based on a photonic crystal delay line. Nat. Commun. 3, 1075.

Schneider, T., Preußler, S., 2014. Quasi-light storage for optical data packets. J. Vis. Exp. 84, 1–8.

Schneider, T., Jamshidi, K., Preubler, S., 2010. Quasi-light storage: a method for the tunable storage of optical packets with a potential delay-bandwidth product of several thousand bits. J. Light. Technol. 28 (17), 2586–2592.

Seeds, A., 2002. Microwave photonics. IEEE Trans. Microw. Theory Tech. 50 (3), 877–887. (URL http://ieeexplore.ieee.org/document/4140112/ http://ieeexplore.ieee.org/document/989971/).

Seeds, A.J., Williams, K.J., 2006. Microwave photonics. J. Light. Technol. 24 (12), 4628–4641. http://ieeexplore.ieee.org/document/4063431/.

Serafino, G., Scotti, F., Lembo, L., Hussain, B., Porzi, C., Malacarne, A., Maresca, S., Onori, D., Ghelfi, P., Bogoni, A., 2019. Toward a new generation of radar systems based on microwave photonic technologies. J. Light. Technol. 37 (2), 643–650.

Shen, B., Lin, H., Merget, F., Azadeh, S.S., Li, C., Lo, G.-Q., Richardson, K.A., Hu, J., Witzens, J., 2019. Broadband couplers for hybrid silicon-chalcogenide glass photonic integrated circuits. Opt. Express 27 (10), 13781.

Shi, T., Chen, Y., 2021. Multiple radio frequency measurements with an improved frequency resolution based on stimulated Brillouin scattering with a reduced gain bandwidth. Opt. Lett. 46 (14), 3460. (URL https://opg.optica.org/abstract.cfm?URI=ol-46-14-3460).

Shin, H., Cox, J.A., Jarecki, R., Starbuck, A., Wang, Z., Rakich, P.T., 2015. Control of coherent information via on-chip photonic-phononic emitter-receivers. Nat. Commun. 6, 6427. http://www.nature.com/doifinder/10.1038/ncomms7427.

Smit, M., Leijtens, X., Ambrosius, H., Bente, E., van der Tol, J., Smalbrugge, B., de Vries, T., Geluk, E.-J., Bolk, J., van Veldhoven, R., Augustin, L., Thijs, P., D'Agostino, D., Rabbani, H., Lawniczuk, K., Stopinski, S., Tahvili, S., Corradi, A., Kleijn, E., Dzibrou, D., Felicetti, M., Bitincka, E., Moskalenko, V., Zhao, J., Santos, R., Gilardi, G., Yao, W., Williams, K., Stabile, P., Kuindersma, P., Pello, J., Bhat, S., Jiao, Y., Heiss, D., Roelkens, G., Wale, M., Firth, P., Soares, F., Grote, N., Schell, M., Debregeas, H., Achouche, M., Gentner, J.-L., Bakker, A., Korthorst, T., Gallagher, D., Dabbs, A., Melloni, A., Morichetti, F., Melati, D., Wonfor, A., Penty, R., Broeke, R., Musk, B., Robbins, D., 2014. An introduction to InP-based generic integration technology. Semicond. Sci. Technol. 29 (8), 083001 (URL http://stacks.iop.org/0268-1242/29/i=8/a=083001?key=crossref.bea5a474059a0a83c3c51c404b579c72).

Song, K.Y., Herráez, M., Thévenaz, L., 2005. Observation of pulse delaying and advancement in optical fibers using stimulated Brillouin scattering. Opt. Express 13 (1), 82–88.

Urick, V.J., Mondich, M.J., Sunderman, C.E., Kozak, D.A., Goetz, P.G., Rabinovich, W.S., Pruessner, M.W., Mahon, R., Williams, K.J., 2016. Microwave phase shifting using coherent photonic integrated circuits. IEEE J. Sel. Top. Quantum Electron. 22 (6), 353–360. http://ieeexplore.ieee.org/document/7479572/.

Van Laer, R., Kuyken, B., Van Thourhout, D., Baets, R., 2015. Interaction between light and highly confined hypersound in a silicon photonic nanowire. Nat. Photonics 9 (3), 199–203. http://arxiv.org/abs/1407.4977. http://www.nature.com/articles/nphoton.2015.11.

Vidal, B., Piqueras, M., a., Martí, J., 2007. Tunable and reconfigurable photonic microwave filter based on stimulated Brillouin scattering. Opt. Lett. 32 (1), 23–25.

Wang, J., Ashrafi, R., Adams, R., Glesk, I., Gasulla, I., Capmany, J., Chen, L.R., 2016. Subwavelength grating enabled on-chip ultra-compact optical true time delay line. Sci. Rep. 6 (1), 30235. URL https://doi.org/10.1038/srep30235 http://www.nature.com/articles/srep30235.

Xu, Y., Yang, Y., Li, X., Wang, X., Zou, W., 2021. Chip-scale Brillouin instantaneous frequency measurement by use of one-shot frequency-to-power mapping based on lock in amplification. Chin. Opt. Lett. 19 (11), 113902. (URL https://www.researching.cn/articles/OJ9ccabe48ea7dabf4/html).

Yao, X.S., 1997. High-quality microwave signal generation by use of Brillouin scattering in optical fibers. Opt. Lett. 22 (17), 1329. https://www.osapublishing.org/abstract.cfm?URI=ol-22-17-1329.

Yao, J., 2009. Microwave photonics. J. Light. Technol. 27 (3), 314–335. (URL http://ieeexplore.ieee.org/document/4785283/).

Zhang, W., Minasian, R.A., 2011. Widely tunable single-passband microwave photonic filter based on stimulated Brillouin scattering. IEEE Photon. Technol. Lett. 23 (23), 1775–1777.

Zhang, W., Minasian, R.A., 2012a. Switchable and tunable microwave photonic Brillouin-based filter. IEEE Photon. J. 4 (5), 1443–1455. (http://ieeexplore.ieee.org/document/6242369/).

Zhang, W., Minasian, R.A., 2012b. Ultrawide tunable microwave photonic notch filter based on stimulated Brillouin scattering. IEEE Photon. Technol. Lett. 24 (14), 1182–1184. (URL http://ieeexplore.ieee.org/document/6208824/).

Zhang, W., Yao, J., 2016. Silicon-based integrated microwave photonics. IEEE J. Quantum Electron. 52 (1), 1–12. (URL http://ieeexplore.ieee.org/document/7331228/).

Zhang, H., Yin, H., Jia, H., Yang, J., Chang, S., 2011. Study of effects of obstacle on non-line-of-sight ultraviolet communication links. Opt. Express 19 (22), 21216–21226. (URL http://www.opticsexpress.org/abstract.cfm?URI=oe-19-22-21216).

Zou, X., Lu, B., Pan, W., Yan, L., Stöhr, A., Yao, J., 2016. Photonics for microwave measurements. Laser Photonics Rev. 10 (5), 711–734.

CHAPTER TEN

Integrated Brillouin lasers and their applications

Daniel J. Blumenthal[a,*], Irina Kabakova[b], Peter T. Rakich[c], and Kerry Vahala[d]

[a]Department of Electrical and Computer Engineering, University of California Santa Barbara, Santa Barbara, CA, United States
[b]School of Mathematical and Physical Sciences, University of Technology Sydney, Broadway, NSW, Australia
[c]Department of Applied Physics, Yale University, New Haven, CT, United States
[d]Watson Laboratory of Applied Physics, California Institute of Technology, Pasadena, CA, United States
*Corresponding author: e-mail address: danb@ucsb.edu

Contents

Semiconductors and Semimetals, Volume 110
ISSN 0080-8784
https://doi.org/10.1016/bs.semsem.2022.05.004

1. Introduction

The miniaturization of Brillouin technology has the potential to bring the robustness, light weight and small size, and low cost associated with integration to a wide range of applications that require the optical spectral purity and signal processing capabilities of Brillouin light–sound interactions (Komljenovic et al., 2019). Precision applications that today demand spectrally pure light sources (Kessler et al., 2012) include atomic clocks (Ludlow et al., 2007; Young et al., 2020), quantum computing and sensing (Degen et al., 2017; Saffman, 2016), microwave photonics (Marpaung et al., 2013), precision metrology (Patel et al., 2018), atomic and molecular spectroscopy (Bohn et al., 2017; Cygan et al., 2013; Jadbabaie et al., 2020), atomic, molecular, and optical (AMO) (National Academies of Sciences and Engineering and Medicine, 2020), gravitational wave detection (Collaboration, 2016), quantum sensing (Fan et al., 2015), and molecular physics (Bohn et al., 2017). Integrating the ultra-low phase noise and signal processing properties of the Brillouin process to the chip-scale will also enable new applications including integrated microwave photonics (IMWP) (Li et al., 2013, 2014; Marpaung et al., 2013; Merklein et al., 2016), energy efficient digital coherent communications (Blumenthal et al., 2020), gyroscopes (Gundavarapu et al., 2019; Lai et al., 2020; Li et al., 2017), space-based quantum and atomic systems (Loriani et al., 1987), and quantum computing and communications (Orieux and Diamanti, 2016; Saffman, 2016; Wang et al., 2020).

In addition to the reduced size, improved stability, environmental robustness, and manufacturability enabled by integration, spectrally pure sources that emit from the visible (VIS) to infrared (IR) are also enabled. Additionally, new insights, for example into the fundamental source of Brillouin noise in cold environments can be gained from integrated Brillouin structures. Integration also promises combination of Brillouin lasers with other integrated optical components to realize systems-on-chip operating in the UV to IR (Blumenthal, 2020). There has been tremendous progress miniaturizing stimulated Brillouin scattering (SBS) and stimulated Brillouin lasers (SBLs) using a range of materials and integration platforms.

Spectral purity is an important aspect of this topic, and is quantified by measures including laser fundamental linewidth, integral linewidth, fractional frequency stability, and drift. Requirements for these parameters depend on the application. Atomic clocks are among the most demanding, requiring integral linewidths on the order of 1 Hz over a period of 1 to 10 seconds and an ADEV of better than 10^{-17} over 1 second. In general, these requirements are defined by the open loop portion of the system or experimental cycle, for example the loading, cooling, and trapping time of atoms where the laser must remain within lock range of the desired atomic transition. These precision laser systems are realized by Pound–Drever–Hall (PDH) (Black, 2000; Drever et al., 1983) locking of a low fundamental linewidth laser to an ultra-stable exotic reference cavity that is table-top scale, with a vacuum cavity and ultra-high finesse semiconductor crystalline vacuum spaced resonators, employing sophisticated athermal designs and other features to mitigate intrinsic thermal fluctuations (Ludlow et al., 2007; Notcutt et al., 2006). Combining these stabilization techniques with the ultra-low fundamental linewidth of the Brillouin laser, and integrated reference cavities that are CMOS compatible will enable precision portable applications as well as systems-on-chip integration.

In terms of integration, the demand for silicon compatible photonics has accelerated the development of SOI-based silicon photonics, a combination of silicon on insulator (SOI) and III–V11 photonics with CMOS electronics. An illustration of silicon nitride SBL laser chip-scale applications is illustrated in Fig. 1, including atomic clocks and sensors, optical gyroscopes, high capacity fiber optic links, and ultra-low phase noise microwave oscillators. Both SOI and III–V photonics deliver solutions from the communication wavelengths to near-infrared (NIR), e.g., 780 to 2500 nm with recent results in the mid-infrared (MIR), e.g., 2 to 20 μm. For applications that operate at wavelengths shorter than 1 μm, wide bandgap materials such as silicon nitride, aluminum nitride, tantala, and chalcogenide-integrated platforms are utilized. The Brillouin gain process in particular has been achieved in silica, silicon nitride, chalcogenide, and silicon.

In this chapter we give a brief history of the Brillouin gain process in optical waveguides and SBLs realized with integrated Brillouin resonators. A brief discussion on the laser fundamental and integral linewidths, the fractional frequency noise, and drift is provided. We discuss progress to date on Brillouin processes in integrated silica, silicon nitride, silicon, and chalcogenide platforms. The underlying Brillouin physics in waveguide materials and the range of waveguide structures and related requirements for photon–phonon phase matching are covered as well as research efforts to implement SBS gain and lasing in these waveguide material systems. Cascaded Brillouin lasing is discussed, including mode engineering approaches to inhibit cascading to enable further linewidth narrowing and higher output powers. Finally, several applications using these integrated Brillouin technologies are presented.

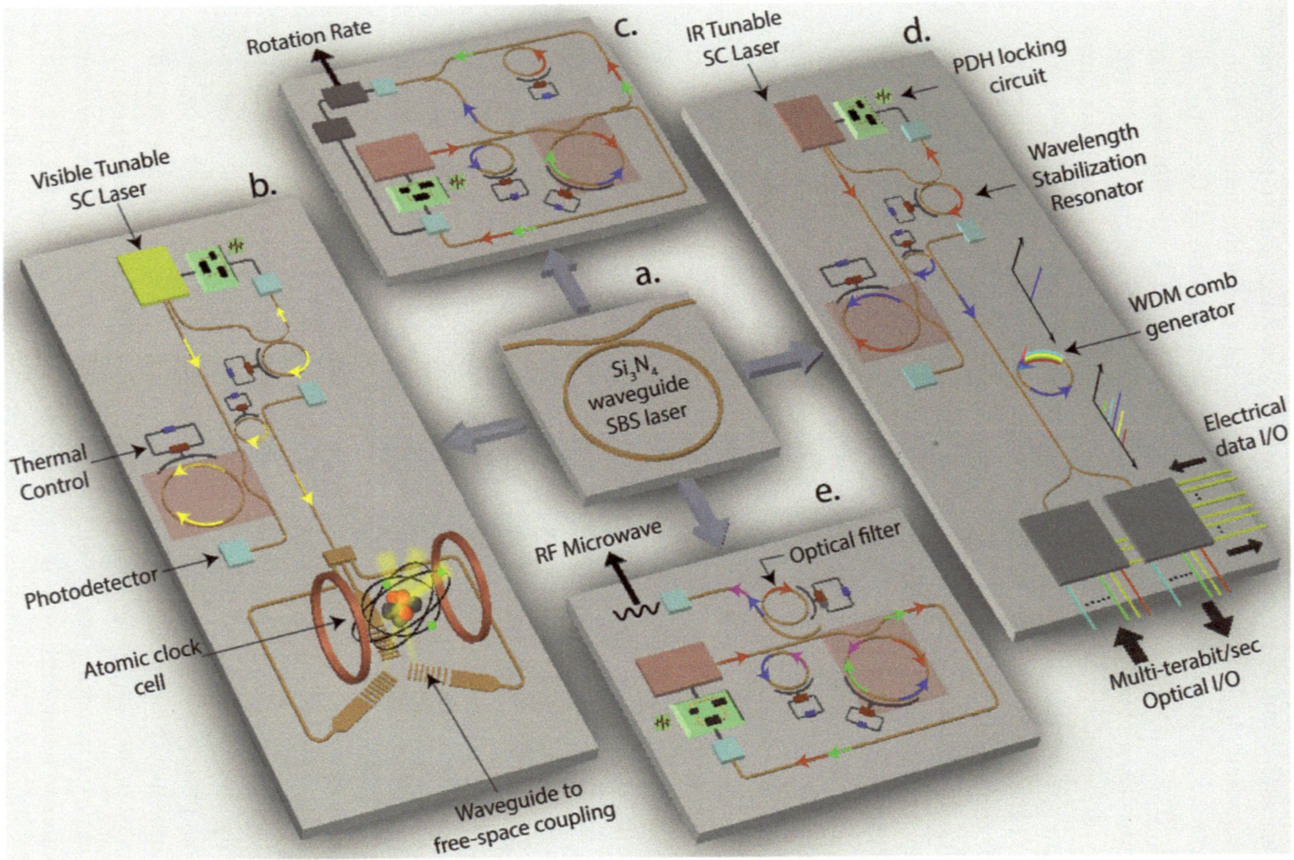

Fig. 1 Si_3N_4 waveguide Brillouin laser and system-on-chip application examples. (A) Si_3N_4 waveguide stimulated Brillouin scattering (SBS) laser. (B) Chip-scale atomic clock. (C) Integrated laser optical gyroscope. (D) WDM multiterabit per second coherent transceiver. *Credit Gundavarapu, S., Brodnik, G. M., Puckett, M., Huffman, T., Bose, D., Behunin, R., Wu, J., Qiu, T., Pinho, C., Chauhan, N., Nohava, J., Rakich, P. T., Nelson, K. D., Salit, M., Blumenthal, D. J., 2019. Sub-hertz fundamental linewidth photonic integrated Brillouin laser. Nat. Photonics 13 (1), 60–67. https://doi.org/10.1038/s41566-018-0313-2. Gundavarapu et al. (2019) is licensed under CC BY 4.0.*

2. Phase noise and linewidth

2.1 Background

The high coherence of fiber Brillouin lasers (Stokes et al., 1982) was recognized decades ago and applied to generate highly coherent microwave signals (Callahan et al., 2011; Yao, 1997). Miniature, Brillouin sources are a more recent development made possible by systematic efforts in boosting optical Q factor (equivalently, reducing optical waveguide loss) of microresonators. While initial demonstrations of these devices used discrete ultra-high-Q resonators (Grudinin et al., 2009; Tomes and Carmon, 2009), steady progress made possible silicon-chip-based devices using dielectrics including chalcogenides, silica, silicon, and silicon nitride (Gundavarapu et al., 2019; Lee et al., 2012; Loh et al., 2015; Otterstrom et al., 2018a; Pant et al., 2011a; Yang et al., 2018). Most recently, ultra-high-Q resonators fabricated on CMOS foundry lines are used to make Brillouin lasers that can be readily integrated with other optical components and electronics, creating the prospect of ultra-high coherence integrated photonic systems. This steady progress has resulted in ultra-narrow linewidth Brillouin lasers (Gundavarapu et al., 2019; Li et al., 2012a). In addition to applications in radio-frequency synthesis (Li et al., 2013, 2014), they are being applied in ring-laser gyroscopes (Gundavarapu et al., 2019; Lai et al., 2020; Li et al., 2017) and high-coherence reference sources (Liu et al., 2022a; Loh et al., 2015). This work will ultimately enable a transition of traditionally tabletop systems to far more compact and rugged integrated ones. It may also make possible entirely new application opportunities for Brillouin-integrated photonics (Eggleton et al., 2013).

2.2 Parametric nature of Brillouin lasers

Brillouin laser action is, at a fundamental level, a parametric process in which a strong pump wave scatters in one of two possible ways with a phonon into two new waves: the Stokes wave resulting from phonon emission, and the anti-Stokes wave resulting from phonon absorption (see chapter "SBS-based fiber sensors" by Zadok et al.). Most microcavity-based Brillouin lasers are of the whispering gallery type, featuring ring or race-track style geometries that support only one or a few transverse spatial modes. In these geometries, phase matching of the scattering process is greatly restricted so that Stokes and anti-Stokes waves propagate opposite to the initial pumping direction.

More generally, when a dense spectrum of transverse modes exists, forward scattering is also possible (Bahl et al., 2012). The switching of propagation direction requires a microwave-frequency phonon that is near microwave X-band rates for optical pumps operating in the telecom band. The Stokes and anti-Stokes waves are accordingly down-shifted (phonon emission) or up-shifted (phonon absorption) in frequency by this Brillouin shift frequency.

Microwave-rate (frequency) phonons experience damping rates that are typically much greater than the optical damping of a cavity mode in an ultra-high-Q resonator. This strong difference or asymmetry in damping rates means that, dynamically, the phonon field amplitude (both emission and absorption) follows the optical fields, and it can be eliminated adiabatically within the equations of motion. As a result, the Brillouin laser dynamics can be understood in terms of optical fields only. A notable feature of this arrangement is that the Stokes wave experiences power build-up that is approximately proportional to the number of photons in the Stokes mode (and also to the pump wave intensity). This situation, which is also characteristic of Brillouin lasers in optical fibers, is responsible for the term "stimulated" appearing in the name stimulated Brillouin scattering (when in fact the process is parametric). Under this strongly damped phonon approximation, the Stokes wave experiences gain responsible for laser action, while the anti-Stokes wave experiences absorption. For efficient stimulated scattering and amplification of the Stokes wave, the frequency of the Stokes mode must lie within the Brillouin gain spectrum, which itself is centered at a frequency equal to the pumping mode frequency less the Brillouin frequency shift (see Fig. 2). Ensuring this spectral arrangement requires precise dimensional control of the resonator, the required tolerances for which are typically a few microns and readily attainable with lithographically defined resonators.

2.3 Pump frequency noise

Even while Brillouin lasers mimic the stimulated process, their underlying parametric nature is critical to understanding the way fundamental sources of noise impact their linewidth. Consider now a Brillouin laser that is pumped above threshold and operating on a single Stokes line. Its phase noise and linewidth are the result of several different noise mechanisms that are now considered individually. First is a technical source of source that occurs on account of the parametric (phase sensitive) nature of the Brillouin scattering process. Here, a coupling of the pump frequency noise

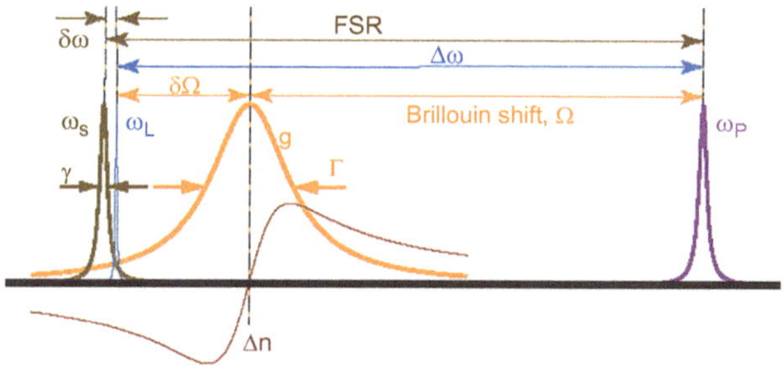

Fig. 2 Brillouin gain process in the frequency domain. A pump wave at ω_P (*purple*) creates Brillouin gain (g, *orange*) for a cavity mode at ω_S (*brown*). A Brillouin laser field (*blue*) oscillates at ω_L. The Brillouin-induced refractive index (Δn, *red*) induces frequency pulling $\delta\omega$ of the laser frequency. Brillouin shift frequency (Ω), gain spectrum linewidth (Γ), and cavity linewidth (γ) are also indicated. Shifts in the pumping frequency cause tuning of the gain spectrum maximum that in turn shifts the refractive index relative to ω_S. This changes the amount of pulling $\delta\omega$ thereby coupling pumping frequency noise to Brillouin laser noise. *Reprinted with permission from Yuan, Z., Wang, H., Wu, L., Gao, M., Vahala, K., 2020. Linewidth enhancement factor in a microcavity Brillouin laser. Optica 7 (9), 1150–1153. https://opg.optica.org/optica/fulltext.cfm?uri=optica-7-9-1150&id=437890. © The Optical Society.*

into the Stokes frequency occurs. This coupling is dramatically suppressed by the quasi-stimulated feature of Brillouin scattering (i.e., resulting from strongly damped phonon modes). The coupling is caused by frequency pulling $\delta\omega_S$ of the Stokes frequency when the pump frequency shifts by $\delta\omega_P$. The physical origin of this pulling is easily explained (here we give a short explanation in the caption of Fig. 2). Its impact on the Stokes wave, is to directly transfer frequency fluctuations (equivalently phase noise) in the pump to the Stokes laser field. Accordingly, a simple relationship connects the phase noise spectral density of the pump wave $W_P(\Omega)$ to that of the Stokes wave $W_S(\Omega)$ as follows (Debut et al., 2000),

$$W_S(\Omega) = \left[\frac{\gamma}{\gamma + \Gamma}\right]^2 W_P(\Omega) \tag{1}$$

where γ and Γ are the photon and phonon damping rates, respectively. Because the ratio Γ/γ is typically a factor of 30–100, the pulling factor suppresses the pump phase noise by 30–40 dB. However, even with this strong suppression, the ability of Brillouin lasers to operate at sub Hertz fundamental linewidths (see below) means that care must be taken in pump selection.

It should also be noted that an important category of applications, which include Brillouin gyroscopes (Gundavarapu et al., 2019; Lai et al., 2020; Li et al., 2017), operate with co-lasing Stokes waves that are pumped by the same pumping laser, but along opposite directions of the whispering gallery resonator. In this case, the above pumping noise becomes common-mode to the two Stokes waves. The relative phase noise of these waves, which sets the sensitivity limit of the gyroscope, is therefore unaffected by the pump frequency noise. This common mode noise cancellation is highly effective and has been verified to offset frequencies as low as 0.1 Hz in lasers having sub-Hertz fundamental linewidths (Lai et al., 2020).

2.4 Fundamental frequency noise (ST linewidth)

Fundamental phase noise of conventional lasers is the result of spontaneous emission into the laser mode. This produces a spectrally flat frequency noise spectrum and a corresponding linewidth given by the modified Schalow–Townes linewidth formula (Henry, 1982; Lax, 1967; Schawlow and Townes, 1958),

$$\Delta\nu_{ST} = \frac{\hbar\omega^3}{4\pi Q_T Q_{ex} P}\left(N_{th} + \frac{1}{2}\right)(1 + \alpha^2) \tag{2}$$

where ω is the laser frequency, Q_T (Q_{ex}) is the total (external) Q-factor ($1/Q_T = 1/Q_0 + 1/Q_{ex}$ where Q_0 is the intrinsic Q factor), P is the laser output power, N_{th} is the thermo-optical quanta in the laser mode (negligible at room temperature), α is the linewidth enhancement factor (Henry factor) (Henry, 1982), and $\hbar$ is the reduced Planck's constant. Key features of this noise are its inverse power dependence and its inverse dependence upon the Q product $Q_T Q_{ex}$.

While these same features are present in Brillouin laser, the underlying driver of fundamental frequency noise is the result of thermal phonons, not spontaneous emission. Physically, this results from participation of a phonon field in the generation of the Stokes laser wave. This mechanical field provides a channel through which thermo-mechanical bath energy couples directly into the Stokes field. At microwave rates, this bath energy results in large thermo-mechanical populations of phonons, greatly increasing the frequency noise and linewidth of the Stokes wave, as well as making its value proportional to temperature (Li et al., 2012a; Suh et al., 2017). The importance of this source of frequency noise, which is absent in conventional lasers, was not recognized in Brillouin lasers until quite recently and results in a fundamental linewidth given by (Li et al., 2012a; Yuan et al., 2020),

$$\Delta\nu_{SBL} = \left(\frac{\Gamma}{\gamma+\Gamma}\right)^2 \frac{\hbar\omega_{\rm L}^3}{4\pi Q_{\rm T} Q_{\rm ex} P_{\rm SBL}} (n_{\rm th} + {\rm N}_{\rm th} + 1)(1+\alpha^2) \tag{3}$$

Comparing this expression with Eq. (2), the factor involving power and optical Q is identical. However, there are several differences. First, and most important, is the appearance of n_{th}, the number of thermal quanta in the phonon mode at the temperature of operation. At room temperature $n_{th} \approx k_B T/\hbar\Omega$ which increases the fundamental Brillouin linewidth by nearly 600× relative to that of a conventional laser having otherwise comparable properties. Other less important factors are the near-unity correction factor $[\Gamma/(\gamma+\Gamma)]^2$ and the appearance of a "1" instead of "1/2" in the parentheses. The former relates to the finite damping rate of the phonons (Yuan et al., 2020), while the latter can be interpreted as the combined zero point contributions of optical and mechanical fields to the linewidth (Li et al., 2012a), the latter of which is absent in Eq. (2).

Finally, it is important to comment on the linewidth enhancement factor α in Eq. (3). As in conventional lasers this factor is associated with detuned operation of the laser frequency relative to the gain maximum. In Brillouin lasers, it has been recently studied and measured to produce enhancements as large as 50× for detuned operation. However, here again the parametric nature of the Brillouin process plays a major role, and α is determined by the degree to which the laser operation is properly phase matched. Specifically, α is given by two equivalent frequency-detuning quantities relative to perfect phase matching (Yuan et al., 2020),

$$\alpha = \frac{2\delta\Omega}{\Gamma} = \frac{2\delta\omega}{\gamma} \tag{4}$$

In the first equality, phonon mode detuning $\delta\Omega \equiv \Omega - \Delta\omega$ is normalized by Γ, the Brillouin gain bandwidth (i.e., phonon decay rate constant). In the second equality, optical mode detuning $\delta\omega \equiv \Delta\omega - \text{FSR}$ (FSR is the unpumped cavity free-spectral-range) is normalized by γ, the photon decay rate constant. Note that the sign of α changes to either side of perfect phase matching. Refer to Fig. 2 for additional details.

Despite the large phonon enhancement of fundamental frequency noise relative to conventional lasers, it is nonetheless possible to achieve very narrow linewidths in Brillouin lasers on account the large optical Q factors possible in high-performance microresonators. Sub-Hertz linewidth performance has been demonstrated in silica and silicon nitride devices including observation of the inverse power dependence (see Fig. 3). These data have also been used

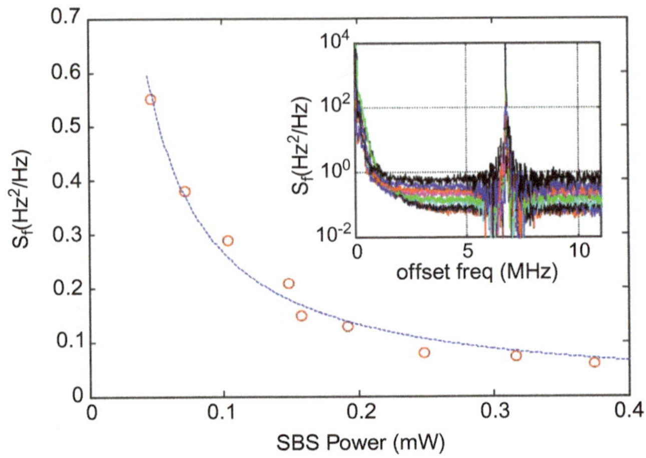

Fig. 3 Fundamental Frequency spectral density vs laser output power. The white noise frequency fluctuation spectral density for single laser line operation is measured vs laser power. A noise level less than 0.1 Hz^2/Hz is measured at less than 0.5 mW output power, corresponding to a short-term linewidth around 0.5 Hz. Inset shows the actual frequency fluctuation spectral densities measured at a series of output powers using a frequency discriminator (interferometer). The singularity in the spectrum near 6 MHz corresponds to the null in the transfer function of the interferometer. *Reprinted with permission from Li, J., Lee, H., Chen, T., Vahala, K. J., 2012a. Characterization of a high coherence, Brillouin microcavity laser on silicon. Opt. Express 20 (18), 20170–20180. © The Optical Society.*

to extract n_{th} for comparison to theory (Li et al., 2012a) as well as to verify the temperature dependence of the fundamental linewidth (Suh et al., 2017). Finally, the α factor has also been measured and compared to theory (see Fig. 4) (Yuan et al., 2020).

2.5 Frequency noise from Brillouin cascade

So far, the frequency noise of single line Brillouin lasers (Fig. 5A) has been discussed. As described earlier there are two phonon scattering processes that occur in Brillouin lasers: the Stokes and anti-Stokes (see green arrows in Fig. 5A). The Stokes process involves phonon emission and is responsible for transferring pump power to the laser field. The anti-Stokes process also performs a critical function by inducing absorption at the pumping frequency. This absorption, which is proportional to the laser field power, clamps the circulating pumping power even as external pump power is increased. This clamping is none other than the familiar gain clamping

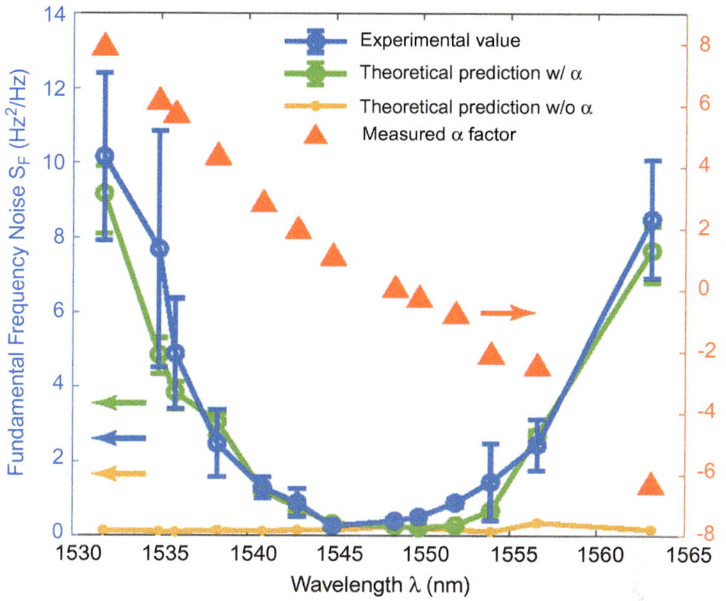

Fig. 4 Brillouin laser linewidth enhancement factor measurement and comparison to theory. Brillouin laser frequency noise (*blue*) measured for a series laser pumping wavelengths. Variation of the pumping wavelength changes the phase matching condition and therefore the α parameter. This causes increases in linewidth by up to 50× for non-phase-matched operation. By measuring the amount of frequency pulling the theoretical α can be determined from Eq. (4). This is plotted as the *red triangles*. Phase matched operation corresponds to $\alpha = 0$. Using these values, the theoretical linewidth prediction (*green*) is given using the nonenhanced linewidth prediction (*yellow*)that was determined by measurement of optical *Q* factors and other parameters. *Reprinted with permission from Yuan, Z., Wang, H., Wu, L., Gao, M., Vahala, K., 2020. Linewidth enhancement factor in a microcavity Brillouin laser. Optica 7 (9), 1150–1153. https://opg.optica.org/optica/fulltext.cfm?uri=optica-7-9-1150&id=437890. © The Optical Society.*

associated with laser operation. In terms of frequency noise, only the Stokes emission process is involved. It creates the large thermal noise contribution described above.

However, with increased pumping, a Brillouin laser field can act as a pumping wave for a second Brillouin laser field. For this to happen, a neighboring mode must exist that satisfies the phase matching condition. This condition is normally satisfied in whispering gallery resonators so that cascading will occur readily (see Fig. 5B and C) (Li et al., 2012a) as well as in silicon nitride waveguide resonator-based Brillouin lasers (Behunin et al., 2018; Gundavarapu et al., 2019). Considering the spectral diagram in Fig. 5B, the cascaded laser 2 will now clamp the circulating power of

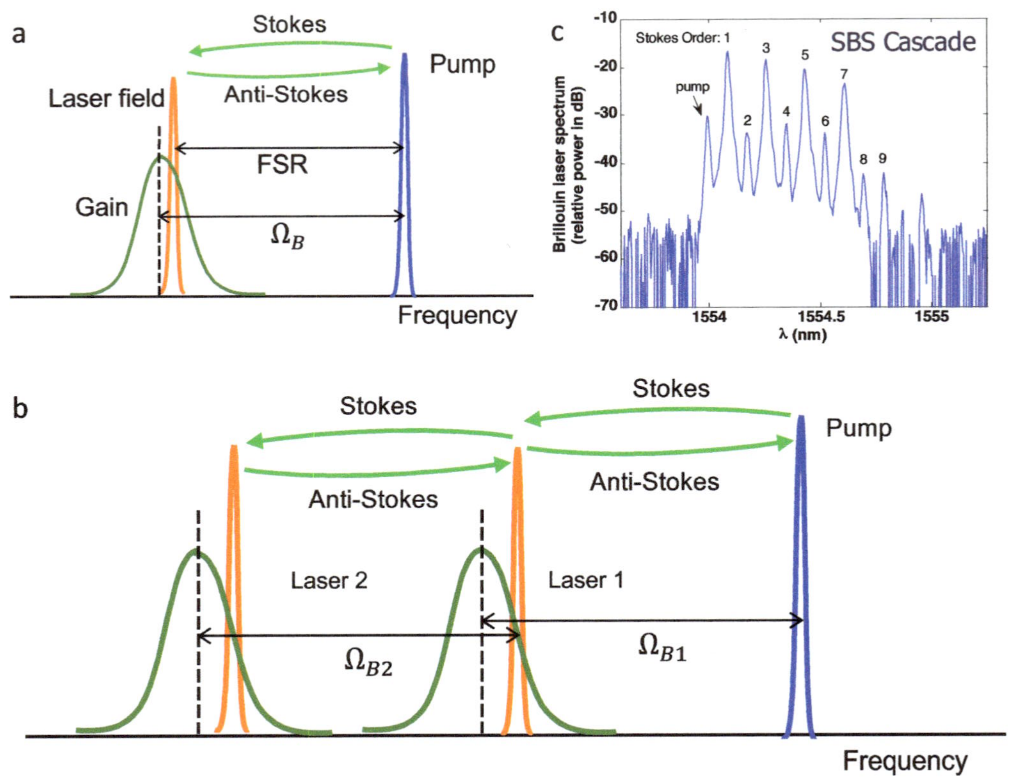

Fig. 5 Description of cascaded Brillouin laser operation and noise mechanisms. (A) Illustration of single line Brillouin laser operation in which a pump laser generates Brillouin gain as described in Fig. 2. Stokes phonon emission and anti-Stokes phonon absorption accompany this process. The former provides a noise channel for thermo-mechanical phonons to couple into the laser field creating the thermal contribution to linewidth given in Eq. (3). (B) Upon cascade, an initial Brillouin laser field (laser 1) will function as a pumping field for generation of a new Brillouin laser field (laser 2). Laser 1 is now subject to an additional noise channel created by the anti-Stokes phonon absorption originating with laser 2. (C) Measured optical spectrum from Li et al. (2012a) showing cascaded operation of a Brillouin laser to nine orders. Also indicated: Brillouin shift frequencies (Ω_B, Ω_{B1}, Ω_{B2}) and free spectral range (FSR). *Reprinted with permission from Li, J., Lee, H., Chen, T., Vahala, K. J., 2012b. Characterization of a high coherence, Brillouin microcavity laser on silicon. Opt. Express 20, 20170–20180. https://doi.org/10.1364/OE.20.020170. © The Optical Society.*

laser 1 (i.e., induce gain clamping) after reaching threshold. And, as before, this clamping occurs by way of the anti-Stokes process. However, it now has an interesting impact on the frequency noise of the laser field 1, which is functioning both as a laser and a pumping field. Specifically, field 1 now receives thermo-mechanical bath energy through two channels: the original channel provided by the Stokes process and a new cascade channel provided by the absorbed phonons of the anti-Stokes process. This additional noise enhances the frequency noise of the laser by adding terms to the linewidth. A specific example follows for a Brillouin laser cascaded to three orders and having equal optical decay rates and Brillouin couplings, in which case the linewidths of the first and third Stokes orders are (Behunin et al., 2018):

$$\Delta\nu_1 = \frac{\hbar\omega_{\mathrm{L1}}^3}{4\pi Q_{\mathrm{T}} Q_{\mathrm{ex}} P_{\mathrm{SBL1}}}\left(3n_{\mathrm{th}} + \mathrm{N}_{\mathrm{th}} + 2\right) \tag{5}$$

$$\Delta\nu_3 = \frac{\hbar\omega_{\mathrm{L3}}^3}{4\pi Q_{\mathrm{T}} Q_{\mathrm{ex}} P_{\mathrm{SBL3}}}\left(n_{\mathrm{th}} + \mathrm{N}_{\mathrm{th}} + 1\right) \tag{6}$$

where α is omitted as well as the near-unity factor associated with finite phonon damping. Also, thermo-mechanical occupancies are taken as equal for phonon populations involved in Stokes and anti-Stokes processes. As expected, the third Stokes order linewidth is unchanged from the non-cascaded case (Eq. 3). However, the first-order Stokes features extra thermo-mechanical noise and quantum noise. To avoid cascade noise, it is possible to operate the Brillouin laser at reduced powers; however, this has the disadvantage of limiting the possible linewidth narrowing. Therefore, methods have been developed to inhibit cascade by introduction of cavity dispersion so as to frustrate the cascade phase-matching condition (Liu et al., 2022b; Merklein et al., 2015; Puckett et al., 2019; Wang et al., 2021). In effect, a mode frequency is shifted away from the spectral location where ideal phase matching would occur.

2.6 Integral linewidth, fractional frequency noise, and drift

As discussed previously, the Brillouin laser fundamental linewidth is set by Eq. (3), which depends on the cavity Q and the number of stored photons in the cavity, as well as fundamental noise from quantum optomechanical modes and linewidth broadening mechanisms. The statistics that contribute to this linewidth can be described by white frequency noise, a memory-less process (Markov) that has a flat power spectrum out to high frequencies

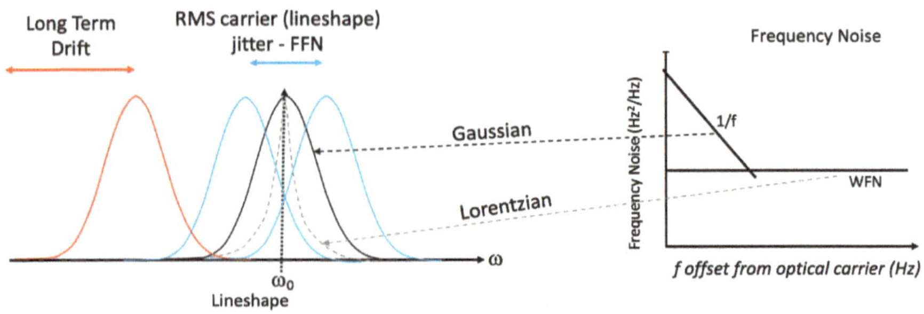

Fig. 6 Relation between laser frequency noise, as a function of frequency offset from carrier (in units of Hz^2/Hz), and laser lineshape. White noise defines Lorentzian fundamental linewidth and energy in wings.

(offset from the optical carrier), which represents random Brownian wander of the laser optical phase over the interval $[0,2\pi]$. Being a Markov process, the fundamental field phase is highly temporally uncorrelated, and can be described by a decaying exponential auto-correlation function. The Wiener–Khinchin theorem relates the laser power spectral density to the Fourier transform of its auto-correlation function (assuming a well-behaved stationary random process), which for white frequency noise is a Lorentzian laser lineshape as illustrated in the gray dashed curve on the left and the flat noise line on the right in Fig. 6. This lineshape is what would be observed if the emission spectrum were captured on a fast time scale; hence it is often referred to as the instantaneous linewidth. The energy in the wings of this lineshape is generated by the high frequency white noise components.

However, only the best stabilized lasers systems in the world are able to emit light approaching a Lorentzian linewidth determined by the fundamental noise sources, and this near perfect output will eventually drift over time. Most laser systems deviate from the fundamental linewidth because the laser cavities and their mirrors are susceptible to internal noise from random index of refraction fluctuations. These fluctuations are driven by random microscopic thermal motion, absorption of random light intensity, and external noise generated by environmental perturbations such as acoustic, mechanical, and ambient temperature fluctuations. The two internal noise sources known as thermorefractive noise (TRN) and photothermal noise (PTN), respectively, are random processes with memory due to various microscopic and macroscopic thermal time constants. In general, these internal and external noise contributions have behavior as a function of frequency that can be described by a polynomial expansion in $1/f^n$. The most well-known of these is the 1/f, or flicker noise, illustrated in the right-hand plot of Fig. 6. This type of noise source generates Gaussian shoulders on the Lorentzian by modulating the fundamental linewidth as depicted in the solid black curve in the left-hand plot of Fig. 6 (Di Domenico et al., 2010). This Gaussian linewidth is essentially filled in by 1/f noise modulation and other technical noise sources described by higher order $1/f^n$ terms (where n can be an integer or noninteger). At very low frequencies, e.g., 10 Hz and lower, the Gaussian lineshape itself will have jitter in carrier frequency, for example the $1/\sigma$ of the jitter distribution is called the fractional frequency noise (FFN). The FFN determines (dominates) the linewidth over a certain observation interval, and the FFN is a percentage of the carrier frequency itself. For example a laser operating at 1550 nm (approximately 193 THz) with a 5.2×10^{-15} FFN describes a 1 Hz linewidth laser. This noise term and its observation interval can also described in terms of the Allan deviation (ADEV) (Allan, 1987). An application will specify the FFN and observation interval, for example an ADEV of 5.2×10^{-15} at 1 s, for the laser to remain within the lock range of a desired atomic transition during open-loop operation when atoms are being loaded and trapped such as the case with atomic clocks. This linewidth jitter is illustrated as the light blue curves of Fig. 6. Over longer time intervals, the carrier, and its lineshape and jitter will drift as the ambient temperature and other long term influences affect the laser cavity and mirrors, unless the laser is locked to an absolute reference such as an atom.

By locking the laser to a reference cavity using a PDH loop (Drever et al., 1983), such as done in precision laser systems (Kessler et al., 2012), the laser takes on the noise characteristics and stability of that cavity. The internal noise contributions of such a cavity down to several Hz is to a large part dominated by the TRN and PTN that are reduced by enlarging the optical mode volume of the cavity and the long-term drift is mitigated by athermalizing and environmentally isolating the reference cavity. In recent years, significant progress has been made with implementing low-thermal noise reference cavities using large mode integrated silicon nitride waveguide resonators. Shown in Fig. 7A is a 1.2-m spiral silica waveguide resonator used to stabilize a fiber laser with relative ADEV of 3.9×10^{-13} at 400 μs averaging time with an integral linewidth measuring less than 100 Hz (Lee et al., 2013). More recently an integrated 4-m long silicon nitride-integrated waveguide resonator was used to lock a semiconductor laser with a relative ADEV of 1.8×10^{-13} at 10 ms with an integral linewidth measuring 36 Hz (Liu et al., 2022a).

Fig. 7 Large mode volume photonic integrated silica and silicon nitride stabilization cavities. (A) Photograph of spiral silica waveguide resonators (Lee et al., 2013). *Left*: 1.2 m spiral resonator. *Upper right*: 4.5 cm spiral resonator. *Lower right*: quarter shown to provide scale. Scale bar, 1 cm. (B) Photograph of 4-m long coil waveguide resonator Liu et al. (2022b) with dimensions shown.

3. Integrated Brillouin platforms

Integration of Brillouin gain and lasing into a photonic platform carries multiple benefits and challenges. Benefits include robust operation, insensitivity to environment disturbances, compact, lightweight design, and lower cost, as well as combination with other components to realize Brillouin-based systems-on-chip. Challenges include creating waveguides and resonators that ensure large Brillouin gain and phase matching between photons and phonons at the wavelengths of interest, properties that are important for linewidth narrowing including low waveguide loss and broad Brillouin gain bandwidth, and integration of the SBS pump, filters and isolators, and modulators. In this section we cover methods and platforms investigated over the last couple of decades, which have enabled translating Brillouin lasing from the bulk- and fiber-based systems to the chip-scale; namely silica, silicon nitride, chalcogenide, and silicon. We break down each platform into key components: a basic discussion of the materials and waveguides, realizing Brillouin gain, forming SBS resonators, fabrication of SBS lasers, and performance measures including phase noise and linewidth.

SBS in silica: Because Brillouin scattering is a signal impairment mechanism in fiber optic communication systems, its properties in silica are well known. It has also gained attention on account of the exceptional coherence properties of fiber-optic-based Brillouin lasers. The earliest Brillouin microcavity lasers used small high-Q spherical resonators (Tomes and Carmon, 2009) or crystalline resonators (Grudinin et al., 2009). These evolved to include silica-on-silicon devices microfabricated as both discrete (Lee et al., 2012) and waveguide-coupled whispering-gallery microcavities (Yang et al., 2018). The ultra-high optical Q factors of these silica structures, now in excess of 1 billion (Wu et al., 2020), have enabled highly efficient Brillouin lasers with fundamental linewidths below 1 Hz (Li et al., 2013). Such narrow linewidths are used in Brillouin ring laser gyroscopes that measure the Earth's rotation (Lai et al., 2020), as well as high performance microwave synthesizers (Li et al., 2013, 2014).

SBS in silicon nitride: The silicon nitride (Si_3N_4) waveguide platform has enabled a broad class of low-loss planar-integrated devices and chip-scale solutions that benefit from the wide bandgap, the resulting transparency over a broad wavelength range (400–2350 nm), and its CMOS wafer-scale foundry process (Blumenthal et al., 2018). The Si_3N_4 platform can be used to realize ultra-low loss waveguides, handle high optical power, be engineered for linear and nonlinear optical functions, and supports a variety of passive and active building blocks (e.g., ultra-high Q resonators, optical filters, highly coherent lasers, optical signal processing circuits, nonlinear optical devices, and frequency comb generators).

The first demonstration of SBS lasing in an all-waveguide silicon nitride waveguide-based photonic integrated circuit (PIC) was reported in 2019 (Gundavarapu et al., 2019). In this work, the large mode volume and low loss resulted in a fundamental linewidth measured at 720 mHz in a 28 million Q resonator, and leveraged optical weakly confining waveguides without acoustic (phonon) guiding. The fundamental noise dynamics in cascaded-ordered operation was analyzed in 2018 (Behunin et al., 2018) and in 2019 application as a microwave oscillator yielded a phase noise of -84 dBc Hz^{-1} at 10 kHZ offset from the 10.93 GHz S1–S3 beatnote carrier (Gundavarapu et al., 2019). In 2019 a grating modulated ring resonator was employed to suppress the second-order Stokes (S2) and allow the laser to operate in the single mode (S1) regime, further reducing the fundamental linewidth and increasing the output power. Application of this laser as a low noise ring-gyro based on SBS was reported in 2019 (Nelson et al., 2019) as well as its application to energy efficient, high capacity coherent wavelength division multiplexed (WDM) fiber optic links for data center interconnects in 2020 (Blumenthal et al., 2020; Brodnik et al., 2020) and for precision frequency synchronized fiber links in 2021 (Brodnik et al., 2021). This laser was used to pump an integrated optical frequency comb in 2020 to study the phase noise transfer (Harrington et al., 2020). Progress in realizing this SBS laser design in the visible regime, operating at 674 and 698 nm (Chauhan et al., 2021), was enabled by ultra-low loss visible light waveguides (Chauhan et al., 2022). Progress in linewidth reduction and improvements in energy efficiency were achieved with threshold power reduction to 320 uW using record low loss 0.034 dB/m waveguides and fabrication of a 720 million intrinsic Q resonator in a 200-mm wafer-scale CMOS-foundry-compatible platform (Liu et al., 2021). Most recently, a novel photonic molecule design, employing a nested resonator design, was used to suppress S2 and allow linewidth reduction to 71 mHz and 10 mW S1 output power (Liu et al., 2022b).

SBS in chalcogenide: Chalcogenide glasses (ChGs), glasses containing one or more chalcogen elements (S, Se, Te), are well-known for their strong nonlinear response. ChGs have been extensively researched since 1950s (Frerichs, 1953) and become one of the leading platforms for photonic integration and signal processing applications in the past 20 years (Eggleton et al., 2011). The intrinsic transparency window of ChGs, which includes much of the molecular fingerprint region of 2–25 μm, makes them attractive for use in infrared-transmitting optical fibers, and as waveguides for optical sensors and telecommunications. The high Kerr nonlinearity of ChGs ($n_2 = 3 \times 10^{-182}/\text{W}$) for most common As_2S_3 glass composition), combined with the strong confinement and dispersion engineering achievable in fiber and waveguide devices, makes them attractive as fast nonlinear optical devices. Standard photolithography and dry etching have been used to produce polymer-clad, dispersion-engineered As_2S_3 rib waveguides measuring 2μm × 850 nm with losses as low as 0.35 dB/cm (Choi et al., 2010). Interestingly, such waveguide devices have been found to exhibit extremely high Brillouin gain of $g_B = 0.7 \times 10^{-9}$ m/W, which is almost 100 times larger than that of a standard telecommunication silica fiber (Pant et al., 2011b). Such high Brillouin gain can be explained by three reasons: (i) the high linear refractive index of ChGs in general (n is between 2 and 3 depending on the glass composition); (ii) high nonlinear refractive index of ChGs; (iii) almost complete overlap between optical and acoustic modes in both chalcogenide fibers and waveguides (see Fig. 21).

The first demonstration of SBS in a 7 cm-long As_2S_3 chalcogenide waveguide (ChWG) on a chip was reported in 2011 (Pant et al., 2011b). The measured Brillouin shift and its full width at half-maximum (FWHM) linewidth were 7.7 GHz and 34 MHz, respectively. A probe gain of 16 dB was obtained for a CW pump power of 300 mW. Shortly after, in 2013, the first chalcogenide single frequency laser was demonstrated by exploiting high Brillouin gain of a ChWG in a ring-cavity configuration (Kabakova et al., 2013). The lasing slope efficiency of 30% was shown after the power of the pump laser exceeded the threshold value of 360 mW. Relatively high lasing threshold can be explained by the losses incurred by the optical wave in a single round trip around the ring cavity, which was estimated to be close to 15 dB and consisted of propagation losses in the waveguide itself (1.4 dB/cm) and the coupling losses (2dB/facet).

SBS in silicon: Brillouin amplifiers and lasers have also been realized using Brillouin-active waveguides fabricated in silicon on insulator materials systems. Owing to the unique photoelastic and mechanical properties of silicon, these waveguides and lasers have a number of features that are distinct from the aforementioned glass- and nitride-based waveguides. Since the sound velocity is higher in silicon than within the silica core of conventional silicon-on-insulator waveguides, conventional SOI waveguides are anti-guiding for sound, meaning that the Brillouin-active phonon modes are leaky. The rapid escape of these phonons from the core of subwavelength silicon waveguides suppresses Brillouin nonlinearities in conventional SOI waveguides (Poulton et al., 2013). This challenge has been addressed by creating suspended optomechanical waveguide structures that permit guidance of both light and sound within the same low loss waveguide structure (Rakich et al., 2012; Shin et al., 2013; Van Laer et al., 2015b). By engineering these waveguide structures, record Brillouin gains (>6000 $W^{-1}m^{-1}$) have been achieved in silicon. Moreover, this waveguide strategy has enabled net Brillouin amplification for pump powers as low as 5 milliwatts (Kittlaus et al., 2016), opening the door to Brillouin amplifiers (Otterstrom et al., 2019) and lasers (Otterstrom et al., 2018a, 2020) in silicon photonics.

In contrast to glass- and nitride-based waveguide systems, the distinct coupling mechanisms within subwavelength silicon waveguides make forward SBS the dominant interactions in Brillouin-active silicon waveguides (Qiu et al., 2013; Rakich et al., 2012; Shin et al., 2013). Using intramodal FSBS, appreciable net Brillouin amplification (>5 dB) has been demonstrated in silicon (Kittlaus et al., 2016). Additionally, engineerable intermodal scattering processes have been used to produce several dB net amplification within linear waveguides (Kittlaus et al., 2017) and 30 dB of gain within resonant structures (Otterstrom et al., 2018a). These same intermodal scattering process have also permitted the first demonstration of a Brillouin laser in silicon (Otterstrom et al., 2018a) and have been used to realize various forms of nonreciprocity (Kittlaus et al., 2018; Otterstrom et al., 2019, 2020).

Since both the photon and phonon modes are tightly confined within these Brillouin-active waveguides, Brillouin gain and the Brillouin frequency both become highly tailorable with waveguide geometry. Since the intramodal and intermodal forward-SBS interactions are mediated by phonon modes that are near cutoff (i.e., with small longitudinal k-vector and finite frequency) the waveguide geometry can be used to tune the Brillouin frequency over a large range of frequencies (e.g., 1–16 GHz) (Shin et al., 2013). Similarly, control of the acousto-optic overlap between photon and phonon modes in these optomechanical waveguides permits optimization of the Brillouin gain. Moreover, by tailoring the Brillouin-active phonon modes within these structures one is able to design the Brillouin gain spectrum to an unprecedented degree (Gertler et al., 2020).

3.1 Brillouin gain in silicon nitride waveguides and silica resonators

The Brillouin gain and phase-matching process in ultra-low loss silicon nitride waveguide and etched silica resonators has important differences from than that found in waveguides that require both optical and acoustic guiding. The process is more like that found in optical fibers (Ponikvar and Ezekiel, 1981; Yeniay et al., 2002) which utilize only optical guiding. Silicon nitride (Si_3N_4) photonics is an integration platform that is complimentary in characteristics and performance to the silicon-on-insulator (SOI) photonics and group III–V photonics platforms, and is implemented using CMOS compatible wafer-scale foundry processes (Blumenthal et al., 2018). Si_3N_4 PIC technology offers ultra-low optical attenuation, from the visible to beyond the infrared (405–2350 nm) (Blumenthal, 2020), and a wide range of integration building blocks to enable systems-on-chip (Blumenthal et al., 2018). These waveguides employ a core layer of Si_3N_4 embedded in a surrounding silicon dioxide (SiO_2) cladding material and can be designed across a wide range of optical confinement factors for a particular wavelength, from weakly confining to strongly confining, leading to a trade-off between loss and minimum bending radius (Huffman et al., 2018). Combinations of waveguides and other high performance components such as microresonators (Yang et al., 2018) can be integrated on the same platform and compact flexible designs can be realized using multiple layer approaches (John et al., 2012).

Extremely high optical quality factors (Q), an important requirement for Brillouin lasers, have been achieved with photonic integrated all-waveguide bus-coupled silicon nitride resonators (Liu et al., 2022c; Puckett et al., 2020) and silicon nitride bus-coupled etched silica resonators (Yang et al., 2018). For example, an all-waveguide silicon nitride ring resonator with 422 million Q reported in 2020 is shown in Fig. 8A, illustrating a single transverse

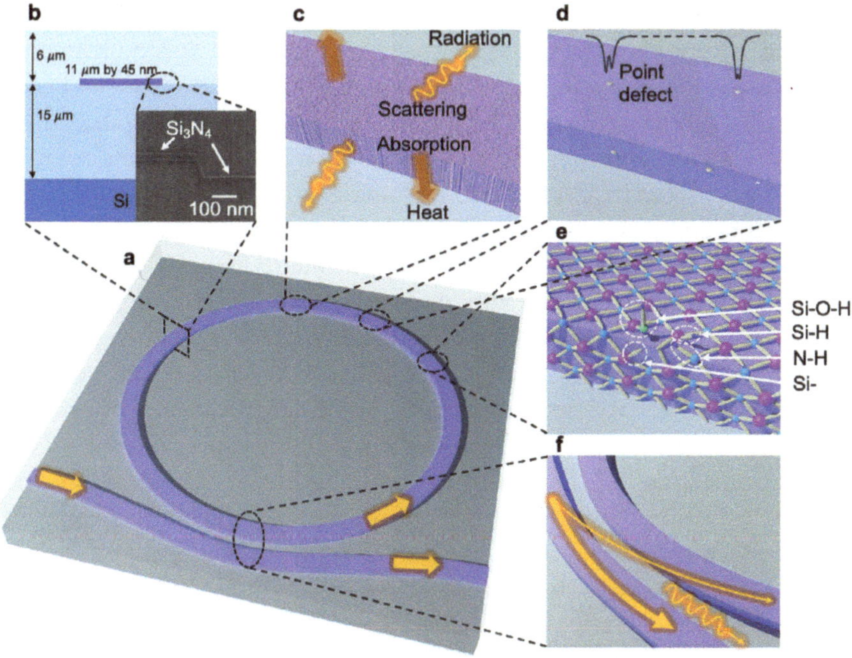

Fig. 8 Ultra-low loss Si_3N_4 waveguide composition and geometry, ring resonator, and waveguide loss mechanisms. *Credit Puckett, M. W., Liu, K., Chauhan, N., Zhao, Q., Jin, N., Cheng, H., Wu, J., Behunin, R. O., Rakich, P. T., Nelson, K. D., Blumenthal, D. J., 2020. 422 Million Q photonic integrated waveguide resonator with a 3.4 billion absorption limit and sub-MHz linewidth. Nat. Commun. 12 (1), 934. https://doi.org/10.1038/s41467-021-21205-4. Puckett et al. (2020) is licensed under CC BY 4.0.*

electric (TE) mode weakly guiding waveguide design. More recently (2021) a transverse magnetic (TM) mode version of this design realized a record 720 million Q and 0.034 dB/m (Liu et al., 2022c). An integrated ultra-low loss, ultra-high Q etched silica resonator coupled to a silicon nitride bus waveguide realized Qs in excess of 200 million in 2018 (Yang et al., 2018) and in 2020 demonstrated Q in excess of 1 billion (Wu et al., 2020).

The Si_3N_4 waveguide core can be designed with a high-aspect ratio thick silicon nitride layer in order to mitigate sidewall scattering losses (Bauters et al., 2011; Liu et al., 2021). For example at 1550 nm operation, a typical resonator waveguide is 11 μm wide by 40 nm thick and the bus waveguide is 7 μm wide by 40 nm. Further details of the process used to achieve this performance can be found in Puckett et al. (2020). Lower losses and higher Q are possible using a transverse magnetic mode (TM) waveguide design to minimize electric-field overlap with the nitride sidewall, realizing

0.034 dB/m loss and 720 million intrinsic Q (Liu et al., 2021). The various loss mechanisms include waveguide surface roughness and bulk material absorption as illustrated in Fig. 8C including point defects that can couple longitudinal modes or the forward and backward propagating modes causing random resonance splitting (Fig. 8D). Additional loss mechanisms can result during the material deposition and waveguide etching processes leading to surface absorption loss (Borselli et al., 2005; Haverlag et al., 1992).

Owing to the wide bandgap of silicon nitride (Bauer, 1977), the wavelength of operation can be shifted down to the visible (approximately 405 nm) by adjusting the nitride core thickness and waveguide width as well as other fabrication parameters. Current state-of-the-art losses and resonator Qs in the 461 to 802 nm wavelength range are 0.01–0.09 dB/cm and associated intrinsic resonator Qs of 60–9.5 million (Chauhan et al., 2022). Waveguides supporting multiple VIS wavelengths on the same wafer using a common nitride core thickness and process flow can be realized using mask-only waveguide-width changes. For example, for operation at $\lambda = 461$ nm, a waveguide width of 1.3 μm minimizes loss and supports a single TE mode. At $\lambda = 674$ nm and $\lambda = 698$ nm a 2.3 μm wide waveguide minimizes loss and supports a quasi-single TE mode with a weakly guided TM mode that is filtered out using appropriate filter bend radii of less than 10 mm. The 2.3 μm width also supports a single mode at 802 nm. The upper and lower cladding thicknesses can vary; typical values are 6 and 15 μm, respectively.

The Brillouin process in both the all-waveguide and etched-silica resonators involves an ultra-low loss propagating optical signal that continuously generates bulk-like phonons as the light propagates in the waveguide (whispering mode in the resonator). In the silicon nitride waveguide, the nitride core and silica cladding form an optimal optical waveguide; however, they are anti-waveguiding for the phonons, leading to generation of a bulk-like phonon continuum rather than discrete energy guided phonons. This is also the case in the airclad etched silica resonator where acoustic absorption is very strong at the interface. Therefore, the Brillouin generated Stokes phonons, coupled into a bulk-like continuum in the oxide for the weakly confining waveguide, have a fast decay rate (on the order of 20 microns). This bulk-like continuum and fast phonon decay lead to a broad Brillouin gain bandwidth on the order of 150 MHz at 1550 nm. These qualities enable operation over a broad range of optical parameters including wavelength as well as device geometries. The resulting Brillouin-generated phonons dampen quickly due to oxide absorption and in the case of silica resonators that of the silica-air interface. The net result is very long-lived photons and very short-lived phonons, which is the desired condition for pump linewidth narrowing (Debut et al., 2000). While the optical mode is guided by the nitride core or the silica air-clad wedge, it predominantly exists in the oxide. As such, the Brillouin nonlinearity leads to an S1 frequency shift similar to that in optical fiber (~11 GHz at 1550 nm pump and ~26 GHz at 674 nm pump).

For the silicon nitride all-waveguide resonator, the predicted and measured Brillouin gain at 1550 nm are shown in Fig. 9. The first-order Stokes, S1, is frequency shifted from the pump by 10.93 GHz and the gain spectral profile is asymmetric with an FWHM of 153 MHz, characteristic of the bulk-like acoustic gratings. This gain linewidth is three to four times broader than that typically found in a guided structure and the Brillouin frequency shift is similar to that of the optical fiber since the shift is determined in part by the material properties of silica and the pump wavelength, since the optical mode propagates predominately in the oxide waveguide cladding, as well as by the acoustic properties of the medium and waveguide. An example of how Brillouin gain is measured using a stimulated pump-probe technique in a Si_3N_4 waveguide spiral is given in Gundavarapu et al. (2019). The gain simulation was performed by incorporating characterization of the nitride and oxide material thin films to extract the optical and acoustic parameters used to populate full vectorial Brillouin gain simulations (Gundavarapu et al., 2019). These values were used to simulate the position, shape and amplitude of the Brillouin gain spectra by constructing fully three-dimensional finite

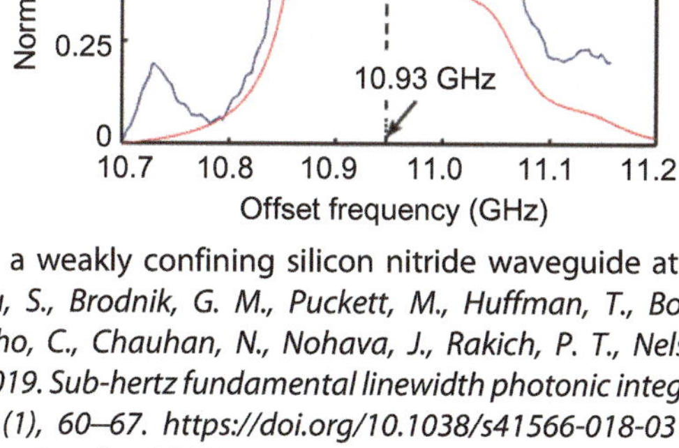

Fig. 9 SBS gain in a weakly confining silicon nitride waveguide at 1550 nm. *Adapted from Gundavarapu, S., Brodnik, G. M., Puckett, M., Huffman, T., Bose, D., Behunin, R., Wu, J., Qiu, T., Pinho, C., Chauhan, N., Nohava, J., Rakich, P. T., Nelson, K. D., Salit, M., Blumenthal, D. J., 2019. Sub-hertz fundamental linewidth photonic integrated Brillouin laser. Nat. Photonics 13 (1), 60–67. https://doi.org/10.1038/s41566-018-0313-2. Gundavarapu et al. (2019) is licensed under CC BY 4.0.*

element method models of the waveguides yielding a gain peak frequency shift of 10.92 GHz. This frequency shift is expected as the optical mode is confined mostly in the oxide. Further details about the Brillouin gain simulation in a nonacoustic guiding Si_3N_4 waveguide are given in Gundavarapu et al. (2019).

The conditions for Brillouin gain in Si_3N_4 waveguides in the IR can be translated to the visible (VIS) light range (Chauhan et al., 2020, 2021) without major changes to the materials or fabrication process. Ultra-narrow linewidth VIS light lasers are critical for AMO physics including atomic clocks, quantum computing, atomic and molecular spectroscopy, and sensing. The ultra-low loss and ultra-high Q components needed for a VIS light SBS laser can be realized using mask only design changes to adjust wavelength on the same wafer and process (Chauhan et al., 2022). For example, Fig. 10 shows the waveguide design and optical field mode profiles for a

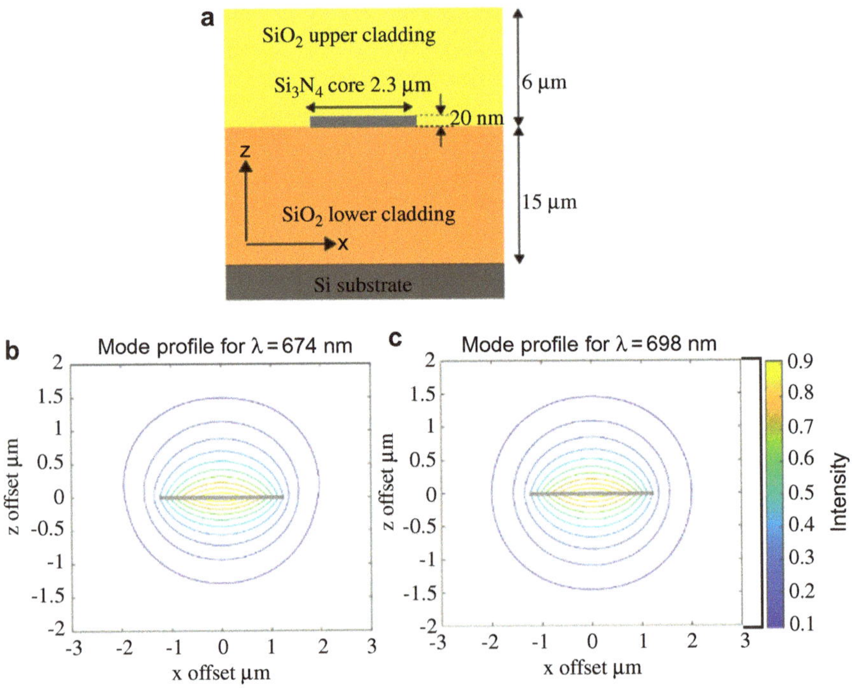

Fig. 10 Waveguide cross section and mode profiles at 674 and 698 nm. *Credit Chauhan, N., Isichenko, A., Liu, K., Wang, J., Zhao, Q., Behunin, R. O., Rakich, P. T., Jayich, A. M., Fertig, C., Hoyt, C.W., Blumenthal, D. J., 2021. Visible light photonic integrated Brillouin laser. Nat. Commun. 12 (1), 4685. https://doi.org/10.1038/s41467-021-24926-8. Chauhan et al. (2021) is licensed under CC BY 4.0.*

single TE mode waveguide designed to operate at both 674 nm light (the clock transition wavelength for the strontium ion) and 698 nm light (the strontium atom clock transition), using on a 20 nm thick, 2.3 μm wide nitride core.

Measurement of Brillouin gain at visible wavelengths presents barriers due to the very weak back-scattered signal. High sensitivity stimulated pump-probe techniques (Gundavarapu et al., 2019) are difficult to implement due to the lack of high-speed visible light modulators. Measuring the extremely weak spontaneous Brillouin signal is also very difficult without knowing the Stokes frequency shift. To predict the Stokes shift, we used a multiphysics vectorial simulation (Chauhan et al., 2021; Gundavarapu et al., 2019) that incorporates measured acoustic and visible wavelength range optical and material parameters. We then employed heterodyne detection between the 674-nm pump and the back-scattered spontaneous signal from a 2-m on-chip spiral waveguide (Fig. 11A and B). The low loss (~1 dB/m) and 2 m length maximizes the spontaneous Brillouin signal, yielding an approximately 10^{-18} (W × spectrum analyzer resolution bandwidth of 100 Hz) heterodyne signal measured with a radio frequency (RF) electrical spectrum analyzer (ESA) (Chauhan et al., 2021). Using time-averaged detection on the ESA we measured a 25.110 GHz peak

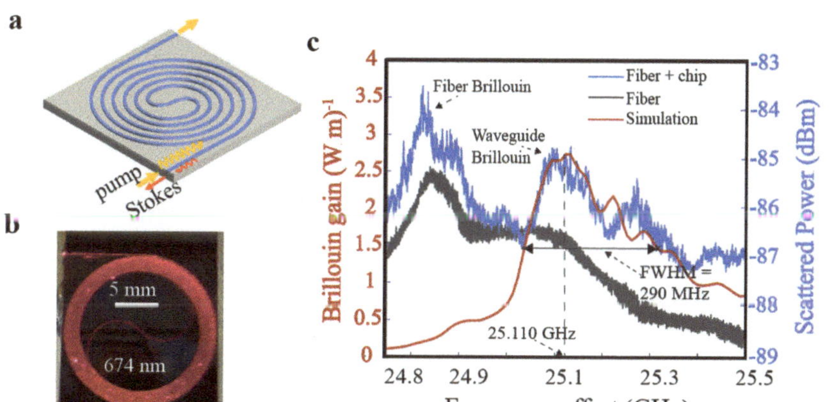

Fig. 11 Heterodyne measurement of 674 nm spontaneous Brillouin gain in a 2 m single-mode waveguide spiral. (A) spiral layout. (B) Photograph of the 2 m spiral illuminated with 674 nm light. (C) Measured and simulated spontaneous Brillouin gain with 25.110 GHz first-order Stokes (S1) frequency shift and 290 MHz bandwidth. *Credit Chauhan, N., Isichenko, A., Liu, K., Wang, J., Zhao, Q., Behunin, R. O., Rakich, P. T., Jayich, A. M., Fertig, C., Hoyt, C.W., Blumenthal, D. J., 2021. Visible light photonic integrated Brillouin laser. Nat. Commun. 12 (1), 4685. https://doi.org/10.1038/s41467-021-24926-8. Chauhan et al. (2021) is licensed under CC BY 4.0.*

frequency shift and 290 MHz gain bandwidth and an SBS gain coefficient of 2.73 $(W\ m)^{-1}$ which agree with numerical simulations (Fig. 11C). The measured broad gain bandwidth and skewed line shape confirms the continuous generation of photons without acoustic waveguiding and coupling to a continuum of bulk acoustic phonon states within the waveguide oxide cladding (Chauhan et al., 2021; Gundavarapu et al., 2019). As shown in Fig. 11C, Brillouin scattering in the optical fiber used to deliver the 674 nm pump laser light is distinguished from the waveguide Brillouin scattering (Fig. 11C blue curve) by decoupling the fiber from the chip and making an independent measurement (Fig. 11C gray curve).

3.2 Brillouin lasing in silicon nitride waveguides and silica resonators

To realize an SBS laser, a bus-coupled on-chip resonator satisfies both the gain conditions described above, and phase matching of a resonance to the Brillouin frequency shift. For the all-waveguide Si_3N_4 laser, a bus-ring resonator with large optical mode volume and ultra-high Q is used (Gundavarapu et al., 2019). Resonators operating in the 1550–1620 nm range today can achieve Q as high as 720 million (Liu et al., 2021), while in the VIS regime, resonators with $Q = 60$ million at 698 nm and 54.4 million at 674 nm have been demonstrated (Chauhan et al., 2022). To minimize losses, the resonator radius is typically larger than the critical bending radius for the fundamental TE mode, while higher order modes are not supported due to large bending losses. An example IR resonator with 720 million intrinsic Q is shown in Fig. 12. This resonator is designed for TM mode operation and is fabricated on a CMOS compatible 200 mm wafer, with highest Q at 1615 nm (Fig. 13). Examples of visible light resonators designed for wavelengths at the strontium neutral and ion clock transitions are shown in Fig. 13.

Ultra-high Q resonators can also be fabricated on-chip by coupling a etched silica resonator to a silicon nitride bus waveguide (Fig. 14A). A scanning electron microscope (SEM) image shows two of the silica ridge resonators connected by a common silicon nitride waveguide (false color red in the image). Zoom-in and cross-sectional views of portions of the resonator are shown in Fig. 14B and C, respectively. Detailed fabrication steps are presented in Fig. 14D. The fabrication process begins by growing a thermal silica layer on a high-purity float-zone silicon wafer (additional details are provided in the figure caption). A phosphoric acid etch is used to define the silicon nitride waveguide. It is important that the silicon nitride is fully

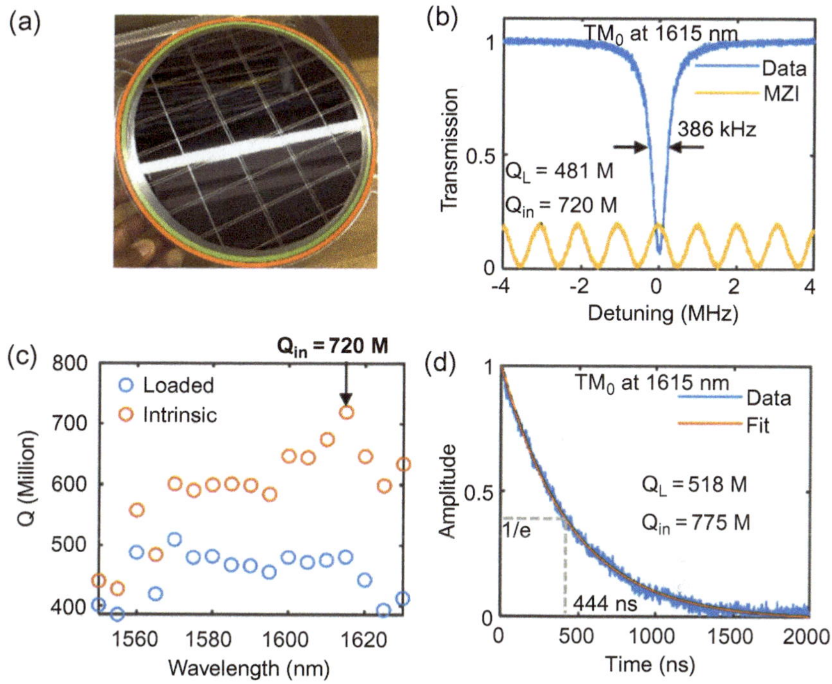

Fig. 12 720 million *Q* Resonator at 1615 nm. (A) 200-mm wafer after the fabrication and dicing process. (B) Resonance spectral scan at 1615 nm with the 1.078 MHz FSR MZI. (C) Measured loaded and intrinsic Qs from 1550 nm to 1630 nm. *M*, million. (D) A ringdown time of 444 ns, corresponding to an intrinsic $Q = 775$ million is measured at 1615 nm. *Reprinted with permission from Liu, K., Jin, N., Cheng, H., Chauhan, N., Puckett, M. W., Nelson, K. D., Behunin, R. O., Rakich, P. T., Blumenthal, D. J., 2022d. Ultralow 0.034 dB/m loss wafer-scale integrated photonics realizing 720 million Q and 380 μW threshold Brillouin lasing. Opt. Lett. 47 (7), 1855–1858. https://doi.org/10.1364/OL.454392. © The Optical Society.*

removed from the silica in this step as otherwise it will adversely impact the optical Q factor. In the penultimate step, a ring aperture is created, as opposed to fully removing the interior silica. This reduces etch loading effects in the final dry-etch step. Not shown in the process flow is the deposition of a thin layer (20 nm) of silica by atomic layer deposition to the etched waveguide. This serves to protect the silicon nitride waveguide during the final dry etch step. The resonator has a maximum thickness of approximately 8.5 μm and the silica layer under the waveguide has a thickness of 2.5 μm. The silicon nitride waveguide has a thickness of approximately 250 nm and is initially 3–3.5 μm in width at the edge of the wafer and as discussed later is

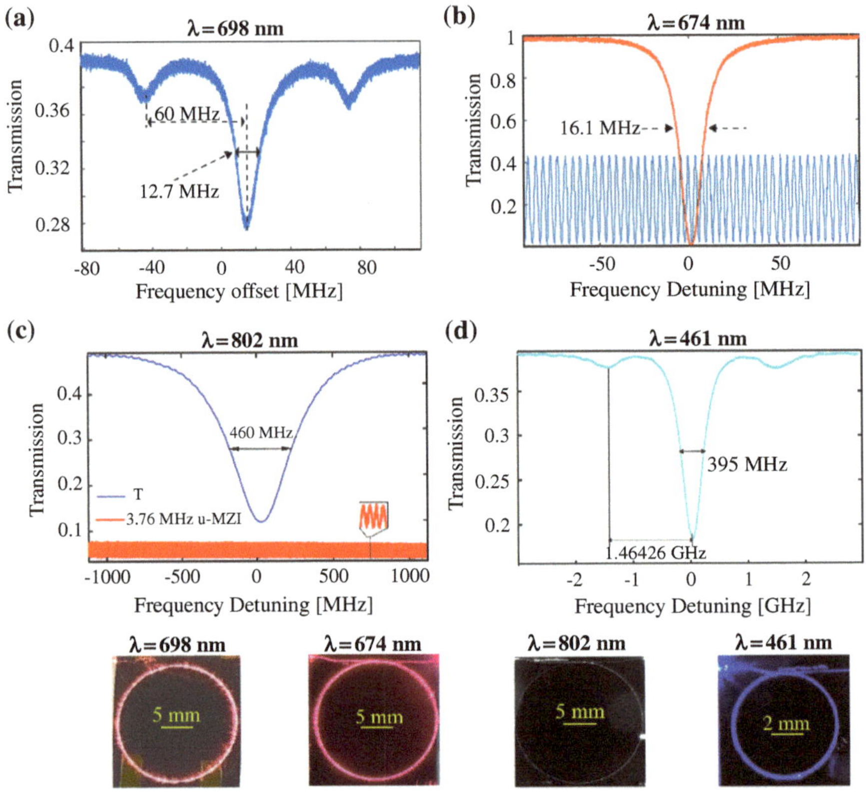

Fig. 13 Visible light SiN resonators. (A) T is resonator transmission spectrum, Q measurement for ring of $R = 9.4$ mm, loaded $Q = 34.4$ million and intrinsic $Q = 60$ million at $\lambda = 698$ nm. (B) Q measurement for ring of $R = 8.9$ mm, loaded $Q = 27.7$ million and intrinsic $Q = 54.4$ million at $\lambda = 674$ nm. (C) Q measurement for ring of $R = 10$ mm, loaded $Q = 0.81$ million and intrinsic $Q = 28$ million at $\lambda = 802$ nm. (D) Q measurement for ring of $R = 3$ mm, loaded $Q = 1.65$ million and intrinsic $Q = 9.5$ million at $\lambda = 461$ nm (E) Resonator images. *Reprinted with permission from Chauhan, N., Wang, J., Bose, D., Zhao, Q., Fertig, C., Hoyt, C., Blumenthal, D. J., 2022. Ultra-low loss visible light waveguides for integrated atomic, molecular, and quantum photonics. Opt. Express 30 (5), 6960–6969. doi:10.1364/OE.448938. © The Optical Society.*

tapered to about 900 nm near the resonator so as to phase match to the resonator optical mode.

SBS lasers can be realized using these resonator designs as shown in Fig. 15A. The relation between the optical resonator resonances, first-order Stokes (S1) frequency shift, Brillouin gain profile, and the optical pump are illustrated in Fig. 15B (not to scale). The S1 (red curve) SBS frequency shift ($\Delta\nu_B$) from the pump (blue curve) is primarily a function of the operating

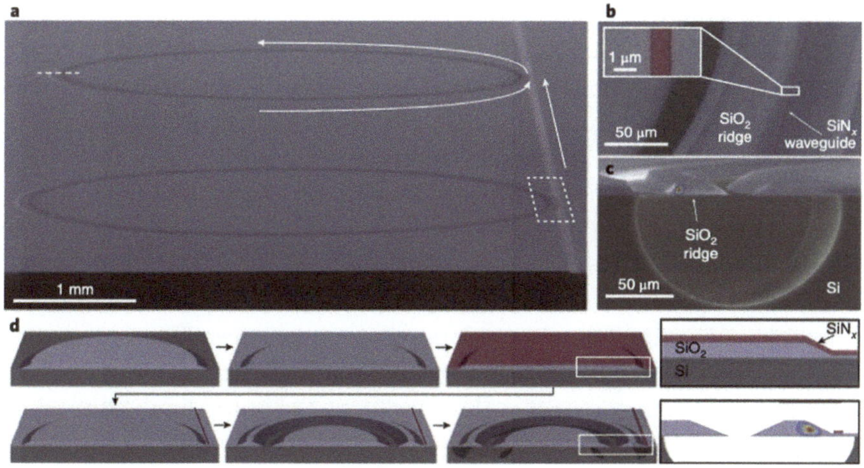

Fig. 14 Ultra-low high *Q* etched silica resonator coupled to a Si_3N_4 bus waveguide. *Credit Yang, K. Y., Oh, D. Y., Lee, S. H., Yang, Q. F., Yi, X., Shen, B., H., W., Vahala, K. J., 2018. Bridging ultrahigh-Q devices and photonic circuits. Nat. Photonics 12, 297–302. https://doi.org/10.1038/s41566-018-0132-5. Yang et al. (2018) is licensed under CC BY 4.0.*

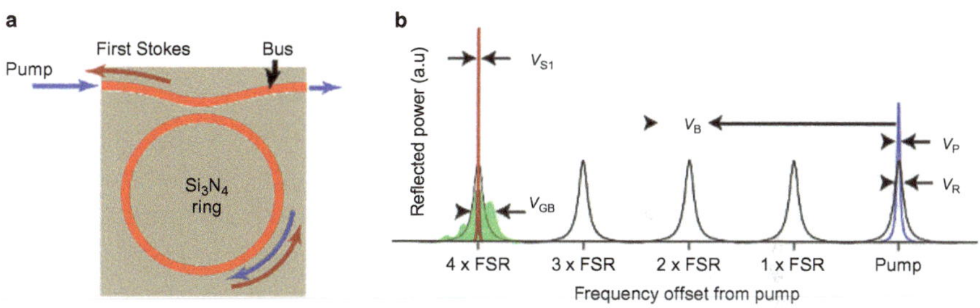

Fig. 15 SBL design in an ultra-low loss silicon nitride resonator. (A) Bus-coupled silicon nitride ring indicating pump and first-order (S1) and third-order (S3) Stokes. (B) Free spectral range (FSR) is engineered to be 1/4 the Brillouin frequency shift. Pump, S1, and Brillouin gain (*green*) are illustrated. *Adapted from Chauhan, N., Wang, J., Bose, D., Zhao, Q., Fertig, C., Hoyt, C., Blumenthal, D. J., 2022. Ultra-low loss visible light waveguides for integrated atomic, molecular, and quantum photonics. Opt. Express 30 (5), 6960–6969. doi:10.1364/OE.448938; Gundavarapu, S., Brodnik, G. M., Puckett, M., Huffman, T., Bose, D., Behunin, R., Wu, J., Qiu, T., Pinho, C., Chauhan, N., Nohava, J., Rakich, P. T., Nelson, K. D., Salit, M., Blumenthal, D. J., 2019. Sub-hertz fundamental linewidth photonic integrated Brillouin laser. Nat. Photonics 13 (1), 60–67. https://doi.org/10.1038/s41566-018-0313-2. Gundavarapu et al. (2019) is licensed under CC BY 4.0.*

wavelength and the acoustic waveguide material properties. The cavity free-spectral range (FSR) is designed to oversample the S1 frequency shift (in this example by a factor of four) providing several advantages: (i) the large cavity optical mode volume reduces the SBL modified Schawlow–Townes fundamental linewidth (see Eqs. 2 and 3) responsible for high frequency offset from carrier noise, (ii) the large mode volume also reduces the thermorefractive noise (TRN) which is responsible for frequency noise in the approximately 10 Hz–1 kHz range, (iii) improvements to fabrication variations and environmental conditions due to the large Brillouin gain bandwidth (153 MHz) overlap with the cavity resonance. The combination of low optical waveguide loss and generation of bulk-like phonons in a critically coupled resonator satisfies the conditions for pump linewidth reduction (Debut et al., 2000).

At 1550 nm, the ring resonator radius is chosen to be 11.83 mm, larger than the critical bend loss limit (approximately 10 mm) for this waveguide design. The resonator physical length, 74.3 mm, resulted in a measured 2.72 GHz FSR. This value differs by only 8 MHz from the target FSR, which combined with the large approximately 153 MHz Brillouin gain bandwidth and resonator linewidth of approximately 7 MHz, highlights the fabrication robustness of a large gain bandwidth relative to the linewidth. Measurements of a 1570 nm silicon nitride waveguide SBS laser are shown in Fig. 16, including the power evolution and S1 linewidth and frequency noise curves. Fig. 16A shows the subthreshold points (1) and (2) that correspond to the spontaneous emission spectra curves in Fig. 16B. Measurements at the threshold of 10.53 mW (point (3)) and above threshold (points (4)–(9)) in Fig. 16A correspond to the frequency–noise curves in Fig. 16C. The dotted lines indicate the far-from-carrier white-frequency-noise level determined by the emission fundamental linewidth. The pump laser frequency noise and its fundamental linewidth are shown in black demonstrating significant pump diffusion phase noise reduction and linewidth narrowing. The linewidth evolution as a function input pump power is shown in Fig. 16C, with a minimum of 0.7 Hz fundamental linewidth observed just below the onset of S2 lasing. The SBL optical threshold power can be reduced by increasing the resonator Q for the same mode volume. A 720 million Q TM mode cavity was used to demonstrate S1 threshold reduction 320 μW (see Fig. 17), an important consideration for energy efficiency.

High-coherence Brillouin laser action at 1550 nm is also achieved using the integrated etched-silica ridge resonator in Fig. 14. To phase match the Brillouin process when pumped near 1550 nm devices with diameters of

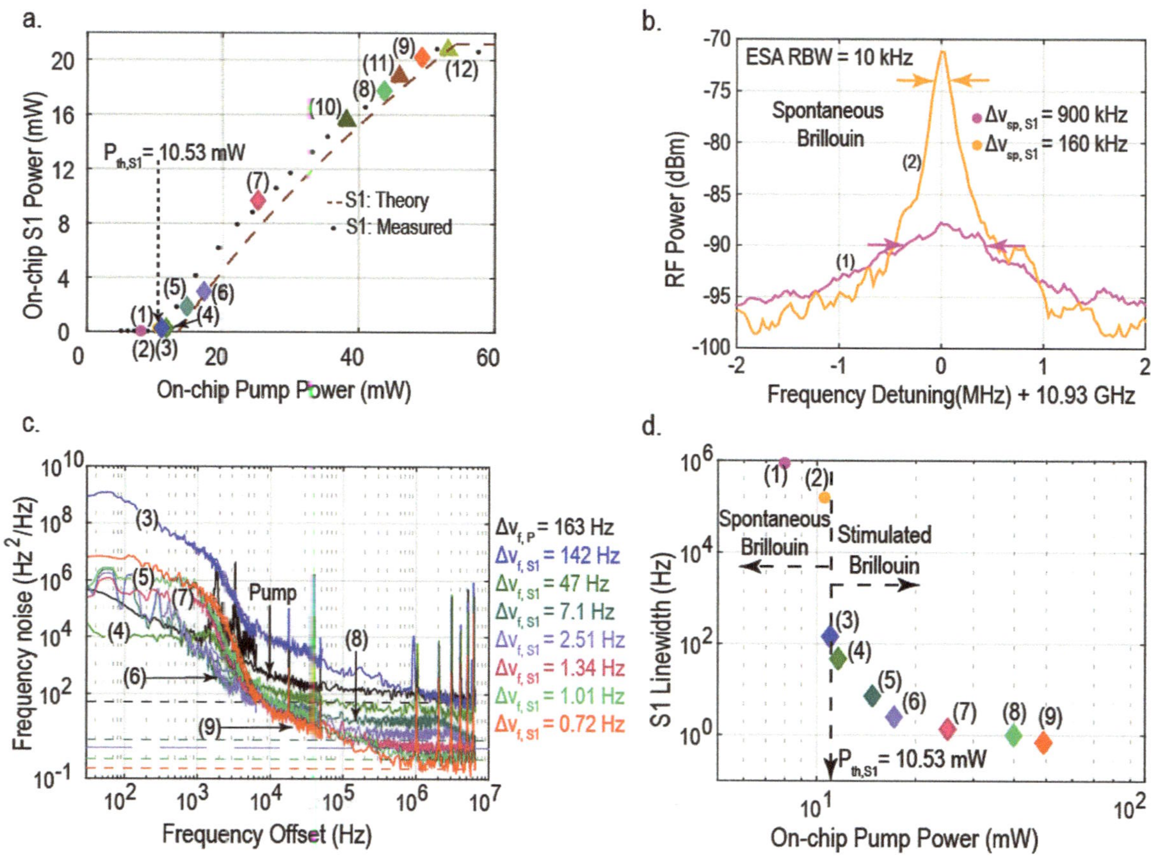

Fig. 16 Sub-Hz fundamental linewidth photonic integrated 1550 nm laser. (A) First-order Stokes power output vs pump power input with threshold. (B) Output power spectrum just below and at threshold. (C) Frequency noise evolution as laser pump power is increased from threshold to just below onset of second-order Stokes (S2) lasing. (D) Summary of output linewidth as a function of on-chip pump power. *From Chauhan, N., Wang, J., Bose, D., Zhao, Q., Fertig, C., Hoyt, C., Blumenthal, D. J., 2022. Ultra-low loss visible light waveguides for integrated atomic, molecular, and quantum photonics. Opt. Express 30 (5), 6960–6969. doi:10.1364/OE.448938; Gundavarapu, S., Brodnik, G. M., Puckett, M., Huffman, T., Bose, D., Behunin, R., Wu, J., Qiu, T., Pinho, C., Chauhan, N., Nohava, J., Rakich, P. T., Nelson, K. D., Salit, M., Blumenthal, D. J., 2019. Sub-hertz fundamental linewidth photonic integrated Brillouin laser. Nat. Photonics 13 (1), 60–67. https://doi.org/10.1038/s41566-018-0313-2. Gundavarapu et al. (2019) is licensed under CC BY 4.0.*

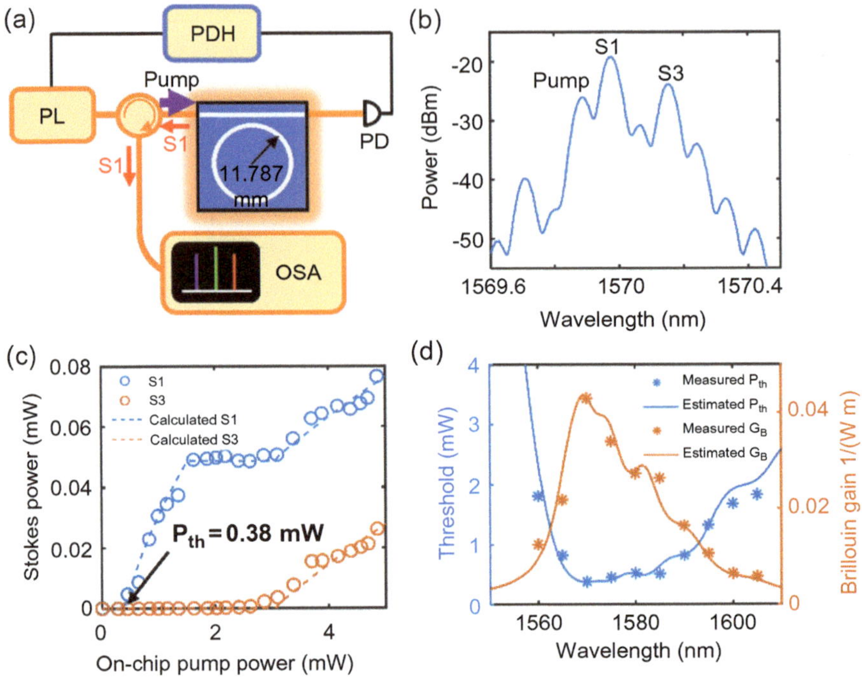

Fig. 17 320 μW threshold 1570 nm silicon nitride waveguide Brillouin laser using a 720 million *Q* resonator. (A) Experimental measurement setup. (B) Cascaded mode operation. (C) First-order (S1) and third-order (S3) Stokes output power vs input pump power, measured and simulated. (D) Measured and simulated first-order Stokes threshold and Brillouin gain as a function of operating wavelength. *Reprinted with permission from Liu, K., Jin, N., Cheng, H., Chauhan, N., Puckett, M. W., Nelson, K. D., Behunin, R. O., Rakich, P. T., Blumenthal, D. J., 2022d. Ultralow 0.034 dB/m loss wafer-scale integrated photonics realizing 720 million Q and 380 μW threshold Brillouin lasing. Opt. Lett. 47 (7), 1855–1858. https://doi.org/10.1364/OL.454392. © The Optical Society.*

approximately 6 mm were fabricated. Fig. 18 shows the optical spectrum of the lasing Stokes wave. The weaker pump signal peak in the spectrum results from the need to collect the lasing Stokes wave in the propagation direction opposite to the pumping direction. The strength of this weaker pump is determined by residual reflection and backscattering in the measurement. The upper left inset in Fig. 18 shows the Stokes power vs pumping and gives a threshold power of approximately 800 μW in the waveguide. The upper right inset shows the microwave beat signal between the pump wave and the Stokes wave. It has a high coherence as evidenced by the resolution bandwidth.

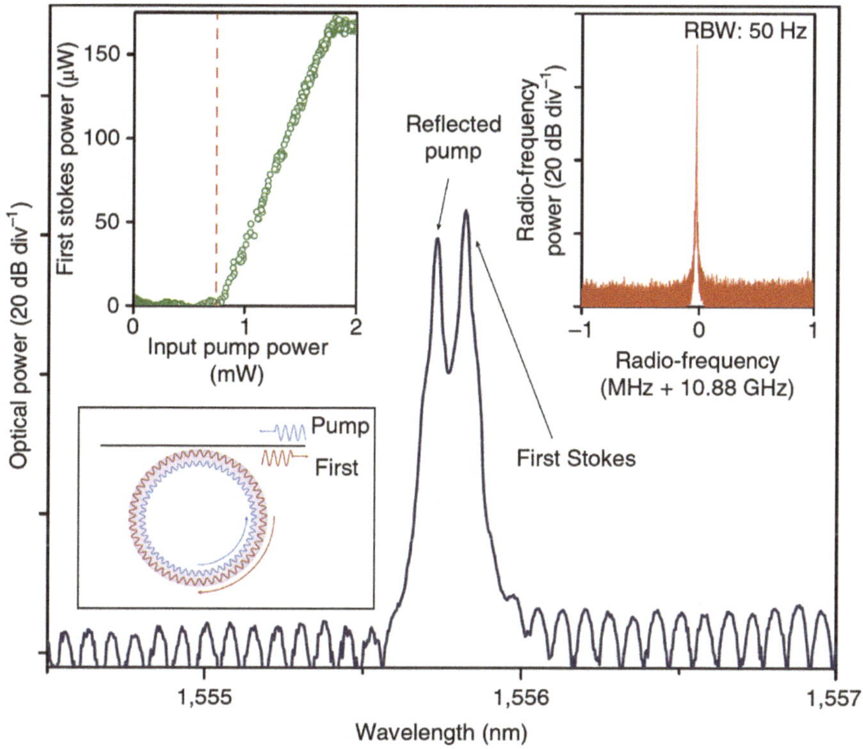

Fig. 18 Demonstration of Brillouin lasing in an integrated optical resonator. Optical spectrum of Brillouin laser with pump and first Stokes waves' spectral peaks indicated. The *upper left inset* is the output laser power vs input coupled power and shows laser threshold behavior (threshold is indicated by the *dashed red line*). All powers are in the waveguide. The clamping of the Stokes power occurs as a result of a second Stokes laser oscillation (i.e., Stokes cascade). The *right inset* shows the radio frequency (RF) power spectrum of the microwave beatnote signal of the back-reflected pump and lasing signal. *RBW*, resolution bandwidth. The *lower left inset* shows pump (*blue*) and Brillouin laser action on first Stokes wave (*red*). *From Yang, K. Y., Oh, D. Y., Lee, S. H., Yang, Q. F., Yi, X., Shen, B., H., W., Vahala, K. J., 2018. Bridging ultrahigh-Q devices and photonic circuits. Nat. Photonics 12, 297–302. https://doi.org/10.1038/s41566-018-0132-5. Yang et al. (2018) is licensed under CC BY 4.0.*

Brillouin lasing in the visible can be achieved with minor design modifications to the 1550 nm all-waveguide resonator design, and the first integrated visible light emission SBS laser was reported in 2021 (Chauhan et al., 2021). The IR laser design was modified by reducing the nitride layer thickness, adjusting the waveguide width, and changing the pump laser to the desired wavelength (offset by the SBS S1 shift). Using the spontaneous SBS gain technique described above, we measured a 25.110 GHz Stokes frequency shift and a 290 MHz Brillouin gain bandwidth, in good agreement with the predicted values. The waveguide consists of a 20 nm tall and 2.3 μm wide silicon nitride core deposited and etched on a lower thermally grown oxide cladding on a silicon substrate, with a TEOS-PECVD deposited upper cladding and a final two-step anneal. Details of the fabrication process are given in Chauhan et al. (2021). The SBS laser is an 8.9509 mm radius bus-coupled ring structure with an FSR designed to be 1/7 of the measured 25.110 GHz peak Stokes shift at 674 nm and a bus-to-ring power coupling coefficient 34 κ_2 of approximately 1.5%. The SBS laser resonator is pumped by an off-chip tapered optical amplifier (TA) that is seeded with 674 nm light from a continuous wave (CW) external cavity diode laser. SBS emission at 674 nm was demonstrated with a 14.7 mW optical threshold, corresponding to a threshold density of 4.92 mW m^2. Above threshold, a 45% slope efficiency is achieved for a 9.28 mW on-chip output power for the first-order Stokes (S1), and a fundamental linewidth of 269 Hz is measured as the pump power is increased to just below the S2 emission threshold. The optical S1 power and linewidth as a function of input pump power are shown in Fig. 19. The laser threshold and various measurement points along the transfer function are shown in Fig. 19A. The panels in Fig. 19B and C show the measured integral linewidth below and just below threshold, and the frequency noise and fundamental linewidth above threshold. The linewidth results are summarized in Fig. 19D.

The versatility of the waveguide laser resonator design was demonstrated by changing its operation to 698 nm, the neutral strontium atom optical clock transition. A resonator was fabricated with modifications to the 674 nm FSR and bus-ring coupling gap to meet the requirements at 698-nm emission. The multiphysics simulation (Chauhan et al., 2021) predicted a 24.243 GHz Stokes shift and 300 MHz Brillouin gain bandwidth for this wavelength as shown in Fig. 20A. As with the 674 nm laser, the FSR is designed to be 1/7 of the S1 frequency shift, resulting in a 9.4-mm radius resonator design. A 3.4 μm bus-ring coupling gap is chosen to operate the resonator in the under-coupled regime with a power coupling coefficient of approximately 1%. For the fabricated devices, a 12.7 MHz cavity resonance width, a 60 million intrinsic Q, a 33.8 million loaded Q, and a 3.421 GHz FSR are measured (see Fig. 20B and C, respectively). Lasing at S1 is observed by pumping the cavity with a Ti:sapphire laser at the expected pump-Stokes frequency offset as measured shown in the pump and S1 optical spectra in Fig. 20D. A pump-S1 heterodyne 23.892 GHz beatnote is measured

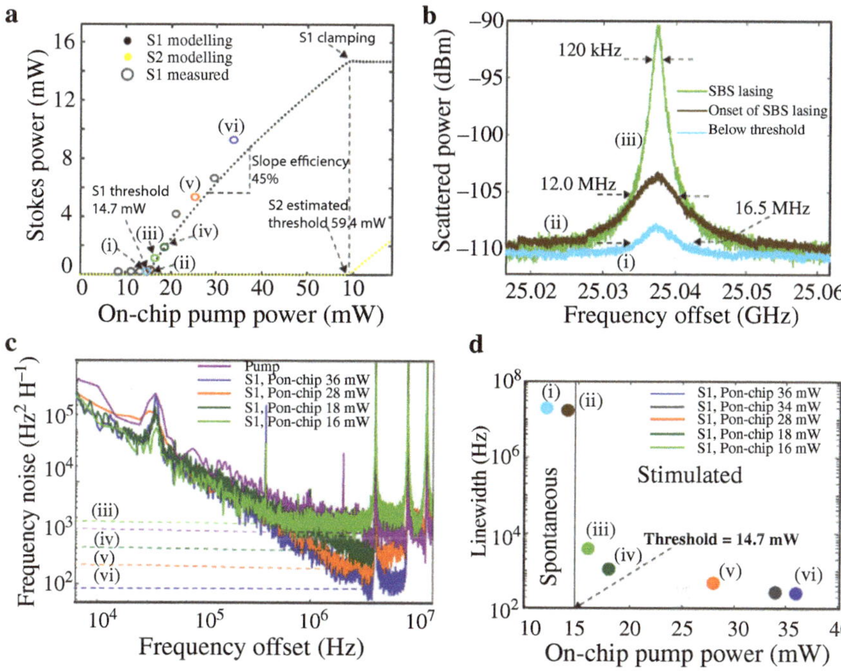

Fig. 19 674 nm SBS laser Stokes threshold, power, and linewidth measurements. (A) Measured and simulated first-order Stokes (S1) output power vs input pump power with threshold and linewidth measurement points indicated. (B) Emission linewidth of S1 below and just below threshold. (C) Frequency noise curves of pump laser and S1 emission indicating reduced fundamental linewidth (indicated by *dashed horizontal lines*) as pump power is increased. (D) Summary of output linewidth (integral below threshold and fundamental above threshold) as a function of input pump power showing linewidth reduction. *From Chauhan, N., Isichenko, A., Liu, K., Wang, J., Zhao, Q., Behunin, R. O., Rakich, P. T., Jayich, A. M., Fertig, C., Hoyt, C.W., Blumenthal, D. J., 2021. Visible light photonic integrated Brillouin laser. Nat. Commun. 12 (1), 4685. https://doi.org/10.1038/s41467-021-24926-8. Chauhan et al. (2021) is licensed under CC BY 4.0.*

as shown in the inset of Fig. 20D. The 351-MHz beatnote offset from our simulated shift is most likely due the offset between the Brillouin gain peak and the cavity resonance. The pump laser is free running (i.e., stabilized neither to the resonator nor a reference optical cavity), and the beatnote drifts on the order of 100 kHz over tens of milliseconds. The on-chip pump power is 108 mW and the measured on-chip pump threshold power (P_{th}) is approximately 75 mW.

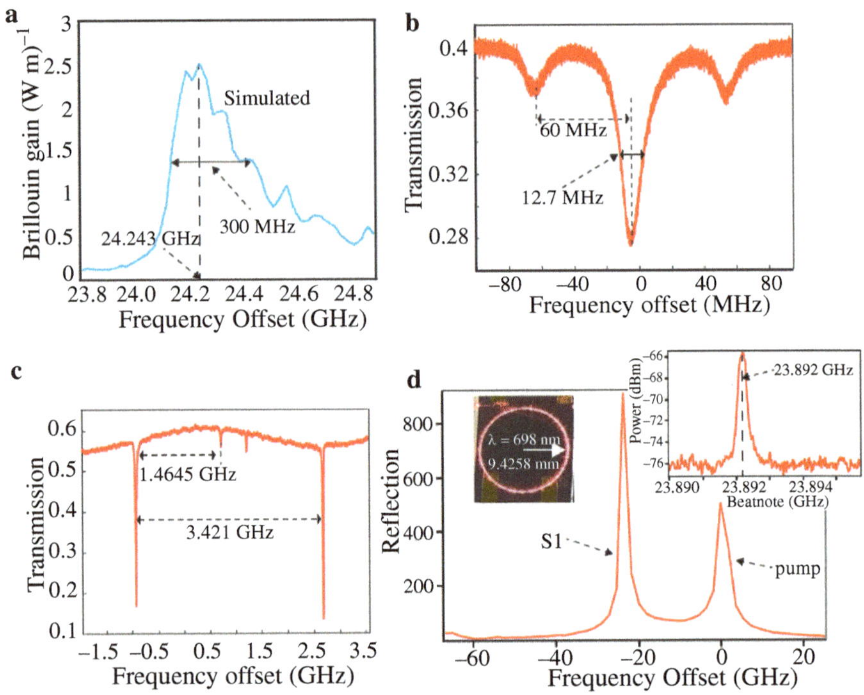

Fig. 20 A 698-nm emission silicon nitride-integrated Brillouin laser. (A) Simulated Brillouin gain at 698 nm. (B) Measured linewidth of 698 nm bus-coupled silicon nitride waveguide SBS resonator. (C) Measured free spectral range (FSR). (D) Measured pump and S1 emission at 698 nm with photograph of device with light coupled from fiber shown in inset. *From Chauhan, N., Isichenko, A., Liu, K., Wang, J., Zhao, Q., Behunin, R. O., Rakich, P. T., Jayich, A. M., Fertig, C., Hoyt, C.W., Blumenthal, D. J., 2021. Visible light photonic integrated Brillouin laser. Nat. Commun. 12 (1), 4685. https://doi.org/10.1038/s41467-021-24926-8. Chauhan et al. (2021) is licensed under CC BY 4.0.*

3.3 Chalcogenide waveguides

3.3.1 Stimulated Brillouin scattering in chalcogenide waveguides integrated on a chip

We have already discussed that the ability to harness SBS in integrated optical chips is key to the realization of many applications in modern photonics that involve light–sound and light–light interactions. Previously we reviewed SBS in material platforms such as silicon and silicon nitrate. In this section, we focus on SBS generation in integrated chalcogenide platforms such wave-guides on a chip. As introduced earlier in the beginning of this section, chalcogenide glasses exhibit both large SBS gain coefficient and Kerr nonlinearity (Eggleton et al., 2011, 2019) and therefore offer an excellent choice for exploiting SBS on a chip.

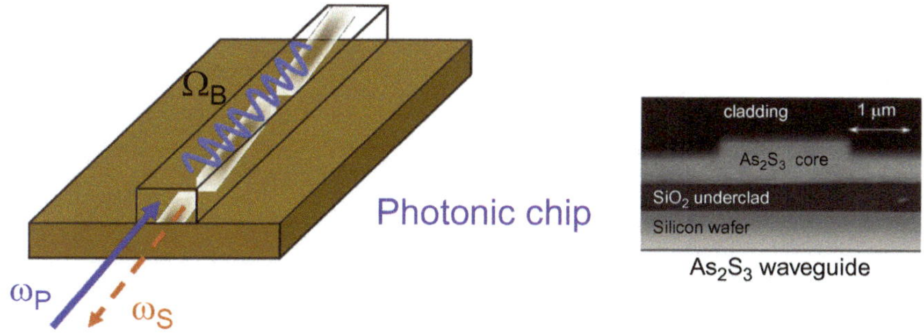

Fig. 21 Concept diagram of on-chip SBS in a chalcogenide waveguide. Interaction of the pump signal (*solid*) of frequency (ω_P) with an acoustic wave of frequency (Ω_B) results in the generation of a backscattered signal at the downshifted frequency $\omega_S = \omega_P - \omega_B$ and SEM image of a typical rib chalcogenide waveguide. *Reprinted with permission from Pant, R., Poulton, C. G., Choi, D.-Y., Mcfarlane, H., Hile, S., Li, E., Thevenaz, L., Luther-Davies, B., Madden, S. J., Eggleton, B. J., 2011a. On-chip stimulated Brillouin scattering. Opt. Express 19 (9), 8285–8290. © The Optical Society.*

The first demonstration of SBS generation on a ChWG was reported in 2011 (Pant et al., 2011b). A 7-cm-long As_2S_3 rib waveguide was used to demonstrate SBS in a back-scattering pump-probe configuration. Fig. 21 shows the principle of chip-based SBS where the interaction of light at frequency ω_P (solid) with a traveling acoustic wave of frequency $\boldsymbol{\Omega}$ generates a backscattered signal (dashed) at the downshifted frequency $\omega_S = \omega_P - \boldsymbol{\Omega}$. A scanning electron microscope (SEM) image of a typical rib waveguide is on the right of Fig. 21.

Unlike silicon-on-silica waveguides which were found to have a negligible acoustic confinement, chalcogenide waveguides show both excellent acoustic and optical guidance and high overlap integral ($\eta = 0.95$) between these two modes (Poulton et al., 2013). This has direct implications for the efficiency of SBS process described by SBS gain parameter and will be discussed next.

3.3.2 Characterization of Brillouin gain in chalcogenide on-chip waveguides

In order to measure the SBS gain spectrum in chalcogenide As_2S_3 waveguides, a pump-probe experiment was performed using the set-up shown in Fig. 22. A 27 dBm Erbium-doped fiber amplifier EDFA1 was used to amplify the laser power before it was coupled to a 50:50 splitter. A continuous wave (CW) pump and a counter-propagating, weak probe signal were used in the experiment. The pump was amplified using a 1 W EDFA. The probe was coupled to the waveguide via port 1 of circulator 3 (C3). The SBS

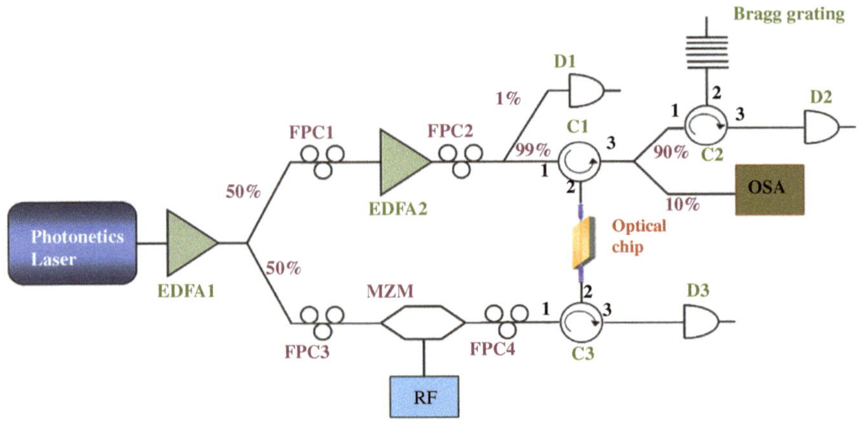

Fig. 22 Pump-probe set-up for measuring the SBS gain spectrum. *Reprinted with permission from Pant, R., Poulton, C. G., Choi, D.-Y., Mcfarlane, H., Hile, S., Li, E., Thevenaz, L., Luther-Davies, B., Madden, S. J., Eggleton, B. J., 2011a. On-chip stimulated Brillouin scattering. Opt. Express 19 (9), 8285–8290. © The Optical Society.*

gain spectrum was obtained using a fixed input pump power of 300 mW and measuring the probe power as the probe frequency was varied using a Mach–Zehnder modulator (MZM) driven by a RF signal generator. The polarization of the pump and the probe signal were adjusted using fiber polarization controllers (FPCs) 1–4 to maximize the gain (Pant et al., 2011b).

Fig. 23A shows the measured SBS gain spectrum (circle markers) and its Lorentzian fit (dash line). The measured Brillouin shift for the chip was found to be 7.7 GHz, which is consistent with the measured Brillouin shift in As_2S_3 fiber, and the measured full width at half maximum (FWHM) line width was 34 MHz. We can use the measured Brillouin shift and FWHM to calculate the gain coefficient for the chip-integrated ChWG using the relation:

$$g_B = \eta \frac{4\pi n^8 p_{12}^2}{c\rho\Omega\Delta\lambda_p^3}, \tag{7}$$

where η is the overlap integral between the optical mode and acoustic mode-induced density change; p_{12} is the longitudinal elasto-optic coefficient; c is the speed of light in vacuum; ρ is the material density; Ω is the Brillouin shift; Δ is the Brillouin linewidth and λ_P is the pump wavelength. Using the measured values of Ω and Δ, the Brillouin gain g_B for the chalcogenide photonic chip was calculated to be 0.74×10^{-9} m/W. The predicted value of g_B was confirmed by measuring power transfer from the pump wave to the Stokes wave directly as illustrated in Fig. 23B.

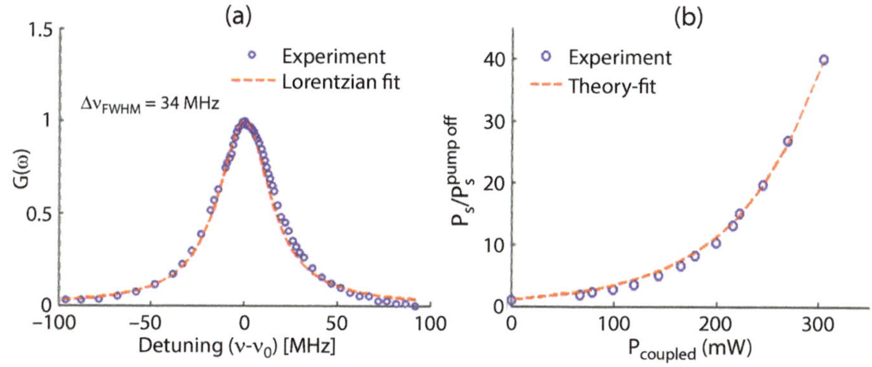

Fig. 23 Pump-probe measurements. (A) the Brillouin gain spectrum obtained using the input pump power of 300 mW while the probe frequency is varied and (B) output probe power, at the Brillouin shift, normalized to the output probe power when the pump is off. *Reprinted with permission from Pant, R., Poulton, C. G., Choi, D.-Y., Mcfarlane, H., Hile, S., Li, E., Thevenaz, L., Luther-Davies, B., Madden, S. J., Eggleton, B. J., 2011a. On-chip stimulated Brillouin scattering. Opt. Express 19 (9), 8285–8290. © The Optical Society.*

3.3.3 Narrow linewidth Brillouin laser based on chalcogenide photonic chip

The high gain reported for chalcogenide waveguides (approximately 100 times larger compared to standard telecommunication silica fibers) is advantageous for exploiting chalcogenide integrated platform for creating Brillouin lasers. This idea was demonstrated in 2013 by Kabakova et al. (2013) where a 7-cm long As_2S_3 waveguide was used a gain medium within otherwise fiber ring resonator. The coupling between the fiber loop and the waveguide was achieved by means of ultra-high numerical aperture (UHNA-4) fiber tapers. The schematic of the experimental setup for a Brillouin single-frequency laser is shown in Fig. 24A, with the waveguide micrograph to follow in Fig. 24B and the results for numerical simulation of optical and acoustic waveguide modes in Fig. 24C. As the power threshold for SBS generation in the ChWG is significantly lower compared to that for silica fibers, SBS gain only occurs within the waveguide while optical feedback for the Stokes wave is provided by the fiber loop. The optical feedback leads to the optical field build-up inside the As_2S_3 waveguide. Lasing occurs when the gain at the Stokes frequency exceeds the cavity loss. The cavity mode nearest to the peak of the Brillouin gain (free spectral range of the waveguide/fiber loop cavity) is FSR= 8.4 MHz and the gain bandwidth is Δ=34 MHz. This mode experiences maximum amplification and hence, it lases preferentially.

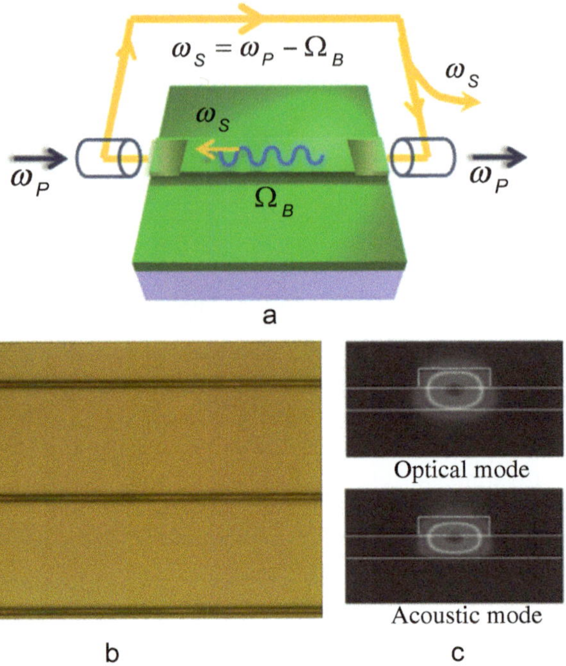

Fig. 24 Brillouin laser on-chip. (A) Schematic of a Brillouin laser based on photonic chip. (B) Micrograph of the As_2S_3 waveguides (*top view*) with vertical tapers (Collaboration, 2016). (C) Numerically calculated acoustic and optical modes in the waveguide. *Reprinted with permission from Kabakova, I. V., Pant, R., Choi, D.-Y., Debbarma, S., Luther-Davies, B., Madden, S. J., Eggleton, B. J., 2013. Narrow linewidth Brillouin laser based on chalcogenide photonic chip. Opt. Lett. 38 (17), 3208–3211. https://doi.org/10.1364/OL.38.003208. http://opg.optica.org/ol/abstract.cfm?URI=ol-38-17-3208. © The Optical Society.*

The pump made a single transit of the cavity before being stripped out by the circulator. The Stokes signal generated in the waveguide, however, was redirected back to the waveguide, and thus made to oscillate inside the cavity. The Stokes wavelength could be aligned to the cavity resonance by tuning the pump wavelength. The number of round trips for the Stokes wave in the ring cavity was limited by the optical losses. These included: (i) propagation loss in the waveguide (−1.4 dB), (ii) the coupling loss (due to vertical tapers this was reduced to 2 dB per facet), and (iii) other losses in optical components such as couples and circulators (1 dB in total). The power in the Stokes signal vs the coupled pump power is plotted in Fig. 25. Just above a pump power of $P_L = 0.36$ W, the Stokes power increased by 37 dB, from −30 dBm to +7 dBm indicating that the lasing threshold was near P_L. The slope efficiency was estimated to be 30%.

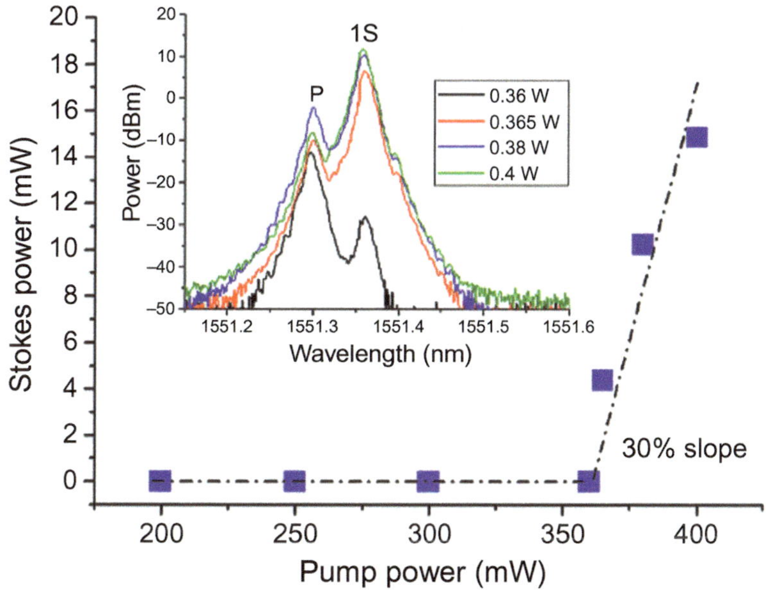

Fig. 25 Power transfer curve between the driving pump laser to the Stokes signal. Shows Brillouin lasing occurs above the power threshold of 0.36 W with the slope efficiency of 30%. *Reprinted with permission from Kabakova, I. V., Pant, R., Choi, D.-Y., Debbarma, S., Luther-Davies, B., Madden, S. J., Eggleton, B. J., 2013. Narrow linewidth Brillouin laser based on chalcogenide photonic chip. Opt. Lett. 38 (17), 3208–3211. https://doi.org/10.1364/OL.38.003208. http://opg.optica.org/ol/abstract.cfm?URI=ol-38-17-3208. © The Optical Society.*

The inset in Fig. 25 shows spectra of the backscattered light circulating in the ring-cavity, measured using an optical spectrum analyzer. The different colored lines correspond to the back-scattered signals for varying pump power close to the lasing threshold. Two wavelengths can be identified: the pump signal at 1551.3 nm and the generated Stokes signal at 1551.37 nm, red-shifted from the pump.

The linewidth of the Brillouin single-frequency laser was measured using a self-heterodyne technique (Richter et al., 1986) with the spectral resolution of 3 Hz. Fig. 26 shows the result of this measurement: the blue line corresponds to the pump, whereas the red curve is the measurement of the lasing signal. The 3 dB bandwidth of Stokes signal was found to be 100 kHz, that is, 15 times smaller than that of the pump laser. A small delta function-like peak on the top of Lorentzian pedestal for the Stokes measurement results from beating of the residual coherent components, which is caused by imperfect suppression of the main carrier in the Mach–Zehnder interferometer used for

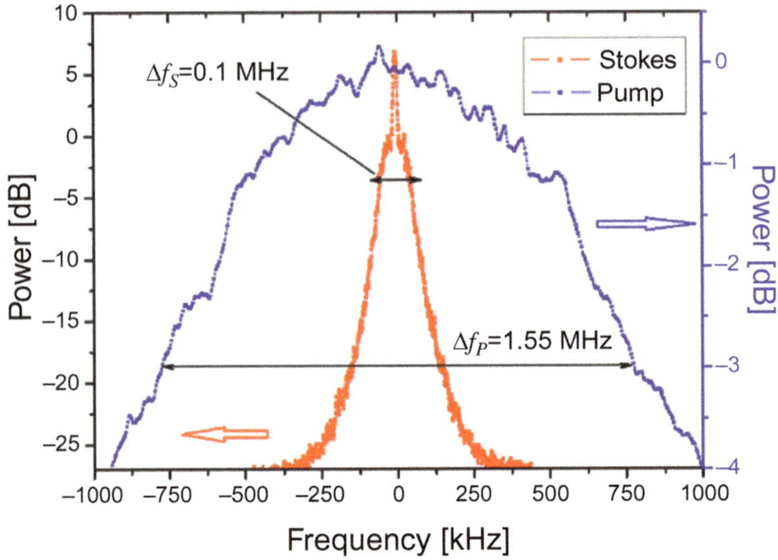

Fig. 26 Spectral output emission. The FWHM of the Brillouin laser measured using RF analyzer. The linewidth is measured at a 3 dB level of a Stokes pedestal. *Reprinted with permission from Kabakova, I. V., Pant, R., Choi, D.-Y., Debbarma, S., Luther-Davies, B., Madden, S. J., Eggleton, B. J., 2013. Narrow linewidth Brillouin laser based on chalcogenide photonic chip. Opt. Lett. 38 (17), 3208–3211. https://doi.org/10.1364/OL.38.003208. http://opg.optica.org/ol/abstract.cfm?URI=ol-38-17-3208. © The Optical Society.*

heterodyne measurements. It is, therefore, a measurement artifact and the Stokes linewidth should be calculated as the full width of half maximum of the Lorentzian pedestal.

3.4 Silicon

3.4.1 Brillouin processes in silicon waveguides

While both Kerr and Raman nonlinearities are radically enhanced by tight optical-mode confinement in nanoscale silicon waveguides (Rong et al., 2004), it is perhaps counterintuitive that these same waveguides do not produce appreciable Brillouin nonlinearities. While silicon on insulator (SOI) waveguides produce low-loss optical waveguidance through total internal reflection, these same structures do not produce total internal reflection for acoustic waves. Since silicon has a higher sound velocity than the silica cladding, SOI waveguides support leaky acoustic modes (Poulton et al., 2013). As a result, the Brillouin-active phonon modes typically radiate before appreciable light–sound scattering can occur, leading to a suppression of Brillouin interactions in conventional SOI waveguides (Rakich et al., 2012).

To realize large Brillouin nonlinearities in silicon it was necessary to develop new optomechanical waveguide structures with simultaneous confinement of optical and acoustic waves that enhance photon–phonon coupling (Rakich et al., 2012). Using suspended waveguide structures that reduce or eliminate the leakage of sound into the cladding, large and engineerable Brillouin nonlinearities have been realized in a variety of optomechanical waveguide structures (Kittlaus et al., 2016, 2017; Shin et al., 2013, 2015; Van Laer et al., 2015a, b). To distinguish these structures from conventional silicon waveguides, we refer to them as Brillouin-active silicon waveguides. Examples of Brillouin-active waveguide geometries, that have been used to produce simulated Brillouin scattering, are seen in Fig. 27; these systems comprised a silicon-on-insulator waveguide whose oxide cladding has been removed using a wet etch. Light is confined to the waveguide core of these waveguide structures through total internal reflection while the phonon mode is confined using either boundary reflection (Fig. 27A and B) or Bragg scattering (Fig. 27C).

In contrast to glass- and nitride-based waveguide systems, which produce backward-SBS interactions, forward Brillouin scattering interactions are dominant within these Brillouin-active silicon waveguides. This is because, the photoelastic tensor component (p_{12}) that ordinarily mediates backward-SBS coupling—through the coupling between transverse electric fields and longitudinal acoustic displacement—is exceptionally small in silicon, effectively stifling the backward-SBS process. Instead, the dominant photoelastic tensor component (p_{11}) of silicon mediates coupling between transverse electric fields and elastic waves with transverse displacement fields. This type of photoelastic coupling permits efficient forward-SBS interactions within these optomechanical waveguide systems.

Simulations and experiments have demonstrated that the optical forces produced by these photoelastic couplings constructively interfere with boundary-induced radiation pressure within subwavelength scale waveguides, contributing to an enhancement of the forward-SBS gain (Rakich et al., 2012; Shin et al., 2013; Van Laer et al., 2015b). By optimizing the overlap between guided acoustic and optical fields within nanoscale silicon waveguides it has been possible to realize record Brillouin gain coefficients (~1000–6000 $W^{-1}m^{-1}$) within these systems (Kittlaus et al., 2016; Shin et al., 2013; Van Laer et al., 2015a,b). However, to realize net optical amplification, the Brillouin gain must exceed the linear and nonlinear losses of the waveguide system. Robust net amplification has been achieved using low-loss (<0.2 dB cm^{-1}) large mode area waveguides that more effectively manage nonlinear losses (Kittlaus et al., 2016). Through forward-SBS interactions,

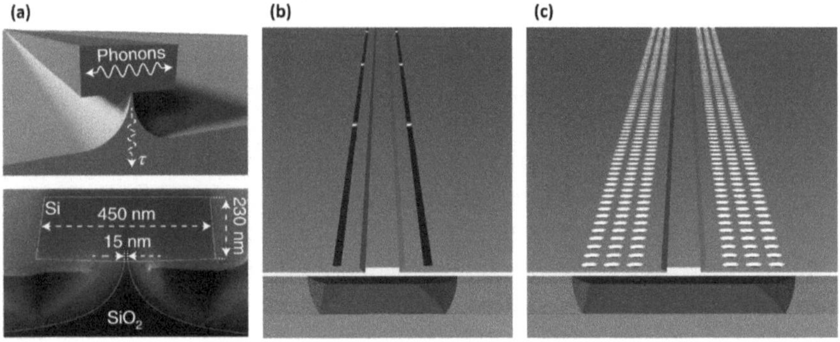

Fig. 27 Different Brillouin-active silicon waveguide structures. (A) nanowire silicon waveguide structure from Van Laer et al. (2015b). (B) suspended waveguide structure of the type reported in Kittlaus et al. (2017) and Kittlaus et al. (2016). (C) phononic crystal clad waveguide of type reported in Shin et al. (2015) and Gertler et al. (2020). *Panel (A) credit Van Laer, R., Kuyken, B., Van Thourhout, D., Baets, R., 2015b. Interaction between light and highly confined hypersound in a silicon photonic nanowire. Nat. Photonics 9 (3), 199–203. https://doi.org/10.1038/nphoton.2015.11. Adapted from Van Laer, R., Kuyken, B., Van Thourhout, D., Baets, R., 2015b. Interaction between light and highly confined hypersound in a silicon photonic nanowire. Nat. Photonics 9 (3), 199–203. https://doi.org/10.1038/nphoton.2015.11; Kittlaus, E. A., Otterstrom, N. T., Rakich, P. T., 2017. On-chip inter-modal Brillouin scattering. Nat. Commun. 8, 15819. https://doi.org/10.1038/ncomms15819; Kittlaus, E. A., Shin, H., Rakich, P. T., 2016. Large Brillouin amplification in silicon. Nat. Photonics 10 (7), 463–467. https://doi.org/10.1038/nphoton.2016.112; Shin, H., Cox, J. A., Jarecki, R., Starbuck, A., Wang, Z., Rakich, P. T., 2015. Control of coherent information via on-chip photonic-phononic emitterreceivers. Nat. Commun. 6, 6427. https://doi.org/10.1038/ncomms7427; Gertler, S., Kittlaus, E. A., Otterstrom, N. T., Rakich, P. T., 2020. Tunable microwave-photonic filtering with high out-of-band rejection in silicon. APL Photonics 5 (9), 096103. https://doi.org/10.1063/5.0015174. Van Laer et al. (2015b), Kittlaus et al. (2017), Kittlaus et al. (2016), Shin et al. (2015), and Gertler et al. (2020) are licensed under CC BY 4.0.*

these waveguides have been used to produce net optical amplification at modest (5 mW) optical powers and appreciable (6.9 dB) of on/off gain corresponding to 5.2 dB of net amplification (Kittlaus et al., 2016).

3.4.2 SBS amplification in silicon waveguides

Two distinct types of forward-SBS scattering processes have been used to produce optical amplification within Brillouin-active silicon waveguides. These include intramodal forward-SBS (Kittlaus et al., 2016; Van Laer et al., 2015a) and intermodal forward SBS (Kittlaus et al., 2017). Through intramodal scattering processes, a phonon with vanishing longitudinal wave vector mediates coupling between guided optical waves having the same spatial character. Through intermodal scattering a phonon with finite longitudinal wave-vector mediates coupling between guided optical waves with distinct spatial character. It should be noted that these scattering processes produce dynamics that are distinct from backward-SBS, as was previously established through studies of optical fiber systems (Kang et al., 2009, 2010, 2011).

The unusual gain dynamics produced through intramodal scattering arises from the fact that the same phonon mode mediates coupling to Stokes and anti-Stokes photons (Kang et al., 2009; Kharel et al., 2016). Consequently, if one seeks to create a laser using such interactions it is necessary to create an asymmetry between the Stokes and anti-Stokes processes (Kang et al., 2009; Kharel et al., 2016). This might be accomplished by using an optical resonance to enhance the Stokes scattering while ensuring that anti-Stokes scattering process is suppressed (i.e., nonresonant). Hence, somewhat more sophisticated resonator designs are necessary to realize a practical laser oscillator based on intramodal forward-SBS scattering. For example, a ring resonator with regularly spaced optical modes, matching the Brillouin frequency, would not meet this criterion, as resonant pump excitation would also yield resonant enhancement of both stokes and anti-stokes scattering process. To break this symmetry it is necessary to make the mode spacing irregular using one of a variety of mode engineering strategies (Liu et al., 2022b; Puckett et al., 2019).

Since the Stokes and anti-Stokes processes are decoupled through intermodal scattering processes (i.e., they are mediated by distinct phonons), the task of laser design becomes more straightforward using intermodal forward-SBS. Additionally, because intermodal processes produce coupling between two distinct spatial modes, these interactions naturally lend themselves to the realization nonreciprocal multiport systems (Kang et al., 2011; Yu and Fan, 2009). For example, using a mode-selective coupler to address the separate spatial modes of a Brillouin-active waveguide (as illustrated in Fig. 28A), it is possible to produce nonreciprocal wave propagation as the basis for nonmagnetic isolators (Kittlaus et al., 2018; Otterstrom et al., 2019; Poulton et al., 2012). Intermodal coupling also opens the door to new Brillouin laser designs (Otterstrom et al., 2018a) that can also exploit nonreciprocal wave propagation to improve immunity to backscatter (Otterstrom et al., 2020).

3.4.3 Silicon SBS laser resonators

To provide context for the flexibility offered by intermodal coupling it is useful to first review the criteria for efficient Brillouin laser operation when using Backward SBS interactions. When using conventional (intramodal) backward-SBS processes to design a Brillouin laser, it is necessary to match

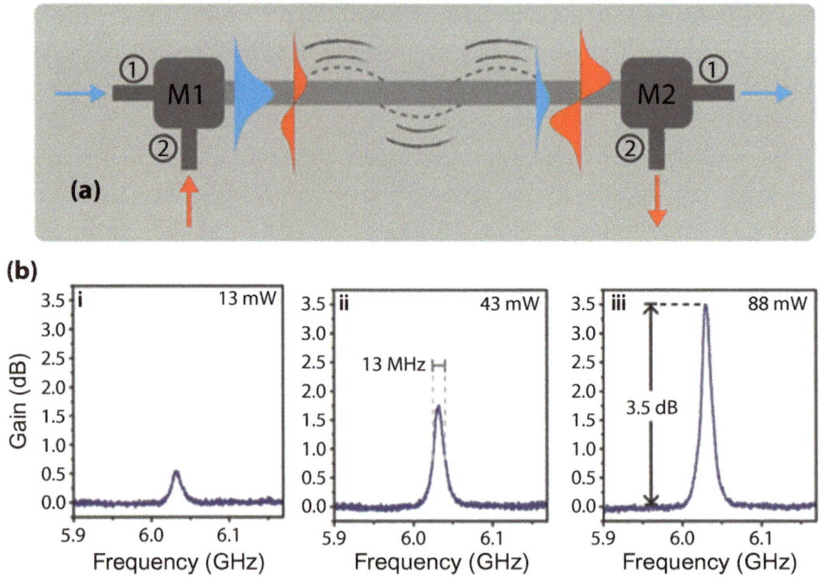

Fig. 28 Intermodal Brillouin Gain in silicon. (A) schematic of multimode Brillouin active waveguide interfaced with mode-selective couplers to address the distinct spatial modes that participate in the intermodal scattering process. (B) Brillouin gain spectrum produced by intermodal waveguide structure pictured in Fig. 29. Through these experiments, the symmetric (*blue*) mode serves as the pump while antisymmetric spatial mode (*red*) is the Stokes wave that is amplified when traversing the Brillouin active waveguide segment. Further details of Brillouin gain measurements found in Kittlaus et al. (2017). *From Kittlaus, E. A., Otterstrom, N. T., Rakich, P. T., 2017. On-chip inter-modal Brillouin scattering. Nat. Commun. 8, 15819. https://doi.org/10.1038/ncomms15819. Kittlaus et al. (2017) is licensed under CC BY 4.0.*

the resonator FSR to the Brillouin frequency to meet the condition for triply resonant Brillouin laser oscillation. Additionally, efficient and low-noise laser oscillation requires that the resonator dimension very precisely matches this condition. These conditions typically constrain the laser size and require very careful dimensional trimming of the resonator to ensure good alignment of the cavity resonances with the Brillouin frequency. However, when designing a laser based on intermodal scattering we now consider frequency separation between distinct spatial modes within a multimode waveguide resonator. In contrast to lasers based on the intramodal backward-SBS scattering process, any number of resonator geometries can readily satisfy the resonance condition, which requires that the frequencies of two distinct spatial modes be separated by the Brillouin frequency as illustrated in Fig. 29B.

In contrast to Brillouin lasers that utilize backward-SBS, the injected pump-wave and the emitted and Stokes-waves copropagate within the

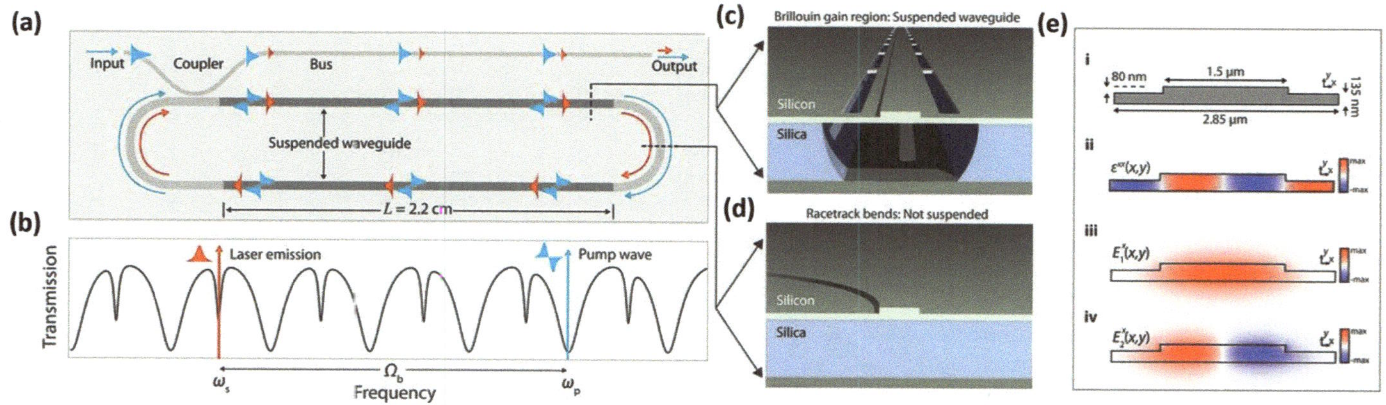

Fig. 29 Schematic of silicon Brillouin laser and basic operation. (A) The Brillouin laser consisting of a multimode racetrack cavity with two Brillouin-active regions (*dark gray*). Pump light (*blue*) is coupled into the antisymmetric spatial mode of the racetrack resonator. Intermodal Brillouin scattering mediates energy transfer from the pump wave (antisymmetric; *blue*) to the Stokes wave (symmetric; *red*). (B) The idealized transmission spectrum for the racetrack cavity. Narrower and broader resonant features correspond to the symmetric and antisymmetric resonances, respectively. Brillouin lasing occurs when the resonance conditions for the pump (antisymmetric) and Stokes (symmetric) waves are simultaneously satisfied. (C and D) Cross sections of the suspended Brillouin-active region and the racetrack bend, respectively. (E) (i) Dimensions of the Brillouin-active waveguide. (ii) Strain profile of the 6-GHz acoustic mode that mediates intermodal scattering. (iii and iv) *x*-directed electric field profiles of the TE-like symmetric and antisymmetric optical modes, respectively. *From Otterstrom, N. T., Behunin, R. O., Kittlaus, E. A., Wang, Z., Rakich, P. T., 2018b. A silicon Brillouin laser. Science 360 (6393), 1113–1116 with permission from AAAS.*

intermodal Brillouin laser pictured in Fig. 29A–D. This laser system (described in Ref. Otterstrom et al., 2018a) consists of a Brillouin-active silicon waveguide that is wrapped into the racetrack configuration seen in Fig. 29. The waveguide cross-section and the optical and elastic fields that participate in intermodal scattering process are shown in Fig. 29A–D. Through operation of this laser, an incident pump-wave is coupled into the anti-symmetric cavity mode (blue) from a bus-waveguide yielding amplification for the symmetric (red) Stokes wave. Above threshold, the self-oscillating Stokes-wave exits the cavity through the bus waveguide. In contrast to conventional lasers based on backward-SBS, this intramodal forward-SBS laser does not produce cascaded Stokes emission (Otterstrom et al., 2018a). Hence, the basic operation this intermodal laser design is unusual in that (1) the Stokes emission copropagates with the pump wave, (2) cascaded scattering is inhibited, and (3) the condition for triply resonant Brillouin laser oscillation can be met at numerous wavelengths for practically any cavity dimension.

3.4.4 SBS lasing and performance

This silicon Brillouin laser also produces line-narrowing dynamics that are distinct from conventional Brillouin lasers. Since the optical decay rate is larger than the acoustic decay rate, this silicon Brillouin laser produces phonon line-narrowing rather than optical line-narrowing (Otterstrom et al., 2018a). When the triply resonant condition is satisfied, the phonon linewidth becomes $\Delta\nu_{ph} = \frac{\Gamma}{4\pi\beta^2}(N_{th} + n_{th} + 1)$, where β^2 is the coherently driven phonon occupation number (Otterstrom et al., 2018a). In limit of vanishing phonon linewidth ($\beta \rightarrow \infty$), the emitted Stokes wave becomes a frequency shifted replica of the incident pump wave. In this limit, the operation of the Brillouin laser is analagous to that of an idealized acousto-optic frequency shifter. Hence, a heterodyne beatnote produced by interference between the pump and Stokes waves would produce a perfectly monochromatic beatnote. However, for finite line-narrowing, the phonon field imparts phase noise onto the Stokes wave, which is detected as phase noise on the heterodyne beatnote. In his way, the heterodyne beatnote between pump and Stokes waves provides a window into the spectrum of the distributed phonon field produced within the Brillouin laser oscillator. The spectrum of Fig. 30B shows heterodyne spectrum produced by pump-Stokes interference, revealing dramatic compression of the phonon linewidth when the Brillouin laser is driven well above threshold. The phonon linewidth narrowing was shown to produce good agreement with the Schawlow–Townes line narrowing expression above.

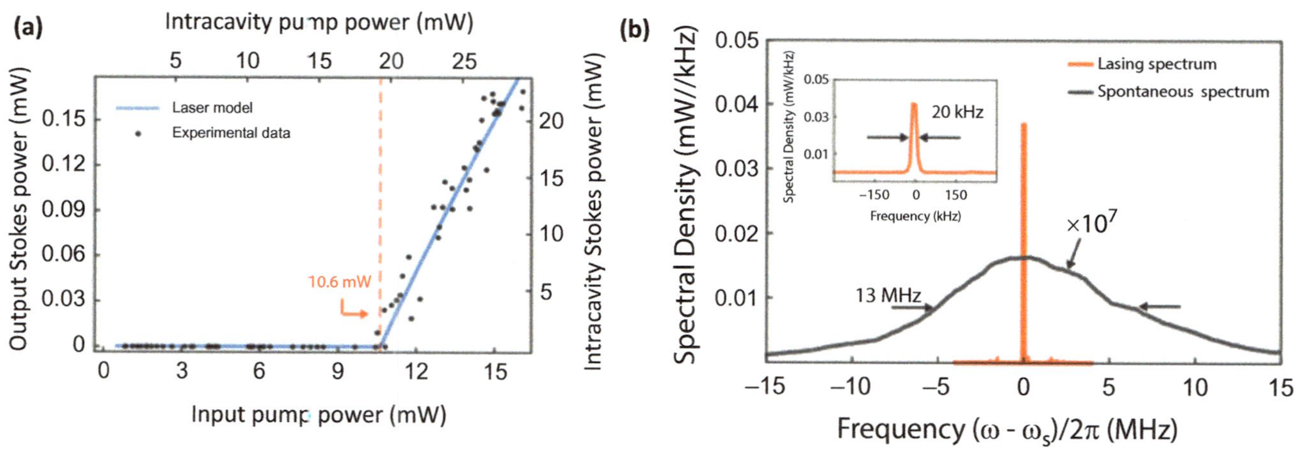

Fig. 30 Threshold and line narrowing characteristics. (A) plot showing emitted Stokes-wave power as a function of input pump power. Threshold for laser oscillation observed at 10.6 mW. (B) comparison of spontaneous Stokes linewidth (without feedback) and the emitted laser linewidth; spectra obtained by heterodyne measurement. *From Otterstrom, N. T., Behunin, R. O., Kittlaus, E. A., Wang, Z., Rakich, P. T., 2018b. A silicon Brillouin laser. Science 360 (6393), 1113–1116 with permission from AAAS.*

4. Cascaded mode operation

Brillouin lasers can be run in cascaded-order mode, with successively higher order Stokes emission line emitted alternately in the forward and backward directions. Operation as a cascaded order source has both advantages and disadvantages, related to common mode noise between orders, noise contributions from higher to lower orders, and linewidth and output power limitations due to clamping. Photomixing of two Stokes tones can produce a low phase noise microwave source through common mode noise cancellation due to their generation in a common cavity (Gundavarapu et al., 2019; Li et al., 2013; Merklein et al., 2016). High-sensitivity Brillouin-based optical gyroscopes through the Sagnac effect have been realized using counter propagating Stokes orders (Gundavarapu et al., 2019; Li et al., 2017). Tradeoffs exist with Stokes cascaded-order operation include linewidth broadening and increased phase noise through feedback from higher to lower order Stokes (Behunin et al., 2018) and clamping of Stokes power. As described in the next section, higher order inhibition can be realized by introducing cavity dispersion to greatly reduce and even eliminate the phase matching condition for the higher Stokes orders (Liu et al., 2022b; Merklein et al., 2015; Puckett et al., 2019; Wang et al., 2021) splitting the resonance where phase matching for the higher order (e.g., S2) would occur in the resonator.

4.1 Phase lock cascaded

As discussed earlier, SBS can be cascaded in a resonator producing multiple Stokes waves spaced by the Brillouin frequency shift, Ω, and forming a Brillouin frequency comb (Buttner et al., 2014). The phase relationship between the comb lines is complex, and SBS by itself provides no phase-locking mechanism since all Stokes waves grow from noise leading to a random spectral phase. Different mechanisms have been suggested to achieve phase-locking of Brillouin combs. Experimentally phase-locking has been demonstrated by limiting the SBS interaction to a small part of the cavity and by using modal dispersion in a short-multimode fiber to reduce wave de-phasing. Nonetheless, the dependence of the Brillouin frequency shift on the absolute frequency of the driving wave is the fundamental limitation to phase-locking of the cascade of Brillouin waves, since each and every order within the cascade occurs at slightly different frequency shift from the previous order.

On the other hand, Kerr-induced four wave-mixing (FWM) is a nonlinear process that also leads to generation of new frequencies. Unlike

SBS, FWM is phase-sensitive with the spacing between the newly generated waves being carefully preserved. Frequency combs generated through FWM in high-finesse micro-resonators have been the subject of extensive research and been shown to exhibit phase-locking (Kippenberg et al., 2004, 2011). Little was known, however, on the interplay between SBS and FWM in the context of frequency comb generation in nonlinear media until 2014. First, Brillouin cascading and phase-locking of Brillouin combs was approached in a chalcogenide As_2Se_3 fiber, which was cleaved to form a low-finesse resonator (Buttner et al., 2014). The feedback created via multiple reflections from the fiber facets helped to reduce the peak power threshold above which Brillouin cascading occurred (this was established to be approximately 1W). Relatively high threshold peak power as well as the long length of the fiber resonator ($L = 36$ cm), when compared to that in chalcogenide rib waveguides integrated on a chip discussed in Sections 3.3.2 and 3.3.3, can be explained by comparing effective mode areas for the fiber ($A_{eff}^{f} = 22\mu m^2$) and the waveguide ($A_{eff}^{w} = 3\mu m^2$), respectively. It has been concluded, however, that FWM plays an essential role in phase-locking of the Brillouin cascade, offering pathways for the synchronization of the waves' phases and formation of short pulses. Finally, 40 ps pulse formation at the repetition rate of 8 GHz, which is determined by the Brillouin frequency shift in As_2Se_3 fiber, was demonstrated (Buttner et al., 2014).

Building on these fiber results, the same group of authors explored SBS cascading on a chalcogenide chip platform (Merklein et al., 2015). To improve power requirements for generating a cascade of Brillouin waves, the chalcogenide waveguide's dispersion has been modulated through creation of a periodic structure, a waveguide Bragg grating. Bragg gratings have been widely used in the context of communications and fiber lasers as narrow band filters and mirrors. In this demonstration, however, the periodic structure of a grating modulates the density of photonic states and provides frequency bands in which SBS process undergoes enhancement (for frequencies at the edges of the photonic gap) and inhibition (inside the photonic gap). Fig. 31 schematically illustrates this very concept. The Bragg grating periodicity and structure has to be designed carefully to provide enhancement for SBS process not only for the pump wave but also for high order cascaded waves, first- and second-order Stokes waves. This is achieved by photoinducing a superstructure Bragg grating with three separate photonic bandgaps, which are spectrally separated exactly by 8 GHz, the Brillouin frequency shift in As_2S_3 glass. Thus, whenever the pump wavelength is tuned to the low-wavelength edge of the first photonic gap, the first and second Stokes waves become automatically aligned with the edges of the second and third photonic gaps thus providing enhancement to all waves at once (Merklein et al., 2015).

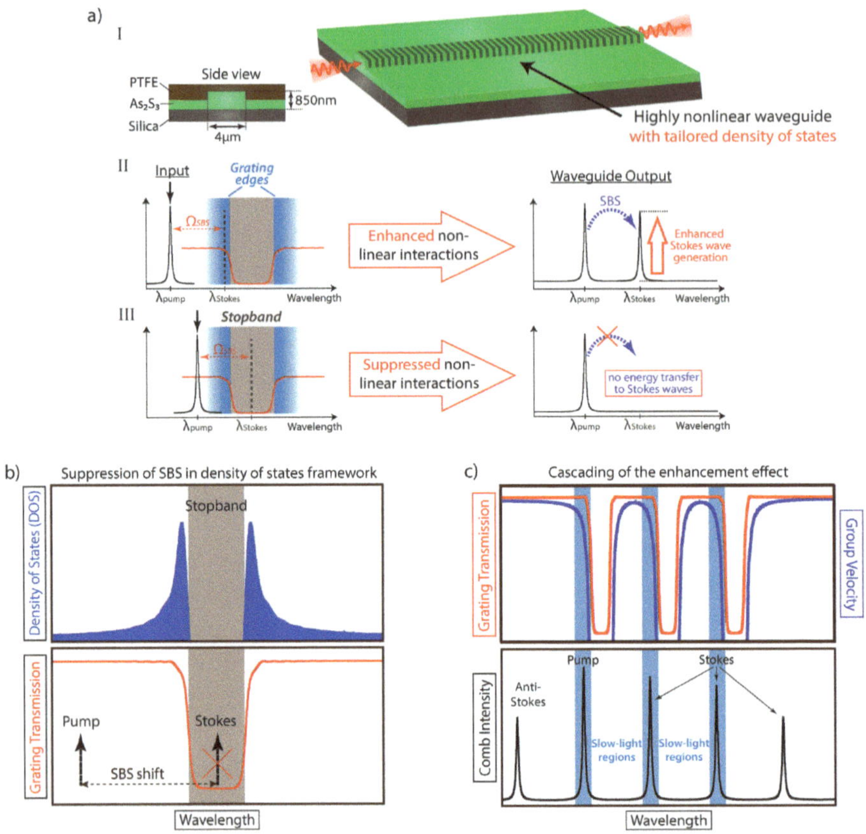

Fig. 31 Basic principle for enhancing or inhibiting SBS by tailoring the density of states in a waveguide. (A) (I) Scheme of the 850 nm by 4 μm chalcogenide (As_2S_3) rib waveguide on a silica substrate with a Polytetrafluoroethylene over-cladding. (II) Tuning the frequency of the optical waves (here shown for the first Stokes wave) into the slow-light region at the band edge of the PBG leads to an enhancement of the nonlinear interactions. (III) Adjusting the pump frequency so that the Stokes frequency falls into the stopband, however, leads to an inhibition of the SBS process. (B) Illustration of the density of states and the stopband for a 1D photonic bandgap structure. The DOS increases toward the band edge of the structure and approaches zero in the stopband, which enables the full suppression of any Stokes wave generation. (C) Shows schematically how the enhancement effect can be cascaded using a multiband grating to form a Brillouin frequency comb. It illustrates the distinctive design of the multiband grating, matching the Brillouin shift and each having narrow bandwidth, so that multiple comb lines are aligned with the grating edges. *From Merklein, M., Kabakova, I. V., Büttner, T. F. S., Choi, D.-Y., Luther-Davies, B., Madden, S. J., Eggleton, B. J., 2015. Enhancing and inhibiting stimulated Brillouin scattering in photonic integrated circuits. Nat. Commun. 6 (1), 6396. https://doi.org/10.1038/ncomms7396. Merklein et al. (2015) is licensed under CC BY 4.0.*

The results of this experiment are demonstrated in Fig. 32 with (A) showing the superimposed spectrum of the Brillouin frequency comb (black line) and the multi-stop-band Bragg grating (red line). For the purpose of the density of states enhancement, the wavelength of the pump laser is chosen on the left, low-wavelength edge of the first photonic bandgap. For the pump power just above the SBS threshold of 0.6 W, the cascade of five high-order Stokes and four anti-Stokes waves are clearly seen and demonstrated in Fig. 32A. While Stokes waves are generated through SBS cascading, anti-Stokes waves are produced via FWM. The first process has much higher gain than the second, as Kerr nonlinearity is orders of magnitude weaker than the effects of electrostriction governing SBS process. The power distributed across Stokes waves is therefore much larger, resulting in stronger Stokes waves compared to their anti-Stokes counterparts.

The broadest Brillouin comb generated in a chalcogenide chip to date is shown in Fig. 32D: 14 lines are clearly seen with fairly symmetric amplitude envelop structure. This result was obtained at the peak pulse power of 1.3 W, far above the SBS threshold. Interestingly, the same waveguide without a Bragg grating supports only four comb lines at similar level of pump power, demonstrating the effectiveness of the approach to enhance nonlinear interactions applied in this work (Merklein et al., 2015).

4.2 Multiple order generation in silicon nitride resonators

In silicon nitride resonators, cascaded order operation is reached efficiently due to the very high Q, low lasing threshold, and precision phase-matched resonance locations with respect to the Brillouin frequency shift for S1 and higher orders. For example, in the silicon nitride cavity described in earlier sections of this chapter, the first-order Stokes S1 is frequency down-shifted from the pump, to a distance in frequency corresponding to four cavity free-spectral ranges (FSRs). The long cavity yields a small FSR and relaxes the fabrication precision required to phase match the optical and acoustic fields for this first and higher orders as shown in the experimental setup in Fig. 33. It is important to pay attention to the fundamental linewidth determined by the photon number in the resonator as well as the noise fed back from higher order Stokes modes, which can be described using a cascaded-order Brillouin laser noise dynamics theory (see Eqs. 2 through 6) (Behunin et al., 2018).

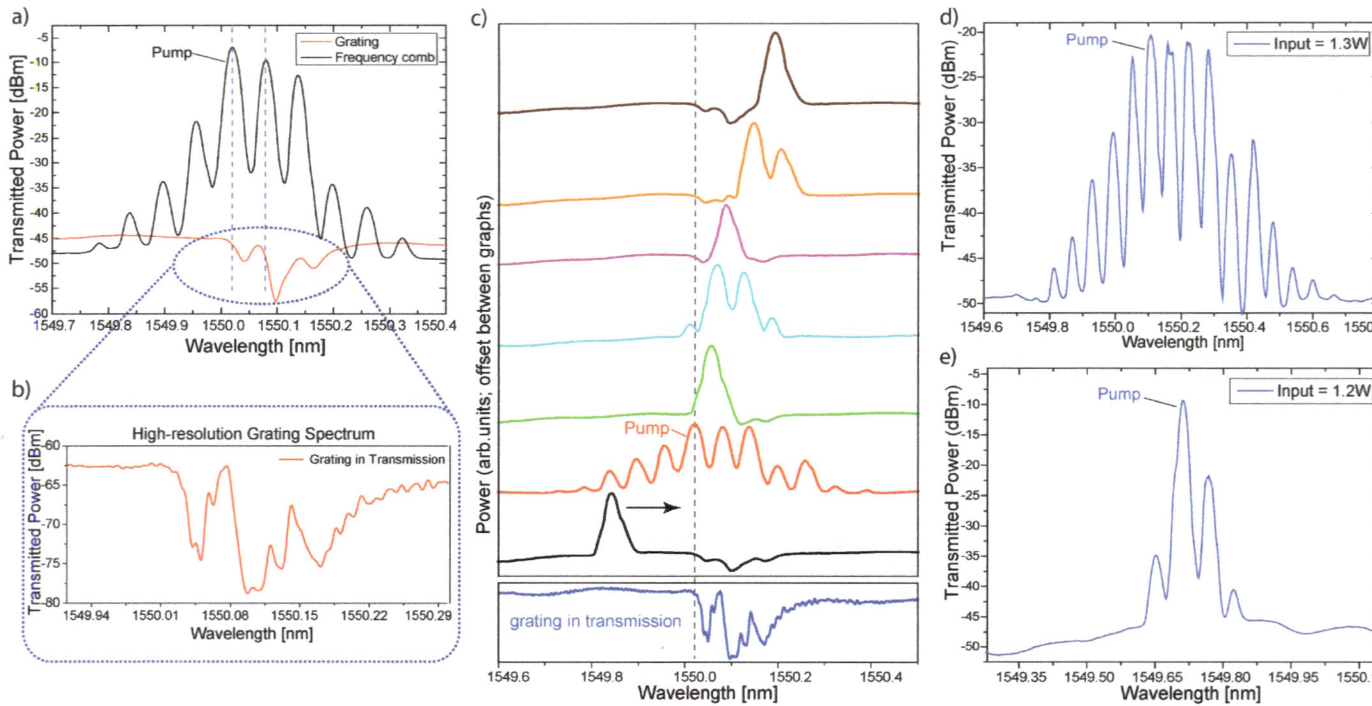

Fig. 32 Slow-light enhancement effect of stimulated Brillouin scattering. (A) Alignment of the Brillouin frequency comb with respect to the multistopband Bragg grating. The pump and lowest order Stokes waves lay at the edges of the individual stopbands, where the density of states was high but the normalized transmission was almost unity. (B) High-resolution spectrum of the multistopband structure. (C) Wavelength scan across the grating spectrum. The SBS process was significantly enhanced in the slow-light regions. (D) Brillouin frequency comb with 15 lines generated by harnessing the tailored density of states in the waveguide. e, Maximum SBS cascading in the waveguide without the grating. *From Merklein, M., Kabakova, I. V., Büttner, T. F. S., Choi, D.-Y., Luther-Davies, B., Madden, S. J., Eggleton, B. J., 2015. Enhancing and inhibiting stimulated Brillouin scattering in photonic integrated circuits. Nat. Commun. 6 (1), 6396. https://doi.org/10.1038/ncomms7396. Merklein et al. (2015) is licensed under CC BY 4.0.*

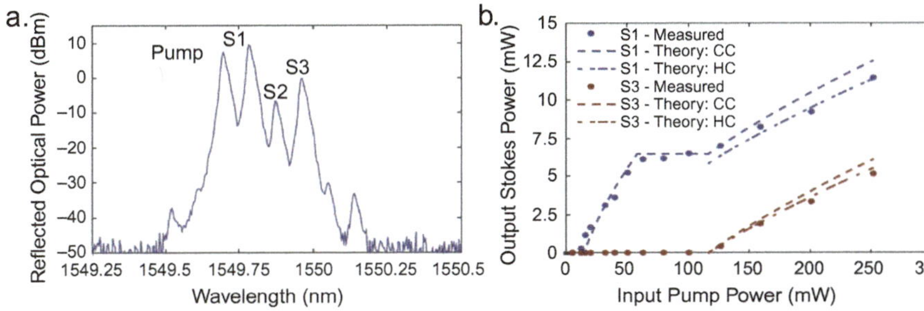

Fig. 33 Si_3N_4 cascaded order operation. Pump light injected into the resonator builds to high intensities, leading to spontaneous Brillouin scattering that populates the Stokes mode (S1). The circulating pump amplifies this incoherent Stokes light, initiating laser oscillation when the round-trip gain compensates for the resonator losses. Brillouin interactions couple counter-propagating optical fields (e.g., pump and Stokes (S1)) through traveling sound waves that provide gain for optical amplification. (A) Further increase in the pump power produces successive cascaded-order lasing where distinct wavelengths at the cavity resonances are emitted. (B) Clamping of S1 at the threshold onset of S2 lasing. *Credit Gundavarapu, S., Brodnik, G. M., Puckett, M., Huffman, T., Bose, D., Behunin, R., Wu, J., Qiu, T., Pinho, C., Chauhan, N., Nohava, J., Rakich, P. T., Nelson, K. D., Salit, M., Blumenthal, D. J., 2019. Sub-hertz fundamental linewidth photonic integrated Brillouin laser. Nat. Photonics 13 (1), 60–67. https://doi.org/10.1038/s41566-018-0313-2. Gundavarapu et al. (2019) is licensed under CC BY 4.0.*

5. Mode engineering

There are cases where fundamental linewidth reduction or increased output laser power is needed. One approach is to design the SBS laser resonator with a filter function designed to inhibit the S2 and higher order Stokes emission and allow the pump laser energy to continue to drive the S1 emission. Overall Stokes inhibition was first demonstrated in 2015 by Merklein et al. (2015). In this work, an optical stop-band was used to inhibit Brillouin all together (see Fig. 34A) and the power remains within the pump well above the SBS threshold. Measurements and simulations are in good agreement as illustrated in Fig. 34.

In 2019, the concept of using mode engineering to produce single Stokes mode lasing, enabling continued increase in S1 with increase pump, while inhibiting S2 was demonstrated in a modified Si_3N_4 SBS laser (Puckett et al., 2019). The resonator was designed as a grating modulated ring resonator waveguide shown in Fig. 35. This modulation was designed to periodically

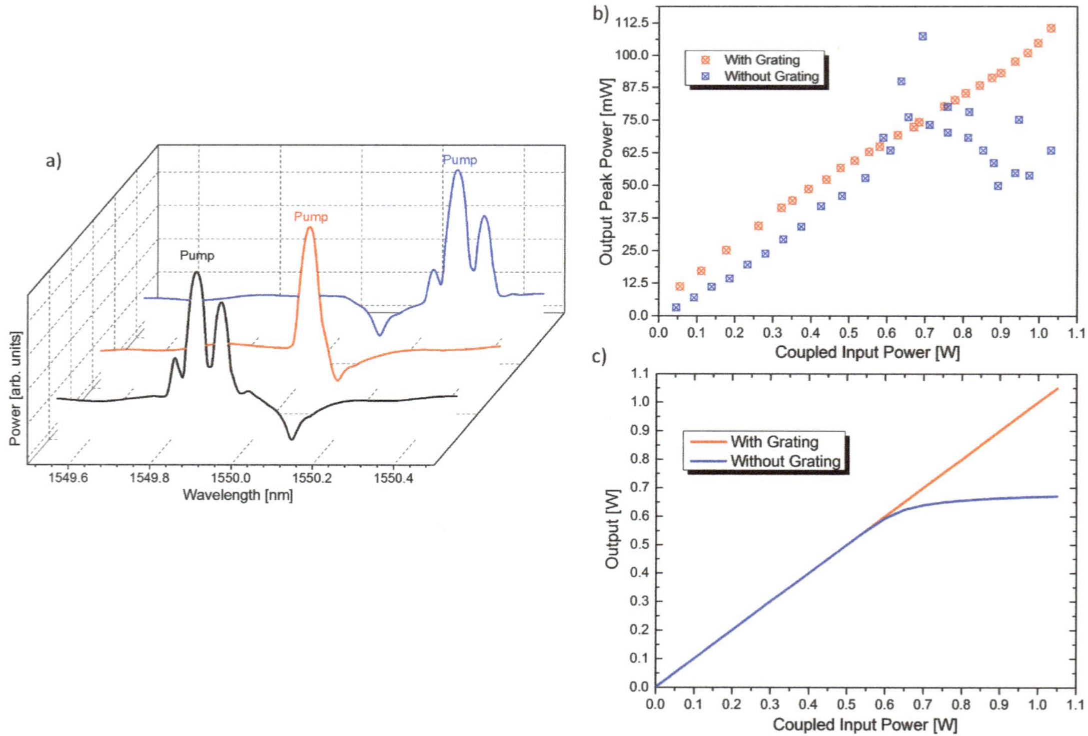

Fig. 34 The demonstration of SBS inhibition in a chalcogenide waveguide. (A) Spectra obtained as the wavelength of the pump ($P_p = 0.8$ W) we scanned relative to the PBG structure. The power was just above the SBS threshold $PSBS = 0.6$ W and a Stokes and anti-Stokes wave was generated at frequencies outside the stopband. By positioning the pump so that the Stokes wave fell into the stopband, SBS was no longer generated. (B) Pump power output with (*red data points*) and without (*blue data points*) the PBG structure. Without the grating the pump was depleted as soon as the SBS threshold was reached while with the grating the generation of SBS was inhibited and no pump depletion was observed. (C) Simulations of the experiment obtained by solving coupled mode equations with and without grating confirming the measurements presented in (B). *From Merklein, M., Kabakova, I. V., Büttner, T. F. S., Choi, D.-Y., Luther-Davies, B., Madden, S. J., Eggleton, B. J., 2015. Enhancing and inhibiting stimulated Brillouin scattering in photonic integrated circuits. Nat. Commun. 6 (1), 6396. https://doi.org/10.1038/ncomms7396. Merklein et al. (2015) is licensed under CC BY 4.0.*

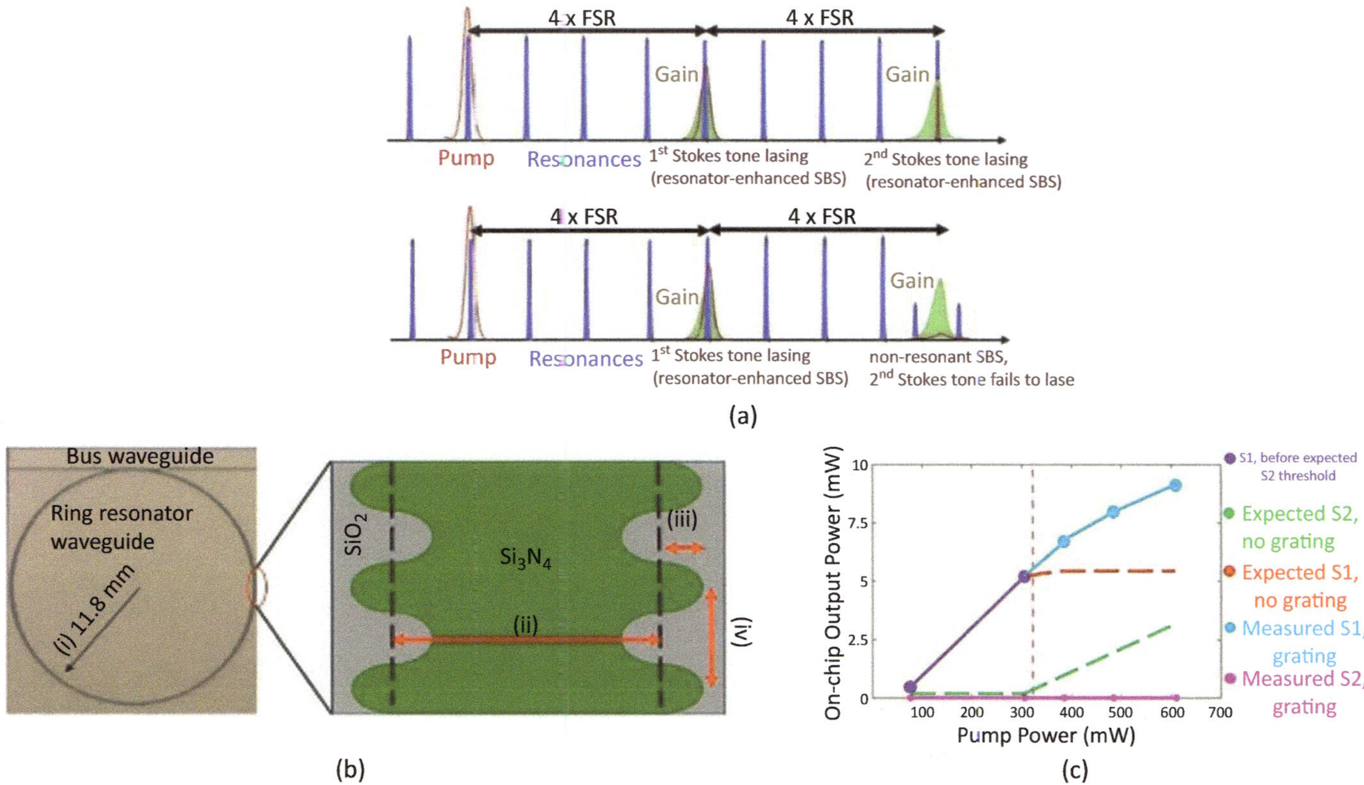

Fig. 35 Grating modulated resonator for second-order Stokes lasing inhibition. (A) Cascaded lasing action in ring resonator without a grating (*top*), no cascaded lasing with grating to split resonance at twice Brillouin frequency shift (*bottom*). (B) Fabricated ring resonator with 11.8 mm radius; Expanded inset to the right—top view of section of grating modulated ring resonator waveguide. (C) Cascaded operation without grating enables first-order Stokes (S1) and second-order Stokes (S2) lasing. Grating modulated resonator prevents lasing of S2 and allows power in S1 to continue to be pumped. *Reprinted with permission from Puckett, M., Bose, D., Nelson, K., Blumenthal, D. J., 2019. Higher order cascaded SBS suppression using gratings in a photonic integrated ring resonator laser. In: Conference on Lasers and Electro-Optics (CLEO), San Jose, CA, SM4O.1. © The Optical Society.*

split the resonance every eight FSRs as shown in Fig. 35B. By locating the pump eight resonances away from the split, S1 emits in the fourth resonance and S2 is blocked. More recently a photonic molecule design (nested ring resonators) was used to inhibit S2 lasing (Liu et al., 2022b). With continued exchange of the pump energy into the S1 mode, the increased photon density and high Q, a 70-mHz fundamental linewidth, and an S1 output power of 10 mW were demonstrated.

6. Applications of on-chip Brillouin

The waveguide-integrated Brillouin laser, shown in Fig. 1 illustrates four example system-on-chip applications, highlighting the advantage of low waveguide loss in the 405–2350 nm wavelength range. In this section we review applications that have been demonstrated or proposed using Brillouin integrated technology and future prospects for systems-on-chip applications.

6.1 Optical gyroscopes

An optical gyroscope can be implemented using Brillouin laser in the cascaded mode, achieving rotation rate sensitivities near the Earth's rotation rate (Gundavarapu et al., 2019; Lai et al., 2020). In this approach, sub-Hertz Brillouin laser linewidths result from the inverse quadratic scaling of fundamental phase noise on Q factor. These narrow linewidths endow the device with high rotation-rate sensitivity. However, the gyroscope performance is also boosted by increased resonator size (increased Sagnac effect) that normally would require higher pumping power. Here, the ultrahigh-Q resonance provides an offsetting effect, enabling gyro operation at low (milliwatt-scale) power levels. A photonic integrated Brillouin laser optical gyroscope was demonstrated using the laser chip and discrete front-end gyro components on a rotation table as show in Fig. 36A. The laser is pumped to emit two Stokes orders. While the gyroscope is at rest under Earth background rotation, the frequency spacing between adjacent optical tones is very close to the Brillouin frequency shift $f_B = \Delta\nu_B$. As the gyroscope rotation is increased, the Sagnac-induced frequency shift (Sanders et al., 2016) results in a relative increase in frequency difference between the counter-propagating first- and second-order Stokes tones, denoted as $\Delta\nu$ in Fig. 36B. The sign of $\delta\nu$ depends on the direction of applied rotation and the value of $\delta\nu$ is used to determine the rotation rate. The measured frequency

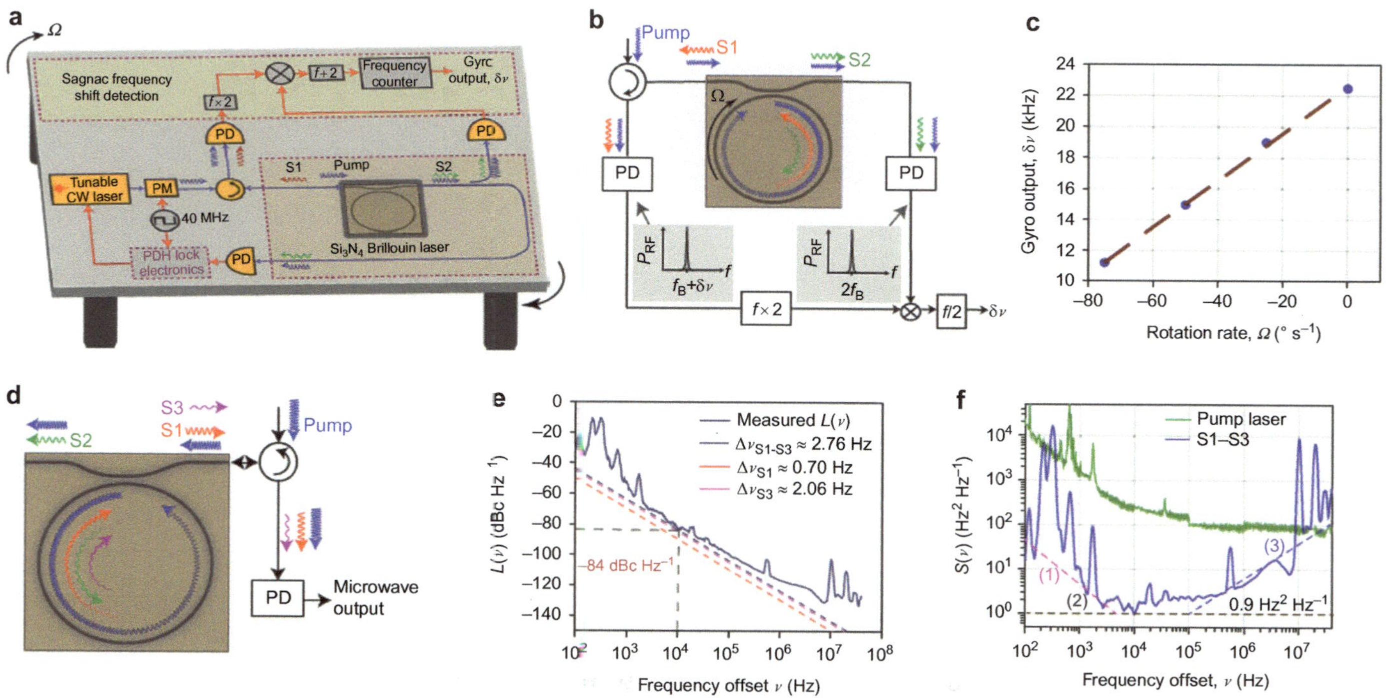

Fig. 36 Integrated laser optical gyroscope and photonic microwave synthesizer demonstrations. (A) Brillouin laser optical gyroscope configured on a rotational stage operated at a rate ω (s–1). (B) Illustration of Sagnac frequency shift ($\delta\nu$) in the Stokes orders circulating in the resonator (pump, *blue*; S1, *red*; S2, *green*). The beatnote between reflected pump and S1 is frequency doubled and mixed with the beatnote between pump and S2 to extract the Sagnac frequency shift. (C) Measured gyro output for an applied rotation rate varying from 0 to 75 s^{-1}. The measured scale factor is 152 Hz per degree per second. (D) Photonic microwave synthesizer configured by photo-mixing S1 and S3. (E) Measured single-sided phase noise of beatnote between S1 and S3 and corresponding theoretical fit lines to indicate the fundamental linewidth of S1 emission. (F) Comparison of pump frequency noise with S1–S3 beatnote frequency noise. *Dashed lines* (1), (2), and (3) show the fits to flicker frequency noise, white frequency noise and flicker phase noise regions, respectively. The common noise peaks at 0.6, 1.7, and 3.9 kHz indicate residual transfer of pump phase noise to the Brillouin laser. *Data from Gundavarapu, S., Brodnik, G. M., Puckett, M., Huffman, T., Bose, D., Behunin, R., Wu, J., Qiu, T., Pinho, C., Chauhan, N., Nohava, J., Rakich, P. T., Nelson, K. D., Salit, M., Blumenthal, D. J., 2019. Sub-hertz fundamental linewidth photonic integrated Brillouin laser. Nat. Photonics 13 (1), 60–67. https://doi.org/10.1038/s41566-018-0313-2.*

shift for an applied rotation rate varying from 0 to 75 s^{-1} in steps of 25 s^{-1} is shown in Fig. 36C and demonstrates the linear operation with a scale factor of 152 Hz per degree per second.

6.2 Low-phase noise microwave oscillators

Photonic microwave frequency synthesizers (known as radio frequency (RF) photonic oscillators) produce a microwave signal with phase noise comparable to the best-available electronic oscillators and exhibit frequency stability orders of magnitude better than electronic solutions (Fortier et al., 2011). Discrete component optical frequency comb generators utilize temperature-stabilized magnesium fluoride (MgF2) microresonators to achieve record-low phase noise of −170 dBc Hz^{-1} at 10 MHz offset from the fixed 10 GHz RF carrier and −90 dBc Hz^{-1} close-to-carrier at 100 Hz offset (Liang et al., 2015). RF carrier tunability can be realized with discrete component optoelectronic-optic photonic oscillators that incorporate chip-scale Brillouin waveguides to achieve better than −100 dBc Hz^{-1} at 100 kHz offset as the carrier is tuned from 5 to 40 GHz (Merklein et al., 2016).

Moving RF photonic oscillators to the chip scale will lower power consumption, weight, and volume. The RF carrier frequency is typically fixed by the device physical dimensions; however, optical frequency comb-based approaches can provide multiple RF carrier outputs. Discrete component microresonators utilizing Kerr comb generation produce microwave signals with measured phase noise as low as −115 dBc Hz^{-1} at 10 kHz frequency offset from a 35 GHz carrier (Liang et al., 2015). Low noise microwave frequency generation was reported using a dual-cavity configuration with an etched-silica microresonator Brillouin laser stabilized to a high-Q microcavity reference with noise performance measuring close to −90 dBc Hz^{-1} at a frequency offset of 10 kHz from a 21.7 GHz carrier (Li et al., 2013). Creating truly tunable microwave photonic oscillator outputs today is accomplished with multiple photonic chip solutions and adaptation of state-of-the-art dual-comb, self-referenced frequency synthesizers (Spencer et al., 2018).

Demonstration of the silicon nitride integrated SBS laser as an RF photonic oscillator is shown in Fig. 36, measures −84 dBc Hz^{-1} at 10 kHz frequency offset from a 21.8 GHz carrier and a close-to-carrier of −30 dBc Hz^{-1} at 100 Hz carrier offset. Fig. 36D shows the oscillator schematic and the measured phase and frequency noise are shown in Fig. 36E and F with the 10-kHz frequency offset denoted. The RF photonic oscillator leverages cancellation of common mode noise sources, such as cavity TRN, between the two Stokes orders in the laser resonator. Employing a large mode volume silicon nitride integrated reference cavity and lock circuit (Liu et al., 2022a) has the potential to further suppress the noise at lower frequency offsets toward the fundamental limit. A more detailed discussion of application of Brillouin to microwave signal generation is given in chapter "Brillouin light scattering in biological systems" by Kabakova et al. in Volume 2 of this book.

6.3 Visible light atom, molecular and quantum applications

Ultra-narrow linewidth visible light lasers provide the spectral purity required for precision atomic, molecular and optical (AMO) physics including atomic clocks (Ludlow et al., 2015) atomic and molecular spectroscopy (Bohn et al., 2017; Cygan et al., 2013; Jadbabaie et al., 2020), and quantum sensing (Degen et al., 2017; Fan et al., 2015). Historically, precision laser systems are locked to lab-scale, environmentally isolated optical reference cavities to obtain the low phase noise and high frequency stability needed to address narrow atomic optical clock transitions in atoms (Fox et al., 2003; Ludlow et al., 2007). These lab-scale systems provide state-of-the-art timing (Nicholson et al., 2015); however they pose challenges for atomic and molecular experiments of ever-growing complexity and for a portable and autonomous optical clocks. Fiber-based Brillouin lasers have been used in trapped ion systems (Loh et al., 2020); however, their emission in the IR requires additional second harmonic generation stages, as well as susceptibility to environmental disturbances. To address new applications, there is a need for ultra-low phase noise visible wavelength lasers that are smaller, more reliable, and can reduce power efficiency and complexity associated with intermediate frequency conversion as well as reduced sensitivity to environmental disturbances. Photonic integrated SBS visible lasers (Chauhan et al., 2021) are a compact "direct-drive" solution as the optical laser oscillator (OLO) for an optical clock. An example vision of a photonic integrated circuit (PIC) that incorporates a visible integrated silicon nitride SBS laser operating at 674 nm for a trapped strontium ion clock application is illustrated in Fig. 37. In this example an 674-nm semiconductor laser pumps a silicon nitride SBS laser and S1 is filtered using a silicon nitride waveguide dual-bus filter (Huffman et al., 2018). A tunable sideband can be generated using a piezoelectric silicon nitride modulator (Wang et al., 2022) and locked to a silicon nitride waveguide optical reference cavity (Lee et al., 2013; Liu et al., 2022a) to reduce the linewidth and provide

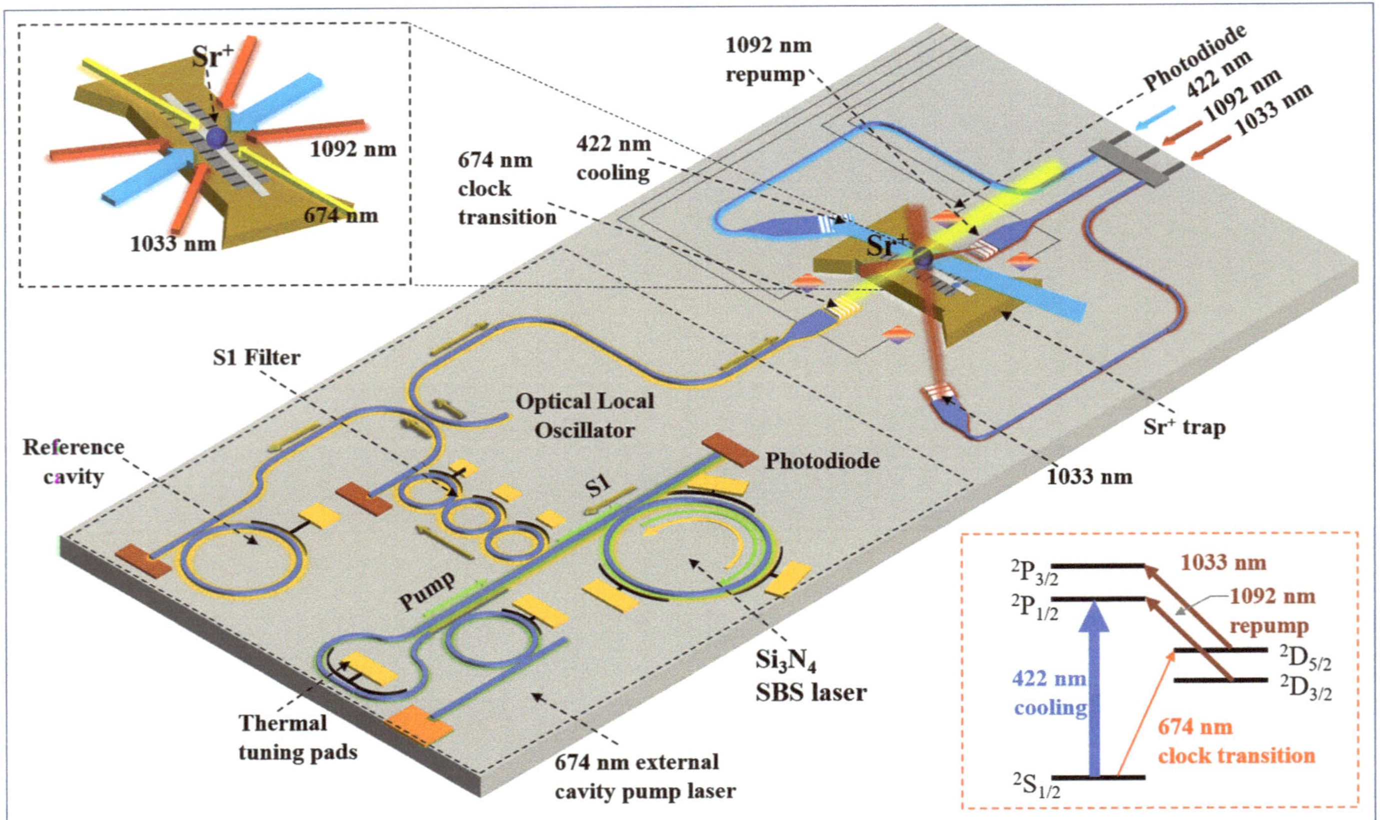

Fig. 37 See figure legend on opposite page.

the stability needed to lock to the optical clock transition. The stabilized 674 nm OLO can be coupled to a $88Sr^+$ ion in an electrostatic trap (Ivory et al., 2020) using, for example, silicon nitride waveguide to free-space grating couplers (Chauhan et al., 2019). Additional cooling and repump beams can be provided via lasers coupled to silicon nitride waveguides and free-space grating couplers.

6.4 Precision frequency synchronized fiber links

Precision optical-frequency and phase synchronization over fiber is important for a variety of applications ranging from timekeeping to quantum optics. These systems historically are built using lab-scale lasers and reference cavities as well as optoelectronic carrier and phase lock control. Chip-scale versions of these systems have the potential to dramatically broaden the application landscape by reducing the cost, size, and power of such exquisite sources. Traditional electronic and optoelectronic chip-scale approaches to carrier and phase synchronization include electronic phase locked loops (EPLLs) (Kikuchi, 2016; Perin et al., 2017), optical phase locked loops (OPLLs) (Balakier et al., 2018; Park et al., 2012; Satyan et al., 2010; Simsek et al., 2018; Zydney et al., 1986), and DSP carrier recovery and phase estimation (Crivelli et al., 2014; Laperle, 2013; Pillai et al., 2014). Several factors drive how well these techniques scale in performance, complexity, and power consumption at optical carrier frequencies, including laser and fiber phase noise and stability, modulation bandwidth, feedback loop noise, bandwidth requirements, and high frequency analogue and digital electronics. Compact, ultra-low linewidth, stabilized lasers with ultra-low phase noise can enable circuitry normally associated with coherent RF and wireless

Fig. 37 Example of integrated Brillouin laser used as the optical laser oscillator (OLO) for strontium ion ($88Sr^+$) clock transition. An integrated 674 nm external cavity Si_3N_4 tunable laser can serve as the SBS OLO, while a silicon nitride optical filter isolates and routes a portion of the SBS first-order Stokes (S1) to a modulator for sideband locking to an on-chip silicon nitride waveguide frequency reference cavity. The reference cavity stabilizes the SBS sideband for locking to the narrow ion transition. A single $88Sr^+$ ion trap is shown as example of chip-integrated trap 5356. Fiber-coupled 1092 and 1033-nm lasers can be converted to free-space beams for state re-pumping and on-chip photodiodes 57 can provide monitoring for signal feedback for locking. *Adapted from Chauhan, N., Isichenko, A., Liu, K., Wang, J., Zhao, Q., Behunin, R. O., Rakich, P. T., Jayich, A. M., Fertig, C., Hoyt, C.W., Blumenthal, D. J., 2021. Visible light photonic integrated Brillouin laser. Nat. Commun. 12 (1), 4685. https://doi.org/10.1038/s41467-021-24926-8. Gundavarapu et al. (2019) is licensed under CC BY 4.0.*

communications to support links at optical frequencies. Examples of low bandwidth feedback techniques used in precision scientific experiments are the PDH (Black, 2000; Drever et al., 1983), method coupled with thermally engineered glass or single crystal high-Q optical reference cavities (Drever et al., 2017; Kessler et al., 2012) housed in sophisticated environmental isolation systems. These state-of-the-art, lab-scale lasers are capable of linewidths below 10 mHz with a carrier instability of 4×10^{-17} at timescales between one and a few tens of seconds.

A fiber link employing independent, mutually highly coherent, SBLs that are frequency stabilized at the chip-scale is illustrated in Fig. 38. A cavity-stabilized SBL consists of a tunable semiconductor pump laser locked to a silicon nitride waveguide SBS laser using a low bandwidth PDH lock optoelectronic loop. The SBS laser output is modulated by an acousto-optic modulator (AOM) to generate the stabilized optical carrier. The AOM is driven by a voltage-controlled oscillator (VCO) that aligns and locks the optical carrier to a resonance of a compact silica reference cavity (Zhang et al., 2020) using a second low bandwidth PDH lock circuit. This reference cavity lock stabilizes the SBS optical carrier and reduce the laser phase noise and integral linewidth within the loop bandwidth of several hundred kHz. The stabilized SBL emission is split to create a transmitted carrier and a receiver-side tunable optical local oscillator (LO). The laser output is coupled to the fiber optic link and transmitted to a receiving node with a second independent stabilized SBL of the same design. The incoming signal is optically mixed with a single sideband modulated local oscillator CS-SBS laser. This photo-mixed signal serves as the error signal for a homodyne optical phase lock. The error signal is filtered by a proportional-integral2-derivative (PI2D) loop filter with integral poles at 50 and 100 kHz and fed back to the local oscillator tuning control to close the phase lock, completing the final lock loop. The high mutual coherence between the transmit and receive stabilized SBLs results in ultra-low residual phase error and improved phase lock capture- and hold-in range over a nonstabilized link. The linewidth reduction properties of the SBS laser significantly reduces the mutual phase noise at frequency offsets outside the bandwidth of the final optical phase lock.

The ability to provide this degree of phase synchronization on a 193 THz optical carrier using low bandwidth electronics typical of RF links opens the door to new applications for precision optical frequency synchronized fiber links. By combining the phase noise reduction dynamics of integrated SBLs with the inherent stability of a compact 1 billion Q reference cavity, a 43 Hz

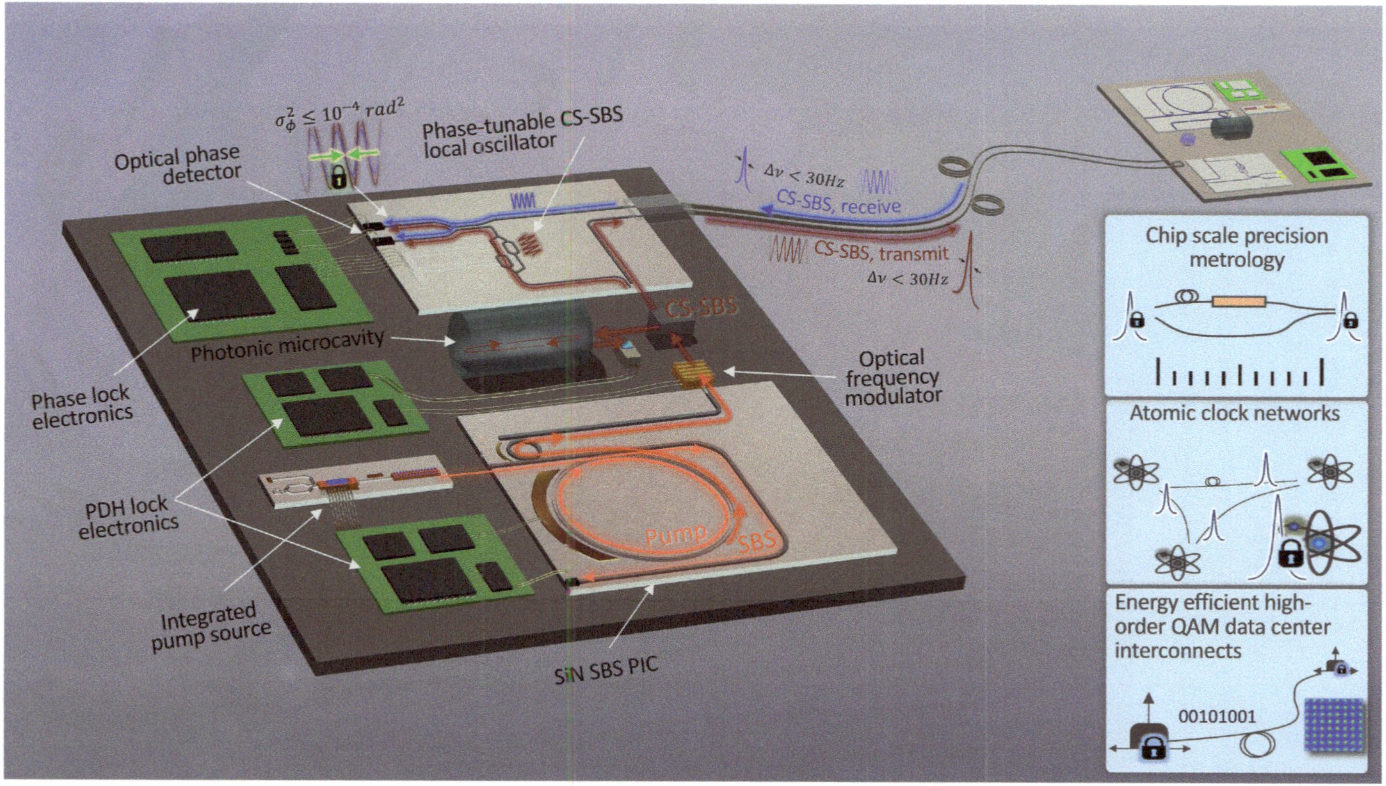

Fig. 38 Brillouin laser Optically synchronized precision fiber link. A precision optical frequency synchronized fiber link can be realized with independent cavity-stabilized stimulated Brillouin scattering (CS-SBS) lasers. The stabilized Brillouin lasers at the transmit and receive nodes provide stable carriers for a low-bandwidth optical frequency stabilized phase-lock loop (OFS-PLL). This type of compact, carrier-stable link will enable applications in precision metrology, distributed atomic timekeeping, and high-capacity coherent fiber optic communications. *Credit Brodnik, G. M., Harrington, M. W., Dallyn, J. H., Bose, D., Zhang, W., Stern, L., Morton, P. A., Behunin, R. O., Papp, S. B., Blumenthal, D. J., 2021. Optically synchronized fibre links using spectrally pure chip-scale lasers. Nat. Photonics 15 (8), 588–593. https://doi.org/10.1038/s41566-021-00831-w. Brodnik et al. (2021) is licensed under CC BY 4.0.*

heterodyne beatnote integral linewidth and fractional frequency instability better than 2×10^{-13} at 50 ms with independent optical sources over a 200 m fiber optic link is achieved. In the future, the individual components used to construct the optical frequency stabilized phase-lock loop (OFS-PLL) can be readily integrated in silicon nitride and heterogeneous silicon photonic platforms with mW-tier BiCMOS electrical circuits.

6.5 Coherent fiber optic communications

The frequency synchronized optical link described in the prior section can be applied to energy-efficient high-order coherent optical communications. The continued growth in data centers will place increased demands on data center fiber optic interconnects and drive underlying electronic application specific integrated circuit (ASIC) switch chips that route DCI ethernet traffic to 100 Tbps and beyond in the future. This astounding growth will push the limits of today's fiber link technologies that connect these packet switches, including power dissipation, density, and engineering practicality. To overcome these limits, high capacity coherent wavelength division multiplexing (WDM), which is traditionally relegated to the metro and long-haul networks, will need to move into the data center. However, implementing coherent WDM into the data center, particularly for link distances less than 2 km, will require elimination of power consuming and costly technologies like the digital signal processor (DSP). Additionally, new photonic integration technologies will be needed to place the coherent fiber optic link interfaces directly with the switch ASICs to alleviate the bandwidth, power, and density limits. The ultra-low phase noise of the Brillouin-stabilized source enables DSP-free Tbps coherent optical links in data center interconnects (Blumenthal et al., 2020; Morsy-Osman et al., 2018; Perin et al., 2017). A DSP-free approach that utilizes the ultra-low phase noise of Brillouin lasers is the ARPAe-funded FRESCO (Frequency Stabilized Coherent Optical Links for Low Energy DCIs) project. FRESCO utilizes spectrally pure, ultra-stable Brillouin light source technology to enable high capacity high-order WDM QAM with low bandwidth, low power electronics normally associated with RF links. Terabits per second FRESCO links based on shared, stabilized sources and high-density coherent WDM silicon photonic coherent transceivers that are colocated with the switch ASIC will pave the way to a DSP-free coherent WDM scalable DCI solution. The frequency and phase lock between transmit and receive nodes can lever power efficient, low bandwidth loop electronics that do not depend on symbol rate. This is in contrast to conventional to digital signal processors (DSPs) (Liu et al., 2014) that scale in operating speed and power as the baud rate increases. Stabilized integrated Brillouin lasers used as the transmit and receive (local oscillator) sources can serve as single low noise pumps for integrated nonlinear optical frequency combs (OFCs) (Harrington et al., 2020). Such techniques enable amortization of complexity, cost, and power consumption across many wavelengths for a fiber link.

References

Allan, D.W., 1987. Time and frequency (time-domain) characterization, estimation, and prediction of precision clocks and oscillators. IEEE Trans. Ultrason. Ferroelectr. Freq. Control 34 (6), 647–654. https://doi.org/10.1109/T-UFFC.1987.26997.

Bahl, G., Tomes, M., Marquardt, F., Carmon, T., 2012. Observation of spontaneous Brillouin cooling. Nat. Phys. 8 (3), 203–207.

Balakier, K., Ponnampalam, L., Fice, M.J., Renaud, C.C., Seeds, A.J., 2018. Integrated semiconductor laser optical phase lock loops. J. Sel. Top. Quantum Electron. 24 (1), 1–12. https://doi.org/10.1109/JSTQE.2017.2711581.

Bauer, J., 1977. Optical properties, band gap, and surface roughness of Si_3N_4. Phys. Status Solidi (A) 39 (2), 411–418. https://doi.org/10.1002/pssa.2210390205.

Bauters, J.F., Heck, M.J.R., John, D., Dai, D., Tien, M.-C., Barton, J.S., Leinse, A., Heideman, R.G., Blumenthal, D.J., Bowers, J.E., 2011. Ultra-low-loss high-aspect-ratio Si_3N_4 waveguides. Opt. Express 19 (4), 3163–3174.

Behunin, R.O., Otterstrom, N.T., Rakich, P.T., Gundavarapu, S., Blumenthal, D.J., 2018. Fundamental noise dynamics in cascaded-order Brillouin lasers. Phys. Rev. A 98 (2), 023832. https://doi.org/10.1103/PhysRevA.98.023832.

Black, E.D., 2000. An introduction to Pound-Drever-Hall laser frequency stabilization. Am. J. Phys. 69 (1), 79–87. https://doi.org/10.1119/1.1286663.

Blumenthal, D.J., 2020. Photonic integration for UV to IR applications. APL Photonics 5, 3020903. https://doi.org/10.1063/1.5131683.

Blumenthal, D.J., Heideman, R., Geuzebroek, D., Leinse, A., Roeloffzen, C., 2018. Silicon nitride in silicon photonics. Proc. IEEE 106 (12), 2209–2231. https://doi.org/10.1109/JPROC.2018.2861576.

Blumenthal, D.J., Ballani, H., Behunin, R.O., Bowers, J.E., Costa, P., Lenoski, D., Morton, P.A., Papp, S.B., Rakich, P.T., 2020. Frequency-stabilized links for coherent WDM fiber interconnects in the datacenter. J. Lightwave Technol. 38 (13), 3376–3386. https://doi.org/10.1109/JLT.2020.2985275.

Bohn, J.L., Rey, A.M., Ye, J., 2017. Cold molecules: progress in quantum engineering of chemistry and quantum matter. Science 357 (6355), 1002–1010. https://doi.org/10.1126/science.aam6299. http://www.sciencemag.org/lookup/doi/10.1126/science.aam6299.

Borselli, M., Johnson, T., Painter, O., 2005. Beyond the Rayleigh scattering limit in high-Q silicon microdisks: theory and experiment. Opt. Express 13, 1515–1530. https://doi.org/10.1364/OPEX.13.001515.

Brodnik, G., Harrington, M., Bose, D., Netherton, A., Zhang, W., Stern, L., Morton, P., Bowers, J., Papp, S., Blumenthal, D., 2020. Chip-scale, optical-frequency-stabilized PLL for DSP-free, low-power coherent QAM in the DCI. In: Conference on Optical Fiber Communications (OFC), San Diego, CA.

Brodnik, G.M., Harrington, M.W., Dallyn, J.H., Bose, D., Zhang, W., Stern, L., Morton, P.A., Behunin, R.O., Papp, S.B., Blumenthal, D.J., 2021. Optically synchronized fibre links using spectrally pure chip-scale lasers. Nat. Photonics 15 (8), 588–593. https://doi.org/10.1038/s41566-021-00831-w.

Buttner, T., Kabakova, I., Hudson, D., Pant, R., Poulton, C., Judge, A., Eggleton, B., 2014. Phase-locking and pulse generation in multi-frequency Brillouin oscillator via four wave mixing. Sci. Rep. 4, 5032. https://doi.org/10.1038/srep05032.

Callahan, P.T., Gross, M.C., Dennis, M.L., 2011. Frequency-independent phase noise in a dual-wavelength Brillouin fiber laser. IEEE J. Quantum Electron. 47 (8), 1142–1150.

Chauhan, N., Bose, D., Puckett, M., Moreira, R., Nelson, K., Blumenthal, D.J., 2019. Photonic integrated Si_3N_4 ultra-large-area grating waveguide MOT interface for 3D atomic clock laser cooling. In: Conference on Lasers and Electro-Optics (CLEO), San Jose, CA, pp. 1–2. STu4O.3.

Chauhan, N., Bose, D., Straus, M., Holliman, C.A., Jayich, A.M., Behunin, R.O., Blumenthal, D.J., 2020. Evidence of visible wavelength spontaneous Brillouin scattering in Si_3N_4 waveguides. In: Frontier in Optics & Laser Sciences, FW1D.5.

Chauhan, N., Isichenko, A., Liu, K., Wang, J., Zhao, Q., Behunin, R.O., Rakich, P.T., Jayich, A.M., Fertig, C., Hoyt, C.W., Blumenthal, D.J., 2021. Visible light photonic integrated Brillouin laser. Nat. Commun. 12 (1), 4685. https://doi.org/10.1038/s41467-021-24926-8.

Chauhan, N., Wang, J., Bose, D., Liu, K., Compton, R.L., Fertig, C., Hoyt, C.W., Blumenthal, D.J., 2022. Ultra-low loss visible light waveguides for integrated atomic, molecular, and quantum photonics. Opt. Express 30 (5). 6960–6969. https://doi.org/10.1364/OE.448938.

Choi, D.-Y., Madden, S., Bulla, D.A., Wang, R., Rode, A., Luther-Davies, B., 2010. Submicrometer-thick low-loss As_2S_3 planar waveguides for nonlinear optical devices. IEEE Photonics Technol. Lett. 22 (7), 495–497. https://doi.org/10.1109/LPT.2010.2040170.

Collaboration, L.S., 2016. Observation of gravitational waves from a binary black hole merger. Phys. Rev. Lett. 116 (6), 061102.

Crivelli, D.E., Hueda, M.R., Carrer, H.S., del Barco, M., López, R.R., Gianni, P., Finochietto, J., Swenson, N., Voois, P., Agazzi, O.E., 2014. Architecture of a single-chip 50-Gb/s DP-QPSK/BPSK transceiver with electronic dispersion compensation for coherent optical channels. IEEE Trans. Circuits Syst. Regul. Pap. 61 (4), 1012–1025. https://doi.org/10.1109/TCSI.2013.2283673.

Cygan, A., Lisak, D., Morzynski, P., Bober, M., Zawada, M., Pazderski, E., Ciurylo, R., 2013. Cavity mode-width spectroscopy with widely tunable ultra narrow laser. Opt. Express 21, 29744–29754.

Debut, A., Randoux, S., Zemmouri, J., 2000. Linewidth narrowing in Brillouin lasers: theoretical analysis. Phys. Rev. A 62 (2), 023803. https://doi.org/10.1103/PhysRevA.62.023803.

Degen, C.L., Reinhard, F., Cappellaro, P., 2017. Quantum sensing. Rev. Mod. Phys. 89 (3). https://doi.org/10.1103/RevModPhys.89.035002. 035002-1–035002-39.

Di Domenico, G., Schilt, S., Thomann, P., 2010. Simple approach to the relation between laser frequency noise and laser line shape. Appl. Opt. 49 (25), 4801–4807. https://doi.org/10.1364/AO.49.004801.

Drever, R.W.P., Hall, J.L., Kowalski, F.V., Hough, J., Ford, G.M., Munley, A.J., Ward, H., 1983. Laser phase and frequency stabilization using an optical resonator. Appl. Phys. B 31 (2), 97–105. https://doi.org/10.1007/BF00702605.

Drever, R.W.P., Hall, J.L., Kowalski, F.V., Hough, J., Ford, G.M., Munley, A.J., Ward, H., 2017. 1.5 um lasers with sub-10 mhz linewidth. Phys. Rev. Lett. 118 (26), 263202. https://doi.org/10.1103/PhysRevLett.118.263202.

Eggleton, B.J., Luther-Davies, B., Richardson, K., 2011. Chalcogenide photonics. Nat. Photonics 5 (3), 141–148. https://doi.org/10.1038/nphoton.2011.309.

Eggleton, B.J., Poulton, C.G., Pant, R., 2013. Inducing and harnessing stimulated Brillouin scattering in photonic integrated circuits. Adv. Opt. Photonics 5 (4), 536–587.

Eggleton, B.J., Poulton, C.G., Rakich, P.T., Steel, M.J., Bahl, G., 2019. Brillouin integrated photonics. Nat. Photonics 13 (10), 664–677.

Fan, H., Kumar, S., Sedlacek, J., Kübler, H., Karimkashi, S., Shaffer, J.P., 2015. Quantum sensing. J. Phys. B At. Mol. Opt. Phys. 48 (20), 202001. https://doi.org/10.1088/0953-4075/48/20/202001.

Fortier, T.M., Kirchner, M.S., Quinlan, F., Taylor, J., Bergquist, J.C., Rosenband, T., Lemke, N., Ludlow, A., Jiang, Y., Oates, C.W., Diddams, S.A., 2011. Generation of ultrastable microwaves via optical frequency division. Nat. Photonics 5 (7), 425–429. https://doi.org/10.1038/nphoton.2011.121.

Fox, R.W., Oates, C.W., Hollberg, L.W., 2003. Stabilizing diode lasers to high-finesse cavities. In: Experimental Methods in the Physical Sciences, vol. 40. Academic Press, pp. 1–46, https://doi.org/10.1016/S1079-4042(03)80017-6.

Frerichs, R., 1953. New optical glasses with good transparency in the infrared*. J. Opt. Soc. Am. 43 (12), 1153–1157. https://doi.org/10.1364/JOSA.43.001153. http://opg.optica.org/abstract.cfm?URI=josa-43-12-1153.

Gertler, S., Kittlaus, E.A., Otterstrom, N.T., Rakich, P.T., 2020. Tunable microwave-photonic filtering with high out-of-band rejection in silicon. APL Photonics 5 (9), 096103. https://doi.org/10.1063/5.0015174.

Grudinin, I.S., Matsko, A.B., Maleki, L., 2009. Brillouin lasing with a CaF_2 whispering gallery mode resonator. Phys. Rev. Lett. 102 (4), 043902.

Gundavarapu, S., Brodnik, G.M., Puckett, M., Huffman, T., Bose, D., Behunin, R., Wu, J., Qiu, T., Pinho, C., Chauhan, N., Nohava, J., Rakich, P.T., Nelson, K.D., Salit, M., Blumenthal, D.J., 2019. Sub-hertz fundamental linewidth photonic integrated Brillouin laser. Nat. Photonics 13 (1), 60–67. https://doi.org/10.1038/s41566-018-0313-2.

Harrington, M., Brodnik, G., Briles, T., Stone, J., Streater, R., Papp, S., Blumenthal, D.J., 2020. Kerr soliton microcomb pumped by an integrated SBS laser for ultra-low linewidth WDM sources. In: Conference on Optical Fiber Communications (OFC), San Diego, CA.

Haverlag, M., Vender, D., Oehrlein, G.S., 1992. Ellipsometric study of silicon surface damage in electron cyclotron resonance plasma etching using CF_4 and SF_6. Appl. Phys. Lett. 61, 2875–2877. https://doi.org/10.1063/1.108062.

Henry, C., 1982. Theory of the linewidth of semiconductor lasers. IEEE J. Quantum Electron. 18 (2), 259–264.

Huffman, T.A., Brodnik, G.M., Pinho, C., Gundavarapu, S., Baney, D., Blumenthal, D.J., 2018. Integrated resonators in an ultralow loss Si_3N_4/SiO_2 platform for multifunction applications. IEEE J. Sel. Top. Quantum Electron. 24 (4), 1–9. https://doi.org/10.1038/s41566-019-0498-z.

Ivory, M., Setzer, W.J., Karl, N., McGuinness, H., DeRose, C., Blain, M., Stick, D., Gehl, M., Parazzoli, L.P., 2020. Integrated optical addressing of a trapped ytterbium ion. arXiv:2011.12376.

Jadbabaie, A., Pilgram, N.H., Klos, J., Kotochigova, S., Hutzler, N.R., 2020. Enhanced molecular yield from a cryogenic buffer gas beam source via excited state chemistry. New J. Phys. 22, 022002. https://doi.org/10.1088/1367-2630/ab6eae.

John, D.D., Heck, M.J.R., Bauters, J.F., Moreira, R., Barton, J.S., Bowers, J.E., Blumenthal, D.J., 2012. Multilayer platform for ultra-low-loss waveguide applications. Photonics Technol. Lett. 24 (11), 876–878.

Kabakova, I.V., Pant, R., Choi, D.-Y., Debbarma, S., Luther-Davies, B., Madden, S.J., Eggleton, B.J., 2013. Narrow linewidth Brillouin laser based on chalcogenide photonic chip. Opt. Lett. 38 (17), 3208–3211. https://doi.org/10.1364/OL.38.003208. http://opg.optica.org/ol/abstract.cfm?URI=ol-38-17-3208.

Kang, M.S., Nazarkin, A., Brenn, A., Russell, P.S.t.J., 2009. Tightly trapped acoustic phonons in photonic crystal fibres as highly nonlinear artificial Raman oscillators. Nat. Phys. 5 (4), 276.

Kang, M.S., Brenn, A., Russell, P.S.t.J., 2010. All-optical control of gigahertz acoustic resonances by forward stimulated interpolarization scattering in a photonic crystal fiber. Phys. Rev. Lett. 105 (15), 153901.

Kang, M.S., Butsch, A., Russell, P.S.t.J., 2011. Reconfigurable light-driven opto-acoustic isolators in photonic crystal fibre. Nat. Photonics 5 (9), 549.

Kessler, T., Hagemann, C., Grebing, C., Legero, T., Sterr, U., Riehle, F., Martin, M.J., Chen, L., Ye, J., 2012. A sub-40-mHz-linewidth laser based on a silicon single-crystal optical cavity. Nat. Photonics 6 (10), 687–692. https://doi.org/10.1038/nphoton.2012.217.

Kharel, P., Behunin, R.O., Renninger, W.H., Rakich, P.T., 2016. Noise and dynamics in forward Brillouin interactions. Phys. Rev. A 93 (6), 063806.

Kikuchi, K., 2016. Fundamentals of coherent optical fiber communications. J. Lightwave Technol. 34 (1), 157–179. https://doi.org/10.1109/JLT.2015.2463719.

Kippenberg, T.J., Spillane, S.M., Vahala, K.J., 2004. Kerr-nonlinearity optical parametric oscillation in an ultrahigh-Q toroid microcavity. Phys. Rev. Lett. 93 (8), 083904. https://doi.org/10.1103/PhysRevLett.93.083904.

Kippenberg, T.J., Holzwarth, R., Diddams, S.A., 2011. Microresonator-based optical frequency combs. Science 332 (6029), 555. https://doi.org/10.1038/nature16536.

Kittlaus, E.A., Shin, H., Rakich, P.T., 2016. Large Brillouin amplification in silicon. Nat. Photonics 10 (7), 463–467. https://doi.org/10.1038/nphoton.2016.112.

Kittlaus, E.A., Otterstrom, N.T., Rakich, P.T., 2017. On-chip inter-modal Brillouin scattering. Nat. Commun. 8, 15819. https://doi.org/10.1038/ncomms15819.

Kittlaus, E.A., Otterstrom, N.T., Kharel, P., Gertler, S., Rakich, P.T., 2018. Non-reciprocal interband Brillouin modulation. Nat. Photonics 12 (10), 613–619. https://doi.org/10.1038/s41566-018-0254-9.

Komljenovic, T., Huang, D., Pintus, P., Tran, M.A., Davenport, M.L., Bowers, J.E., 2019. Brillouin integrated photonics. Nat. Photonics 13 (10), 664–677. https://doi.org/10.1038/s41566-019-0498-z.

Lai, Y.-H., Suh, M.-G., Lu, Y.-K., Shen, B., Yang, Q.-F., Wang, H., Li, J., Lee, S.H., Yang, K.Y., Vahala, K., 2020. Earth rotation measured by a chip-scale ring laser gyroscope. Nat. Photonics 14 (6), 345–349.

Laperle, C., 2013. Advances in high-speed DACs, ADCs and DSP for optical coherent transceivers. In: 2013 Optical Fiber Communication Conference and Exposition and the National Fiber Optic Engineers Conference (OFC/NFOEC), Anaheim, CA, SM4O.1.

Lax, M., 1967. Quantum noise. X. Density-matrix treatment of field and population-difference fluctuations. Phys. Rev. 157 (2), 213–231.

Lee, H., Chen, T., Li, J., Yang, K.Y., Jeon, S., Painter, O., Vahala, K.J., 2012. Chemically etched ultrahigh-Q wedge-resonator on a silicon chip. Nat. Photonics 6 (6), 369.

Lee, H., Suh, M.-G., Chen, T., Li, J., Diddams, S.A., Vahala, K.J., 2013. Spiral resonators for on-chip laser frequency stabilization. Nat. Commun. 4, 2468. https://doi.org/10.1038/ncomms3468.

Li, J., Lee, H., Chen, T., Vahala, K.J., 2012a. Characterization of a high coherence, Brillouin microcavity laser on silicon. Opt. Express 20 (18), 20170–20180.

Li, J., Lee, H., Vahala, K.J., 2013. Microwave synthesizer using an on-chip Brillouin oscillator. Nat. Commun. 4, 2097. https://doi.org/10.1038/ncomms3097.

Li, J., Yi, X., Lee, H., Diddams, S.A., Vahala, K.J., 2014. Electro-optical frequency division and stable microwave synthesis. Science 345 (6194), 309–313.

Li, J., Suh, M.-G., Vahala, K., 2017. Microresonator Brillouin gyroscope. Optica 4 (3), 346–348.

Liang, W., Eliyahu, D., Ilchenko, V.S., Savchenkov, A.A., Matsko, A.B., Seidel, D., Maleki, L., 2015. High spectral purity kerr frequency comb radio frequency photonic oscillator. Nat. Commun. 6, 7957. https://doi.org/10.1038/ncomms8957.

Liu, X., Chandrasekhar, S., Winzer, P.J., 2014. Digital signal processing techniques enabling multi-Tb/s superchannel transmission: an overview of recent advances in DSP-enabled superchannels. IEEE Signal Process. Mag. 31 (2), 16–24.

Liu, K., Jin, N., Cheng, H., Puckett, M.W., Behunin, R.O., Nelson, K.D., Rakich, P.T., Blumenthal, D.J., 2021. 720 million quality factor integrated all-waveguide photonic resonator. In: 2021 Device Research Conference (DRC), June, pp. 1–2.

Liu, K., Chauhan, N., Wang, J., Isichenko, A., Brodnik, G.M., Morton, P.A., Behunin, R., Papp, S.B., Blumenthal, D.J., 2022a. Semiconductor laser stabilized by a photonic integrated 4 meter coil-waveguide resonator. In: Conference on Optical Fiber Communications (OFC), San Diego, CA.

Liu, K., Harrington, M., Nelson, K.D., Behunin, R.O., Papp, S.B., Blumenthal, D.J., 2022b. Photonic integrated cascade-inhibited Brillouin laser with sub-100-mHz fundamental linewidth. In: Conference on Lasers and Electrooptics (CLEO), San Jose, CA.

Liu, K., Jin, N., Cheng, H., Chauhan, N., Puckett, M.W., Nelson, K.D., Behunin, R.O., Rakich, P.T., Blumenthal, D.J., 2022c. Ultra-low 0.034 dB/m loss wafer-scale integrated photonics realizing 720 million Q and 380 μW threshold Brillouin lasing. Opt. Lett. 47 (7), 1855–1858. https://doi.org/10.1364/OL.454392.

Loh, W., Green, A.A., Baynes, F.N., Cole, D.C., Quinlan, F.J., Lee, H., Vahala, K.J., Papp, S.B., Diddams, S.A., 2015. Dual-microcavity narrow-linewidth Brillouin laser. Optica 2 (3), 225–232.

Loh, W., Stuart, J., Reens, D., Bruzewicz, C.D., Braje, D., Chiaverini, J., Juodawlkis, P.W., Sage, J.M., McConnell, R., 2020. Ultra-narrow linewidth Brillouin laser with nanokelvin temperature self-referencing. Nature 588 (7837), 244–249. https://doi.org/10.1038/s41586-020-2981-6.

Loriani, S., Schlippert, D., Schubert, C., Abend, S., Ahlers, H., Ertmer, W., Rudolph, J., Hogan, J.M., Kasevich, M.A., Rasel, E.M., Gaaloul, N., 1987. Atomic source selection in space-borne gravitational wave detection. New J. Phys. 21 (6), 063030.

Ludlow, A.D., Huang, X., Notcutt, M., Zanon-Willette, T., Foreman, S.M., Boyd, M.M., Blatt, S., Ye, J., 2007. Compact, thermal-noise-limited optical cavity for diode laser stabilization at 1×10^{-15}. Opt. Lett. 32 (6), 641–643. https://doi.org/10.1364/OL.32.000641.

Ludlow, A.D., Boyd, M.M., Ye, J., Peik, E., Schmidt, P.O., 2015. Optical atomic clocks. Rev. Mod. Phys. 87, 637–701.

Marpaung, D., Roeloffzen, C., Heideman, R., Leinse, A., Sales, S., Capmany, J., 2013. Integrated microwave photonics. Laser Photonics Rev. 7 (4), 506–538. https://doi.org/10.1002/lpor.201200032.

Merklein, M., Kabakova, I.V., Büttner, T.F.S., Choi, D.-Y., Luther-Davies, B., Madden, S.J., Eggleton, B.J., 2015. Enhancing and inhibiting stimulated Brillouin scattering in photonic integrated circuits. Nat. Commun. 6 (1), 6396. https://doi.org/10.1038/ncomms7396.

Merklein, M., Stiller, B., Kabakova, I.V., Mutugala, U.S., Vu, K., Madden, S.J., Eggleton, B.J., Slavík, R., 2016. Widely tunable, low phase noise microwave source based on a photonic chip. Opt. Lett. 41 (20), 4633–4636. https://doi.org/10.1364/OL.41.004633.

Morsy-Osman, M., Sowailem, M., El-Fiky, E., Goodwill, T., Hoang, T., Lessard, S., Plant, D.V., 2018. Dsp-free coherent-lite transceiver for next generation single wavelength optical intra-datacenter interconnects. Opt. Express 26 (7), 8890–8903.

National Academies of Sciences and Engineering and Medicine, 2020. Manipulating Quantum Systems: An Assessment of Atomic, Molecular, and Optical Physics in the United States. The National Academies Press, Washington, DC, ISBN: 978-0-309-49951-4, https://doi.org/10.17226/25613. https://www.nap.edu/catalog/25613/manipulating-quantum-systems-an-assessment-of-atomic-molecular-and-optical.

Nelson, K.D., Puckett, M.W., Wu, J., Bose, D., Gundavarapu, S.C., Blumenthal, D., 2019. Reducing noise in a ring-laser Gyro based on stimulated Brillouin scattering. In: 2019 IEEE International Symposium on Inertial Sensors and Systems (INERTIAL).

Nicholson, T.L., Campbell, S.L., Hutson, R.B., Marti, G.E., Bloom, B.J., McNally, R.L., Zhang, W., Barrett, M.D., Safronova, M.S., Strouse, G.F., Tew, W.L., Ye, J., 2015. Systematic evaluation of an atomic clock at 2×10^{-18} total uncertainty. Nat. Commun. 6 (1), 6896. https://doi.org/10.1038/ncomms7896.

Notcutt, M., Ma, L.-S., Ludlow, A.D., Foreman, S.M., Ye, J., Hall, J.L., 2006. Contribution of thermal noise to frequency stability of rigid optical cavity via hertz-linewidth lasers. Phys. Rev. A 73 (3), 031804. https://doi.org/10.1364/OL.32.000641.

Orieux, A., Diamanti, E., 2016. Recent advances on integrated quantum communications. J. Opt. 18 (8), 083002.

Otterstrom, N.T., Behunin, R.O., Kittlaus, E.A., Wang, Z., Rakich, P.T., 2018a. A silicon Brillouin laser. Science 360 (6393), 1113–1116.

Otterstrom, N.T., Kittlaus, E.A., Gertler, S., Behunin, R.O., Lentine, A.L., Rakich, P.T., 2019. Resonantly enhanced nonreciprocal silicon Brillouin amplifier. Optica 6 (9), 1117–1123.

Otterstrom, N.T., Gertler, S., Zhou, Y., Kittlaus, E.A., Behunin, R.O., Gehl, M., Starbuck, A.L., Dallo, C.M., Pomerene, A.T., Trotter, D.C., et al., 2020. Backscatter-immune injection-locked Brillouin laser in silicon. Phys. Rev. Appl. 14 (4), 044042.

Pant, R., Poulton, C.G., Choi, D.-Y., Mcfarlane, H., Hile, S., Li, E., Thevenaz, L., Luther-Davies, B., Madden, S.J., Eggleton, B.J., 2011a. On-chip stimulated Brillouin scattering. Opt. Express 19 (9), 8285–8290.

Pant, R., Poulton, C.G., Choi, D.-Y., Mcfarlane, H., Hile, S., Li, E., Thevenaz, L., Luther-Davies, B., Madden, S.J., Eggleton, B.J., 2011b. On-chip stimulated Brillouin scattering. Opt. Express 19 (9), 8285–8290. https://doi.org/10.1364/OE.19.008285. http://opg.optica.org/oe/abstract.cfm?URI=oe-19-9-8285.

Park, H.-C., Lu, M., Bloch, E., Reed, T., Griffith, Z., Johansson, L., Coldren, L., Rodwell, M., 2012. 40-Gbit/s coherent optical receiver using a costas loop. Opt. Express 24 (1), 1–12. https://doi.org/10.1364/OE.20.00B197.

Patel, K., Brubaker, M., Kotlerman, A., Salazar, R., Wolf, E., Weld, D.M., 2018. Geodesy and metrology with a transportable optical clock. Nat. Phys. 14 (5), 437–441.

Perin, J.K., Shastri, A., Kahn, J.M., 2017. Design of low-power DSP-free coherent receivers for data center links. J. Lightwave Technol. 35 (21), 4650–4662. https://doi.org/10.1109/JLT.2017.2752079.

Pillai, B.S.G., Sedighi, B., Guan, K., Anthapadmanabhan, N.P., Shieh, W., Hinton, K.J., Tucker, R.S., 2014. End-to-end energy modeling and analysis of long-haul coherent transmission systems. J. Lightwave Technol. 32 (18), 3093–3111. https://doi.org/10.1109/JLT.2014.2331086.

Ponikvar, D.R., Ezekiel, S., 1981. Stabilized single-frequency stimulated Brillouin fiber ring laser. Opt. Lett. 6 (8), 398–400.

Poulton, C.G., Pant, R., Byrnes, A., Fan, S., Steel, M., Eggleton, B.J., 2012. Design for broadband on-chip isolator using stimulated Brillouin scattering in dispersion-engineered chalcogenide waveguides. Opt. Express 20 (19), 21235–21246.

Poulton, C.G., Pant, R., Eggleton, B.J., 2013. Acoustic confinement and stimulated Brillouin scattering in integrated optical waveguides. J. Opt. Soc. Am. B 30 (10), 2657–2664. https://doi.org/10.1364/JOSAB.30.002657. http://opg.optica.org/josab/abstract.cfm?URI=josab-30-10-2657.

Puckett, M., Bose, D., Nelson, K., Blumenthal, D.J., 2019. Higher order cascaded SBS suppression using gratings in a photonic integrated ring resonator laser. In: Conference on Lasers and Electro-Optics (CLEO), San Jose, CA, SM4O.1.

Puckett, M.W., Liu, K., Chauhan, N., Zhao, Q., Jin, N., Cheng, H., Wu, J., Behunin, R.O., Rakich, P.T., Nelson, K.D., Blumenthal, D.J., 2020. 422 Million Q photonic integrated waveguide resonator with a 3.4 billion absorption limit and sub-MHz linewidth. Nat. Commun. 12 (1), 934. https://doi.org/10.1038/s41467-021-21205-4.

Qiu, W., Rakich, P.T., Shin, H., Dong, H., Soljačić, M., Wang, Z., 2013. Stimulated Brillouin scattering in nanoscale silicon step-index waveguides: a general framework of selection rules and calculating SBS gain. Opt. Express 21 (25), 31402–31419.

Rakich, P.T., Reinke, C., Camacho, R., Davids, P., Wang, Z., 2012. Giant enhancement of stimulated Brillouin scattering in the subwavelength limit. Phys. Rev. X 2 (1), 011008.

Richter, L., Mandelberg, H., Kruger, M., McGrath, P., 1986. Linewidth determination from self-heterodyne measurements with subcoherence delay times. IEEE J. Quantum Electron. 22 (11), 2070–2074. https://doi.org/10.1109/JQE.1986.1072909.

Rong, H., Liu, A., Nicolaescu, R., Paniccia, M., Cohen, O., Hak, D., 2004. Raman gain and nonlinear optical absorption measurements in a low-loss silicon waveguide. Appl. Phys. Lett. 85 (12), 2196–2198.

Saffman, M., 2016. Quantum computing with atomic qubits and Rydberg interactions: progress and challenges. J. Phys. B At. Mol. Opt. Phys. 49 (20), 202001. http://stacks.iop.org/0953-4075/49/i=20/a=202001.

Sanders, G.A., Sanders, S.J., Strandjord, L.K., Qiu, T., Wu, J., Smiciklas, M., Mead, D., Mosor, S., Arrizon, A., Ho, W., Salit, M., 2016. Fiber optic gyro development at honeywell. In: SPIE, 9852. International Society for Optics and Photonics. pp. 9852–9814.

Satyan, N., Sendowski, J.B., Vasilyev, A., Rakuljic, G.A., Yariv, A., 2010. Phase noise reduction of a semiconductor laser in a composite optical phase-locked loop. Opt. Eng. 49 (12), 1–6. https://doi.org/10.1117/1.3518077.

Schawlow, A.L., Townes, C.H., 1958. Infrared and optical masers. Phys. Rev. 112 (6), 1940.

Shin, H., Qiu, W., Jarecki, R., Cox, J.A., Olsson, R.H., III, Starbuck, A., Wang, Z., Rakich, P.T., 2013. Tailorable stimulated Brillouin scattering in nanoscale silicon waveguides. Nat. Commun. 4, 1944.

Shin, H., Cox, J.A., Jarecki, R., Starbuck, A., Wang, Z., Rakich, P.T., 2015. Control of coherent information via on-chip photonic-phononic emitter-receivers. Nat. Commun. 6, 6427. https://doi.org/10.1038/ncomms7427.

Simsek, A., Arafin, S., Kim, S.-K., Morrison, G., Johansson, L.A., Mashanovitch, M., Coldren, L.A., Rodwell, M.J., 2018. Low-loss D-shape silicon nitride waveguides using a dielectric lift-off fabrication process. In: Optical Fiber Communication Conference (OFC), San Jose, CA.

Spencer, D.T., Drake, T., Briles, T.C., Stone, J., Sinclair, L.C., Fredrick, C., Li, Q., Westly, D., Ilic, B.R., Bluestone, A., Volet, N., Komljenovic, T., Chang, L., Lee, S.H., Oh, D.Y., Suh, M.-G., Yang, K.Y., Pfeiffer, M.H.P., Kippenberg, T.J., Norberg, E., Theogarajan, L., Vahala, K., Newbury, N.R., Srinivasan, K., Bowers, J.E., Diddams, S.A., Papp, S.B., 2018. An optical-frequency synthesizer using integrated photonics. Nature 557, 81–85. https://doi.org/10.1038/s41586-018-0065-7.

Stokes, L.F., Chodorow, M., Shaw, H.J., 1982. All-fiber stimulated Brillouin ring laser with submilliwatt pump threshold. Opt. Lett. 7 (10), 509–511.

Suh, M.-G., Yang, Q.-F., Vahala, K.J., 2017. Phonon-limited-linewidth of Brillouin lasers at cryogenic temperatures. Phys. Rev. Lett. 119 (14), 143901.

Tomes, M., Carmon, T., 2009. Photonic micro-electromechanical systems vibrating at X-band (11-GHz) rates. Phys. Rev. Lett. 102 (11), 113601.

Van Laer, R., Bazin, A., Kuyken, B., Baets, R., Van Thourhout, D., 2015a. Net on-chip Brillouin gain based on suspended silicon nanowires. New J. Phys. 17 (11), 115005.

Van Laer, R., Kuyken, B., Van Thourhout, D., Baets, R., 2015b. Interaction between light and highly confined hypersound in a silicon photonic nanowire. Nat. Photonics 9 (3), 199–203. https://doi.org/10.1038/nphoton.2015.11.

Wang, J., Sciarrino, F., Laing, A., Thompson, M.G., 2020. Integrated photonic quantum technologies. Nat. Photonics 14, 273–284. https://doi.org/10.1038/s41566-019-0532-1.

Wang, H., Adamski, N., Mu, S., Van de Walle, C.G., 2021. Piezoelectric effect and polarization switching in $Al_{1-x}Sc_xN$. J. Appl. Phys. 130 (10), 104101. https://doi.org/10.1063/5.0056485.

Wang, J., Liu, K., Harrington, M.W., Rudy, R.Q., Blumenthal, D.J., 2022. Ultra-low loss silicon nitride ring modulator with low power PZT actuation for photonic control. In: Conference on Optical Fiber Communications (OFC), San Diego, CA.

Wu, L., Wang, H., Yang, Q., Ji, Q.-x., Shen, B., Bao, C., Gao, M., Vahala, K., 2020. Greater than one billion Q factor for on-chip microresonators. Opt. Lett. 45 (18), 5129–5131. https://doi.org/10.1364/OL.394940.

Yang, K.Y., Oh, D.Y., Lee, S.H., Yang, Q.F., Yi, X., Shen, B., H., W., Vahala, K.J., 2018. Bridging ultrahigh-Q devices and photonic circuits. Nat. Photonics 12, 297–302. https://doi.org/10.1038/s41566-018-0132-5.

Yao, X.S., 1997. High-quality microwave signal generation by use of Brillouin scattering in optical fibers. Opt. Lett. 22 (17), 1329–1331.

Yeniay, A., Delavaux, J.-M., Toulouse, J., 2002. Spontaneous and stimulated Brillouin scattering gain spectra in optical fibers. J. Lightwave Technol. 20 (8), 1425.

Young, A.W., Eckner, W.J., Milner, W.R., Kedar, D., Norcia, M.A., Oelker, E., Schine, N., Ye, J., Kaufman, A.M., 2020. Half-minute-scale atomic coherence and high relative stability in a tweezer clock. Nature 588 (7838), 408–413. https://doi.org/10.1038/s41586-020-3009-y.

Yu, Z., Fan, S., 2009. Complete optical isolation created by indirect interband photonic transitions. Nat. Photonics 3 (2), 91–94.

Yuan, Z., Wang, H., Wu, L., Gao, M., Vahala, K., 2020. Linewidth enhancement factor in a microcavity Brillouin laser. Optica 7 (9), 1150–1153. https://opg.optica.org/optica/fulltext.cfm?uri=optica-7-9-1150&id=437890.

Zhang, W., Stern, L., Carlson, D., Bopp, D., Newman, Z., Kang, S., Kitching, J., Papp, S.B., 2020. Ultranarrow linewidth photonic-atomic laser. Laser Photonics Rev. 14 (4), 1900293. https://doi.org/10.1002/lpor.201900293.

Zydney, A.L., Bennett, J.S., Shahid, A., Bauer, K.W., 1986. Impact of undergraduate research experience in engineering. J. Lightwave Technol. 4 (2), 182–195. https://doi.org/10.1109/JLT.1986.1074698.

CHAPTER ELEVEN

SBS in optical communication systems: The good, the bad and the ugly

Bill Corcoran[a,*] **and A. Choudhary**[b,*]
[a]Department of Electrical and Computer Systems Engineering, Monash University, Melbourne, Australia
[b]Ultrafast Optical Communications and High-performance Integrated Photonics (UFO-CHIP) Group, Department of Electrical Engineering, Indian Institute of Technology, Delhi, India
*Corresponding authors: e-mail address: bill.corcoran@monash.edu; amol.choudhary@ee.iitd.ac.in

Contents

1. Introduction

1.1 Stimulated Brillouin scattering and optical communications

Stimulated Brillouin scattering and optical communications have had a fraught relationship over the lifetimes of these fields. While in some instances, SBS has been explored as a tool to enable novel functions in optical communication systems, it also imposes power limits in optical fiber system. Power limits in turn limits the achievable signal-to-noise ratio (SNR), and improving SNR is key to increasing data rates.

Signal-to-noise ratio directly relates to the data-rate that can be sent through a communication system, as established by Shannon (1948), a quarter of a century after the discovery of Brillouin scattering. In an optical fiber

Semiconductors and Semimetals, Volume 110
ISSN 0080-8784
https://doi.org/10.1016/bs.semsem.2022.05.005

communication system, there is a certain level of system noise. This could be electrical noise in the receiver, optical noise added into the system through amplification, or noise added in the transmitter. This background noise power can be overcome by simply increasing the power of the wanted signal, which then enables an increase in the data rate that can be sent successfully through the system.

SBS, as a nonlinear optical effect, significantly perturbs propagation through optical fiber as power is increased. This perturbation acts as an additional source of noise or distortion that increases with signal power (Aoki et al., 1988), and then can impose a limit to achievable SNR (Chraplyvy, 1990; Montes and Rubenchik, 1992). As stimulated Brillouin scattering occurs over a bandwidth of a few tens of MHz in optical fiber, for signal bandwidths above this, the threshold power for where SBS significantly increases. This is because for signal bandwidths above the SBS intrinsic linewidth, SBS reduces with the ratio of signal bandwidth to SBS intrinsic linewidth. As such, SBS was a significant issue for systems running with signals at a few 100s of MHz bandwidth in the late 1980s early 1990s but became less of a limiting issue as channel bandwidths increased. Because of this, a major focus of prior work has been to suppress SBS in optical communication systems.

On the other hand, SBS has also been investigated as a unique tool for enhancing the operation of optical communication systems. Optical communication provides a compelling context for the application of novel photonics and optical physics. While communication systems in general draw from a vast body of development, the addition of optical components provides new challenges in terms of optimizing systems to achieve the highest possible data rates. The ability of SBS to provide narrowband gain has been used in several contexts that we will expand on in this chapter, as well as more novel aspects such as optical delay (see chapter "Slow light, dynamic gratings, and light storage" by Stiller et al.) and ultranarrow linewidth lasing (see chapter "Integrated Brillouin lasers and their applications" by Blumenthal et al.). These provide significant opportunities to add photonic functionalities to optical communication systems that complement the powerful capabilities of electronic signal processing.

1.2 The optical communications context

To understand the opportunities for the use of SBS in optical communications, it is important to understand these systems in context. Optical communications systems can be broken down into a few important component parts and concepts. A light source (e.g., a laser) at the transmitter side is modulated to imprint information on a lightwave. This modulation perturbs one or more aspects of the light wave, its amplitude, phase, frequency or polarization. At the other end of the system is a receiver based on one or more photodiodes. This receiver can be set up to measure in one or more of the degrees-of-freedom (sometimes termed as 'dimensions') that a lightwave can be modulated in. In between the transmitter and receiver lies the communication channel, generally an optical fiber, and often other optical processing stages (e.g., wavelength division multiplexing couplers to route different wavelengths in different directions in a network). There are a significant number of textbooks that dive deeply into the makeup, design and operation of optical communication systems (e.g.. Agrawal, 2002; Keiser, 2000; Kumar and Deen, 2014). Here, we provide a short overview of some key concepts that are important to the interaction between SBS and optical fiber communication systems.

One of the simplest sorts of optical communication systems are based on on-off keying (OOK), encoding a binary '1' as a high amplitude state, and a '0' as a low amplitude state. A simplified intensity-modulated, direct detection (IMDD) system that could run OOK is shown in Fig. 1. An OOK transmitter can be achieved either simply by turning a laser on and off, or by putting a 'shutter' in front of the laser, usually a modulated Mach-Zehnder interferometer. The receiver structure is also very simple, as a single photodiode can directly measure changes in optical power, and so can

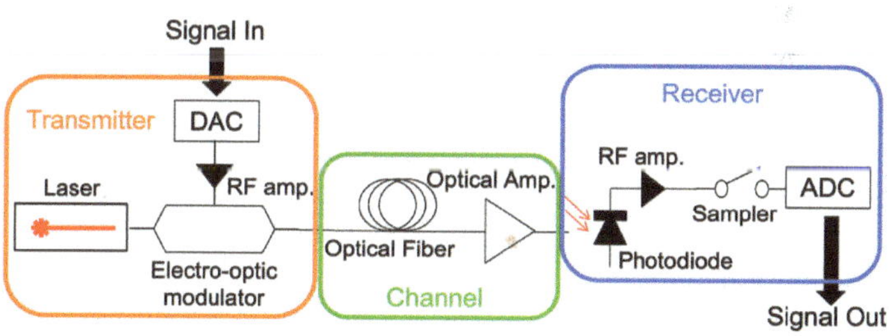

Fig. 1 (A) simple intensity-modulated, direct-detection (IMDD) optical communication system. The transmitter consists of a light source, a method of modulating that light source, and a way to convert from digital information to an analogue waveform—here a laser, modulated with an external electro-optic modulator, driven by a digital-to-analogue converter (DAC). The channel is the transmission medium, and often distorts the signal (adds impairments). The receiver requires a method of translating an optical waveform into an electrical waveform, and back into digital information. Here, intensity is measured using a photodiode, the resulting waveform amplified and gated using a sampler, and then converted to digital information via an analogue-to-digital converter (ADC).

directly convert amplitude modulation into an electrical signal. OOK systems were extensively used commercially and developed in research labs once broadband optical amplifiers (erbium doped fiber amplifiers) became available in the late 1980s, until about 2006–10, when the advent of fast digital electronic signal processing enabled coherent communication systems. Before that stage, a key tension with OOK systems was a mismatch between the very high achievable optical channel bandwidths and low available electronic processing speeds. As such, there were numerous investigations of the potential for optical signal processing techniques to support electronics, including sub-systems based on SBS.

Coherent optical communications systems have provided a big shift in the way fiber optic links are designed and operated. Coherent optical communications systems rely on the beating of signal lightwave with a reference lightwave in a photodetector. In direct detection systems, like those using OOK, the generated photocurrent (I) is related to the optical power (P) and field (E) as $I \propto P = |E|^2$, measuring the beating of the signal lightwave with itself (signal-signal beat). In coherent systems, the incoming electric field (E) is made up of a signal component (E_{sig}) and a reference wave component (E_{ref}). In this case, the generated photocurrent can be represented as $I \propto |E_{sig} + E_{ref}|^2$, which produces an electrical signal proportional to the beating of the reference and signal fields, as $|E_{sig} + E_{ref}|^2 \propto E_{sig}E_{ref}$. Early coherent systems focused on the advantages of using a powerful reference wave at the receiver to increase the overall photocurrent, as a drop in signal power ($P_{sig} = |E_{sig}|^2$) could be compensated by increase in reference wave power ($P_{ref} = |E_{ref}|^2$), such that the detected photocurrent remained high (i.e., the product $E_{sig}E_{ref}$ remains high). This was an important strategy to improve receiver sensitivity (i.e., the minimum useful detectable optical signal power), before the advent of the erbium doped fiber amplifier.

Over the past 15 years or so, coherent detection has become important more in its ability to measure the optical field. If the field of the reference wave is known (or, more accurately, stable enough to be tracked and removed from measurement), then the photocurrent in a coherent detector is directly related to the signal field—if the reference field (E_{ref}) can be treated as a known constant, then the photocurrent is proportional to the signal field, i.e., $I \propto E_{sig} = |E_{sig}| e^{i\phi_{sig}}$. As such, the signal can have both amplitude ($|E_{sig}|$) and phase (ϕ_{sig}) measured simultaneously. Importantly, as the photocurrent I is the product of optical interference between the signal and reference waves, it is also sensitive to polarization. This means that if there are two reference fields by measuring the signal against two reference fields (e.g., $E_{ref,x}$ and $E_{ref,y}$) on orthogonal polarization states (e.g., 'x' and 'y'), the signal polarization can also be simultaneously measured with phase and amplitude.

What this means in practice is that arbitrary transforms in amplitude, phase and polarization can then be applied to the detected optical field, which can help to compensate for distortions accumulated during transmission through optical fiber. Because of this property, distortions such as chromatic dispersion or polarization mode dispersion can be compensated in the electronic domain. In addition to this, the ability to measure the full optical field enables the ability to modulate on all dimensions of the optical field. By enabling multi-level modulation in phase (ϕ_{sig}), amplitude ($|E_{sig}|$) and polarization ($E_{sig,\ x}$, $E_{sig,\ y}$), this allows optical communications systems to move closer to data rates limited only by signal-to-noise ratio, as predicted by Shannon.

In modern coherent systems, signal processing in the optical domain is largely limited to wavelength routing, with most processing occurring in the digital domain. While digital signal processing remains an exceptionally powerful and flexible tool, there are opportunities for photonic approaches to complement digital processing capabilities. Generally, digital signal processing consumes a significant amount of energy, and can also introduce latency in systems due to the need for many computational steps. Photonic methods in these systems are then often judged on their potential to reduce the energy consumption of digital processing, or to help relax computational requirements to reduce latency.

This then sets up some criteria for whether SBS can meaningfully help enhance optical communication systems. In each case, there are trade-offs in terms of overall system complexity, energy consumption of the additional SBS sub-system, and the potential for off-loading time-consuming signal processing steps out of the digital domain. These could be summarized as trade-offs in cost, power and speed. The other aspect to consider is one of the size of the SBS sub-system. We find that this is harder to estimate in research, as there is considerable scope for the integration of various components of photonic sub-systems, but the platforms for these are not necessarily directly accessible by all research groups working in this field. As such, we'll refer to these metrics of cost, power and speed as appropriate when assessing the potential for SBS to contribute to future optical communication systems.

In optical communication systems, the key overriding metric is the bit error rate (BER), which defines how many errors one can expect for a given number of sent bits. In a given communication system, there will be a threshold at which the system is judged to generate too many errors to be viable. Below this threshold BER, the system will be judged to be viable. This threshold will change, depending on the system. For example, a maximum BER of 10^{-12} or 10^{-9} was often considered to be required for systems in the late 1980s or 1990s. This threshold can be relaxed using a family of signal coding techniques known as forward error correction (FEC), in which a number of data bits are replaced with a known set of bits, to enable the receiver to know which bits are in error and correct for these. Common FEC codes enable BERs in the region of 10^{-3}–10^{-2} to be corrected down to a post-FEC BER of 10^{-15}–10^{-12}. Regardless of the BER threshold used or assumed for a system, the performance of a new signal processing sub-system can be judged by the required power or (optical) signal-to-noise ratio ((O)SNR) required at the receiver, with or without the new sub-system in place. In some cases, the new sub-system will be required to improve required power or OSNR. In other cases, the sub-system may be introducing a new functionality that may improve the capabilities of the system overall but come at the cost of a slightly degraded required power or OSNR. This overarching metric for performance provides context for the potential trade-offs in cost, power and/or speed when adding a new sub-system into a larger communications system. While system performance degradation may be allowable in some limited cases where a new functionality is added, a sub-system should on balance improve the achievable BER in order to be considered for implementation.

1.3 Chapter outline

In short, there are a number of situations in which an SBS system can be explored to improve optical communication systems, and we have metrics to judge potential performance on. This chapter reviews the multiple faces SBS presents to optical communications: The good, the bad, and the ugly. We review the early use of SBS as a unique tool for optical signal manipulation, its emergence as a limitation, and recent research progress that places SBS as an interesting capability to enhance modern coherent communications technologies. This is not intended as an exhaustive review of literature over this time, but to highlight some of the interesting areas of study with a few examples.

Section 2 reviews the early investigation of SBS in optical communication systems, taking in the early coherent communications era of the 1980s, up through the 1990s and into the early 2000s during the dominance of on-off keying and direct detection systems.

Section 3 looks at the application of SBS over the past 15 years or so as coherent communication technologies have become dominant in optical communication systems. The unique challenges in coherent optical communication systems set the stage for a revival in interest in SBS in this period and expanded the potential applications of SBS sub-systems.

Finally, Section 4 provides a forward-looking view of the potential for SBS to contribute to future optical communication systems. This includes a review of some recent work within a year or so of the time of writing, as well as a look at some of what we think are outstanding requirements for SBS-based sub-systems before they may become practical for widespread use in optical communication systems.

2. Early use of SBS in communications

Stimulated Brillouin scattering has been investigated in optical communication systems since the 1970s. In this section, we provide a brief overview of various themes of study of SBS in optical communications systems, in the period before the advent of practical of coherent communication systems (i.e., up to around 2005).

Although stimulated Brillouin scattering had been observed in different media earlier than the 1970s, the exploration of this effect in optical fibers bought a new focus of the impact of SBS on the field of optical communications.

In 1972, Ippen and Stolen investigated Brillouin scattering in glass optical fibers with a loss of 20 dB/km (Ippen and Stolen, 1972). These experiments were some of the first to observe SBS in fibers and provided the basis for understanding the significance of SBS in what would eventually become optical fiber communication systems. In their study, a pulsed pump laser in the visible (~535 nm) was injected into this early type of fiber, and the backscattering analyzed with photodetectors and a Fabry-Perot etalon. This study showed that SBS stands out as a nonlinear effect with a low threshold—in these experiments, 40 mW was shown to induce significant SBS. The Stokes shift was also shown to match well with theory, with the observed shift of 32.2 GHz coinciding well with the acoustic velocity of fused silica, which was to be expected in these early fibers. One of the important conclusions that these authors made was that SBS would seem to provide a limit to power spectral density able to be transmitted in optical fibers. These power limits would become a major focus of research.

In the same year, a theory paper outlined SBS as a power limiter in optical fibers (Smith, 1972). The theory outlined in this paper backed up the experimental observations of km (Ippen and Stolen, 1972), providing a rule-of-thumb metric for when optical power into a fiber hit a critical level to generate significant SBS. The theoretical estimate of around 35 mW for a 20 dB/km fiber closely matched the experimental conditions and observations in km (Ippen and Stolen, 1972). This paper also noted the 'thresholding' behavior of SBS, in that a small reduction of power (1 dB) away from thresholds can dramatically reduce the magnitude of SBS.

As fiber losses and fabrication techniques improved, and single mode fibers for lightwaves in the infrared became more common, the limits on optical fiber communications systems given by SBS became clearer.

In 1988, Aoki, Tajima and Ito (Aoki et al., 1988) investigated the effects of SBS in transmission fibers on various binary signal modulation formats, including amplitude shift keying (ASK, i.e., OOK), phase shift keying (PSK) and frequency shift keying (FSK). In single mode fibers, with 0.4–0.5 dB/km loss, using signals modulated on semiconductor lasers working in the 1300 nm waveband, the threshold power for SBS was able to be mapped out for fiber lengths over several kilometers. Thresholds of around 10 mW were observed for fibers >10 km long when launching CW laser light, increasing to 15–30 mW when modulating at 200–400 Mb/s. They showed that modulation at the rates used increased the SBS thresholds by 2× to 4×, by moving 2× to 4× beyond the intrinsic SBS linewidth.

In 1990, Chraplyvy provided a review of the effect of the major fiber optic nonlinearities on the performance of optical communications systems. This provided a guide for what nonlinearities are important for different WDM system configurations. Broadly, at the channel bit rates relevant in the early 90s, Brillouin scattering was important for systems with low channel counts, where there was a high-power spectral density. This is an interesting conclusion, given that power spectral densities in modern systems are lower, as per-channel bit rates are increased and gaps between WDM channels decreased. Moreover, SBS is also a limiting factor in systems that are designed to be robust to non-linear perturbations. An analysis of soliton-based transmission systems (Montes and Rubenchik, 1992) showed that achievable information rates could be capped by SBS, depending on the phase correlations between modulated soliton pulses. These themes of research in the late 1980s and early 1990s very much set SBS up as an impairment to be reckoned with in fiber transmission systems.

However, during the same period, SBS was also recognized for its potential as a unique tool for photonic signal processing in optical communication systems. Applications such as signal conversion, clock recovery, short-pulse generation, wavelength division multiplexing, self-heterodyne detection, and narrow linewidth lasing, among others, were all proposed and investigated.

Optical clock recovery is an application of the frequency selectivity of SBS to high spectral density features. An OOK signal has significant 'clock tones,' which are spectrally dense peaks at the clock rate of the signal, with significantly higher power spectral density than the broader band signal. These occur due to the signal transitioning from high power to low power regularly. By taking an incoming signal and power splitting it, a portion of the incoming power can be shifted up in frequency to match the Stokes shift, amplified, and then counter-propagated in an SBS gain medium (e.g., standard optical fiber). Because the spectral power density of the spectrally narrow clock tones is much greater than that of the broad bandwidth signal, it is only the up-shifted clock tones that provide significant SBS gain. As such, SBS gain is only applied over a limited bandwidth (a few 10s of MHz) at the clock tone frequencies—an example of an experimental set-up for this is shown in Fig. 2A. For a signal forward propagating through the SBS gain medium, these clock tones are enhanced, changing the signal from a high extinction ratio OOK signal, to a low extinction ratio signal that resembles a pulse train with some amplitude variation. This pulse train can then be used to extract a clock signal. In Butler et al. (1995), a clock was extracted from a single channel, pseudo-random OOK signal running at 5 Gb/s (a symbol rate of 5 Gbaud). This type of system can also be run for multiple wavelength channels (as shown in Johnson et al., 1999), due to the self-selecting nature of the SBS gain—as the SBS interaction is pumped by a frequency shifted version of the signal, it is in principle agnostic to the wavelength of the signal. The selectivity of SBS gain also allows for multiple channel operation without interaction between pumps via SBS, although other nonlinear mixing could become an issue (e.g., four-wave mixing). The authors here noted that the change of the Stokes shift with wavelength did provide a practical limit to their demonstration, although they suggested that approaches to broadening the SBS gain would help. For an in depth look at the specifics behind operating such pump recovery systems, a theoretical/numerical study by Zhou et al. (2000) covers many of the practical factors and limitations to this scheme.

The selective amplification of narrowband tones carried with a broader signal can also be used to enable optical carrier recovery for coherent communication systems. An early proposal of this technique was provided by

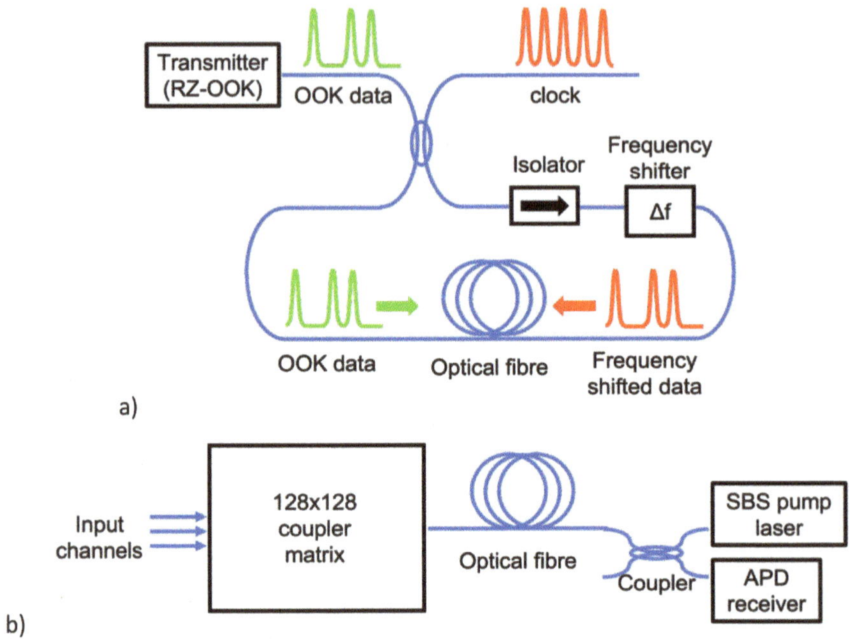

Fig. 2 Sub-systems for SBS-based optical signal processing, representing to major approaches to SBS pumping. (A) A sub-system for optical carrier recovery (Butler et al., 1995) where the SBS pump is derived from the incoming signal. In such a set-up, the incoming optical signal is frequency shifted by a modulator to provide gain for high spectral density components of the signal. (B) A sub-system for channel selection via SBS (Tkach et al., 1989). In this set-up, an independent laser is used to pump the SBS interaction, and is here broadened to allow for drifts in frequency between the incoming signal and the independent pump laser. The matrix to the left is a cascaded coupler system to emulate a 128 × 128 star coupler, to enable proof-of-concept demonstration of a 128-channel optical system. The right-hand side shows the SBS amplifier system and receiver.

Smith et al. in 1986 (Smith et al., 1986). In this proposal, SBS gain was used to amplify a residual, low power optical carrier that was transmitted alongside a data-bearing signal at the receiver. This amplified carrier could then be used as a frequency- (and maybe optical phase-) matched local oscillator for coherent detection, which potentially solved issues with analogue phase locked loops in early coherent receivers. In the proposal by Smith et al., an independent laser was used at the receiver side to pump the SBS interaction, which they note can cause some issues with precise tracking and noise (Smith et al., 1986). Demonstrations building on this work solve this issue by using the same self-seeding approach used for the clock recovery demonstrations we highlighted above.

Optical carrier recovery was also envisaged early as a useful tool for sub-carrier multiplexed systems (Lee et al., 1991). In sub-carrier multiplexed systems, multiple frequency channels are defined on a single laser via a single external modulator. This can be viewed as using single laser to support multiple wavelength channels, by modulating signals at different frequencies around the laser line. In the system presented in Lee et al. (1991), SBS was used to enhance the power of a residual sub-carrier modulated laser line before the receiver, to then enable homodyne detection. The sub-carrier of interest could then be extracted out through microwave filtering. In this way, SBS can boost the carrier power of the desired channel, which increases the signal/carrier beat term for that wavelength channel after photodetection. Microwave band-pass filtering then enables the channel of interest to be selected. This approach is interesting in that it parallels the 'colorless' receiver idea of modern coherent receivers, where a tuneable local oscillator provides the selection of the wavelength channel, and microwave (or digital) filtering then can provide fine-grained filtering of the selected channel (Kumar and Deen, 2014). In the self-homodyne presented in Lee et al. (1991), instead of using a laser local to the receiver as a local oscillator, it is used to provide SBS amplification to select the channel of interest through optical amplifying the incoming optical carrier of that channel.

The optical carrier and optical clock recovery examples highlighted show that SBS can provide an interesting narrowband optical 'gain' filter, where frequency selectivity is through gain, as opposed to loss in a common optical filter. This also then provided the opportunity for this active, selective filtering to be used for wavelength selection in wavelength division multiplexed systems. In 1989, Tkach, Chraplyvy and Derosier demonstrated a 128×128 network, selecting 150 Mb/s channels with channel spacings down to 1.5 GHz. In this case, the SBS pump was an independent laser, which was frequency modulated to broaden the SBS gain to 600 MHz. The set-up for this demonstration is shown in Fig. 2B. This level of frequency selectivity was remarkable for the ability to pack channels closely together, which reduced the amount of optical spectrum that would be wasted as 'guardbands' between channels in a WDM system by a factor of over 100. For this demonstration, it was also noted that there was good scope for scalability, with channel counts of up to 1000 being viewed as feasible, given the noise constraints in the demonstrated system.

Developments of SBS technologies for applications outside of communications have become important for potential applications in modern communication systems. For example, laser cavities using SBS for gain have enabled very high coherence lasing. An example from the early 1990s is Smith et al. (1991), where a laser with an intrinsic linewidth of <30 Hz and a spectral width of about 2 kHz. In this paper, the applications of such narrow linewidth lasers were in sensing or microwave signal synthesis. However, in recent years such lasers have been targeted to provide a better optical reference for coherent detection, which we'll discuss in more detail in later sections.

Stimulated Brillouin scattering is, however, a complex nonlinear interaction. Mapping the properties and features of SBS in a useful way for optical systems can be problematic. An example of this was highlighted in 1994 by van Deventer and Boot (Van Deventer and Boot, 1994). In that work, the polarization properties of SBS were investigated, and found to differ significantly from what might be assumed for other optical amplifiers. In particular, how the noise properties of SBS amplifiers spread over polarization states was found not to be as intuitively expected, following a 2:1 distribution instead of a 1:1 distribution between polarizations. This is where the 'ugly' part of SBS starts to show, in terms of application to optical communications. The ability to predict the individual behavior of components in an optical communication system is key to being able to model how the system will operate as a whole. As such, SBS technologies that are to be included in larger communication systems need to operate in very predictable and reliable ways, and things get ugly for communication systems when they don't operate as planned. This is a common theme in further demonstrations of SBS-based technologies for optical communication systems over the last 15 years or so, and going forward.

3. Resurgence of SBS in optical communications

In the previous section we have seen that while SBS was explored for enhancing optical communications, especially as an amplifier, most of the results were not very promising preventing further interest in these 'difficult to use and model' devices. At the turn of the 21st century, however, there has been a renewed interest in SBS. This has been primarily guided by advances in the understanding of sub-system designs using SBS for microwave photonics and sensing which has given the community a renewed confidence to implement SBS-based devices for optical communications applications. In this section, we will provide an overview of the recent advances in the area of SBS-enhanced optical communication systems.

3.1 Non-coherent systems

The main strength of Stimulated Brillouin scattering lies in its narrowband response of ~30 MHz, which has motivated several applications in filtering of optical and RF signals. One such application has been in optical spectral analysis (Domingo et al., 2005). Here the Brillouin gain is swept through the optical signal and the output response is measured. This is equivalent to scanning a 30 MHz optical filter and measuring the discretized spectral response of the system. This system thus provides an alternate solution to high-resolution optical spectral analysis. However, due to the complexity of such systems, they find very limited use in current coherent optical communication systems. It is expected that such a setup with high resolution could be of interest for future systems due to the reduction of the guardbands.

In addition to exploiting the amplitude response of SBS, there was a strong focus to harness the phase (and thus the delay response, as delay is nothing but the derivative of the phase with respect to the frequency) of SBS for making delay lines. This has been referred to as 'Slow light' (Zadok et al., 2011; Chapter "Slow light, dynamic gratings, and light storage" by Stiller et al.) in the community and was a very popular research topic in optical communications systems as a potential for nanosecond delay lines and optical buffers. SBS-based continuous delay lines based on dynamic gratings were also used for contention resolution between asynchronous optical packets (Porzi et al., 2013). However, there are several limitations which have prevented the deployment of SBS-based delay lines in commercial optical communication systems: (1) limited gain bandwidth of SBS which limits the pulse width putting a low bound to the data rate, and (2) the large power requirements for large delays. The first challenge can be mitigated to an extent by increasing the gain-bandwidth by pump modulation, for example in Cabrera-Granado et al. (2008), however this increases the complexity of the system to minimize the distortions. The second challenge is an inherent disadvantage of SBS, since it is an active process and larger pump power means a larger gain response and hence a larger delay. Furthermore, in modern coherent systems the delays can be handled with advanced digital signal processing (DSP) blocks.

SBS was also explored as a solution to several other problems in optical communication systems. One such problem is the monitoring of the optical signal to noise ratio (OSNR) (Fu et al., 2010; Iredale et al., 2007). The SBS-based ONSR monitoring works on the principle that whenever a spectral component exceeds the Brillouin threshold, the power transfer to the Stokes wave depends on the noise present in the signal. Thus, by measuring the back-reflected power, the OSNR can be calibrated. This concept was used for Non Return to Zero (NRZ) On Off Keying (OOK) systems and was also extended to dual polarization systems (Dahan et al., 2010). Another aspect where SBS was used as a solution was in the encryption and decryption of signals (Yi et al., 2013). In this technique, the authors 'encrypt' 10.83 Gbps NRZ-OOK data using SBS gain filters during transmission. The encryption is achieved by distorting the amplitude and phase of the signal. At the receiver, the same SBS response is used to 'decrypt' the signal, thus recovering the distorted waveform. The proposed keys are the SBS parameters such as the gain, the spectral bandwidth and the shape of the gain response. However, this technique requires the exact matching of the transmission and receiver SBS gain medium. Furthermore, a DSP block can recover the encrypted information by removing the SBS gain response digitally.

Another early implementation of SBS was for demodulating Differential Phase Shift Keying (DPSK) signals at the receiver (Marazzi et al., 2011). In this demonstration a DPSK signal was filtered using a broadened SBS pump of 6.7 GHz bandwidth which was achieved by passing an amplified spontaneous emission (ASE) noise signal through a fiber Bragg grating. This SBS filter converts the phase modulation to amplitude modulation. Typically, such a transformation is done using Mach Zehnder interferometers (MZIs), however these required very precise length difference between the two arms and were complicated to realize with fibers. However, with the advancement of integrated photonic technology, MZIs can now be made very precisely on a photonic chip.

Modulation format conversion is important at network edge nodes, since different applications are serviced through different modulation formats. A typical implementation of the format conversion would be through demodulation to the electrical domain and re-modulation in the new modulation format and re-transmission. However, this increases complexity and therefore all-optical format conversion can be of interest to overcome these challenges. The carrier can be removed using notch filters (Gatti et al., 2008), however due to a large bandwidth of such filters, useful signals can also be attenuated. Format conversion between NRZ-OOK and binary phase shift keying (BPSK) was also realized using SBS (Zhang et al., 2013) which enabled high-resolution filtering. The main difference in the frequency domain between OOK and BPSK signals is the absence of the carrier for BPSK signals. The OOK signal was converted to the BPSK signal by attenuating the carrier through the SBS Loss response arising due to the

Anti-Stokes response (Choudhary et al., 2017b), while the BPSK signal was converted to the OOK signal by amplifying the carrier using the SBS gain response. Using this technique, the authors were successfully able to demonstrate format conversion for 10.86 Gbps signals. A challenge for such a realization is the existence of the gain (loss) response when converting from OOK to BPSK (BPSK to OOK), which limits the total single sideband bandwidth of the system to the SBS shift of ~10.8 GHz in optical fibers.

3.2 Coherent systems

As coherent optical communications made a comeback with the advancement in digital signal processing, researchers also explored SBS for coherent optical communication systems. In current coherent communication systems, flex-grid networking enables flexible channel spacings for more flexible networks at the network edge. This means that the channel spacing need not be 50 GHz, for example channel 1 can have a bandwidth of 40 GHz, while channel 2 can have a bandwidth of 25 GHz depending on the user demand. Such a flexible and efficient network architecture requires reconfigurable optical add drop multiplexers (ROADMs). A common technology for achieving ROADMs is based on Liquid-crystal on silicon (LCoS) (Frisken et al., 2011). However, these filters have a minimum bandwidth of 10 GHz limiting high-resolution signal processing. SBS can alleviate this challenge as bandwidth reconfiguration can be demonstrated from as low as 30 MHz to 3 GHz (Wei et al., 2015). This was achieved using multiple SBS pumps generated using an electronic waveform generator. However, due to the nonlinear response of the modulator and four wave mixing between the pump lines a large ripple was caused in the passband of the filter. Using an iterative feedback loop, the ripple in the passband was reduced to <1 dB. Using this filter, add-drop multiplexing was demonstrated for a 2.2 GHz SBS filter and OFDM signals, as illustrated in Fig. 3. The SBS gain can be used to selectively amplify one of the bands, which performs the drop functionality. The SBS loss can also be used to attenuate a band, removing it completely and thus performing the through functionality. Using a directional coupler another signal can be added, thus performing the add functionality. The advantage of using SBS as filters is that it has a very sharp transition and enables very narrow passbands of <0.5 GHz. The authors demonstrated QPSK with a penalty of 0.7 dB, while the penalty was >2 dB for 16 QAM when using 20 dB of SBS gain and increased to >4 dB when using 30 dB of SBS gain.

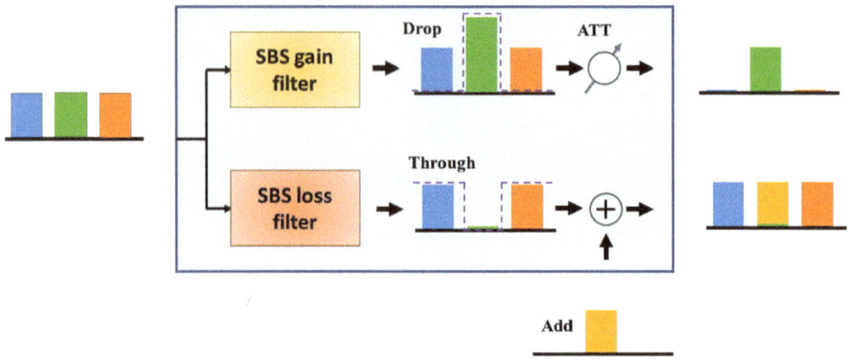

Fig. 3 Add drop multiplexing using SBS. Here SBS gain blocks can amplify a specific band (shown in green) which acts as the drop port. An SBS loss filter can attenuate the signal (green) thus emulating a through port. Another signal can be added at the drop port thus enabling add-drop multiplexing.

Another application for SBS in coherent optical communication systems was demonstrated by Cao et al. (2017). Here the authors implement a self-homodyne detection scheme by transmitting a pilot tone adjacent to the signal. At the receiver, a narrowband SBS amplifier block is inserted to amplify the weak pilot. The amplified pilot is then frequency converted to the centre of the signal using a phase locked optical frequency comb and nonlinear frequency conversion. For each channel, two comb lines were selected and then mixed with the amplified pilot and signal in a nonlinear crystal. The phase relation between the comb lines was controlled to enable detection of I (phase difference $=0°$) and Q (Phase difference $=90°$) sequentially on a photodiode and the average BER was calculated. In this application, the SBS is used as a narrowband amplifier for the pilot and the homodyne detection is done using an additional multi-wavelength laser source, making the system complex.

Four wave mixing (FWM) is a common technique for frequency conversion (Liu et al., 2014) from the signal frequency to an idler frequency. However, the conversion efficiencies strongly depend on the polarization alignment between the signal and the pump waves, which can be improved using SBS (Huang et al., 2016). Another example of SBS-enhanced enhanced nonlinear optical signal processing for coherent optical communications was presented in Huang et al. (2017). In this work, the authors used the SBS induced phase shift to improve the phase matching condition for polarization insensitive four wave mixing. Two SBS pumps are used:

the first induces a gain response on the FWM pump and the other induced SBS loss. This results in a nearly flat amplitude response, but with a strong phase response at frequencies downshifted from the pumps by the Brillouin shift and the phase response is added. By controlling the frequency separation of the two SBS pumps, the phase of the pump is controlled enabling a versatile technique to control the phase response of the pump. This enabled a conversion efficiency improvement of ~6 dB and increase in the bandwidth by 1.1 THz, which resulted in an improvement of 240 km in the reach of a QPSK signal. However, this system has the drawback of being complex.

Phase sensitive regeneration is typically used to reduce the phase noise of a signal using four wave mixing. This employs a binary phase shift keying (BPSK) modulated signal and two strong pumps equidistant in frequency (one blue shifted and the other red shifted) from the signal to be transmitted through a nonlinear fiber. This results in the addition of the signal (S) with its conjugate (S*), causing the phase noise to be reduced for the BPSK signal. However, this technique requires the pumps to be mutually coherent with the signal which typically requires either frequency comb lines or phase-locked loops. Almaiman et al. (2016) used a two-stage nonlinear mixing process with SBS to alleviate the requirement for phase locked loops (PLL). In the first nonlinear mixing process, the signal and a pump are used to generate an idler. In a following SBS stage, the idler is amplified using a narrowband SBS amplifier stage to ensure the idler is amplified to a similar power level as the pump. This signal is then sent through another nonlinear optical fiber for generating the S + S* signal thus reducing the phase noise of the received BPSK signal. This technique resulted in a phase noise reduction of >50% for a 10 Gb/s signal. This system has a complex architecture and is limited to BPSK signals only.

The demonstrations discussed so far have presented very interesting concepts and exciting results, however the applicability in modern systems is challenging due to the large complexity. A simple manifestation of SBS for coherent communications was in self-coherent systems (Banchi et al., 2010), where the local oscillator (LO) is extracted from the transmitted signal, as opposed to using a separate LO laser at the receiver for extracting the phase. This reduces the receiver complexity and reduces the load on the DSP. The authors transmitted BPSK signals and extracted the carrier using selective Brillouin amplification at the receiver, before detecting the signal using a PIN diode thus converting phase fluctuations to intensity fluctuations. High-pass filtering was required to eliminate the low-frequency

Brillouin intensity noise. The received bit error rates for 10 Gbps BPSK signals had a 3-dB penalty with respect to the theoretical limit. This setup required the simultaneous transmission of the pump and the signal through the fiber which would limit the total signal capacity due to the separation of the pump and the signal by ~10 GHz. This effectively renders a significant fraction of the bandwidth unusable, thus reducing spectral efficiency.

As we have already discussed, coherent optical communication systems require sophisticated narrow linewidth lasers at the receivers to act as an LO to extract the phase information from the received signal. Furthermore, digital signal processing blocks are required to eliminate the relative frequency drift between the received signal and the LO. The requirement of an extra laser and additional DSP blocks increases the cost and complexity of the receiver. Another modulation scheme known as self-coherent orthogonal frequency division multiplexing (Self-Co-OFDM) circumvents these requirements by transmitting a carrier with the signal, that could be extracted at the receiver and used as an LO. This is typically done using filters, however most optical filters have a large bandwidth thus necessitating the use of a large guardband between the carrier and the signal sidebands on both sides. This unfortunately removes the signal carriers thus limiting the total data rates. Here again, SBS's narrowband amplifying capability enables efficient filtering at the receiver. SBS can be harnessed in an optical fiber (Giacoumidis et al., 2017) or on a chip (Giacoumidis et al., 2018), however chip-based implementation allows better phase and polarization stability to environmental conditions and SWAP advantages. Chip-based implementation has been made possible due to the large on-chip gains made possible by advancement in fabrication technologies (Choudhary et al., 2016). 116.82 Gb/s Self-Co-OFDM signals with 16 QAM modulation were transmitted over 80 km of transmission fiber and the use of on-chip SBS enabled a record guardband of ~265 MHz. The architecture is shown at the bottom part (1) of Fig. 4. Here the receiver employs a chalcogenide chip which imparts SBS gain and selects the transmitted carrier and suppresses the signal, thus emulating a high carrier to noise power ratio LO. This architecture enables a higher capacity as compared to the benchmark Co-OFDM systems. Fig. 4 also shows the comparison of chip-based SBS carrier recovery, fiber-based SBS carrier recovery and standard carrier-suppressed OFDM in terms of DSP requirements. It is apparent that on-chip SBS enables the lowest electronic signal processing and has a potential for implementation in future communication networks.

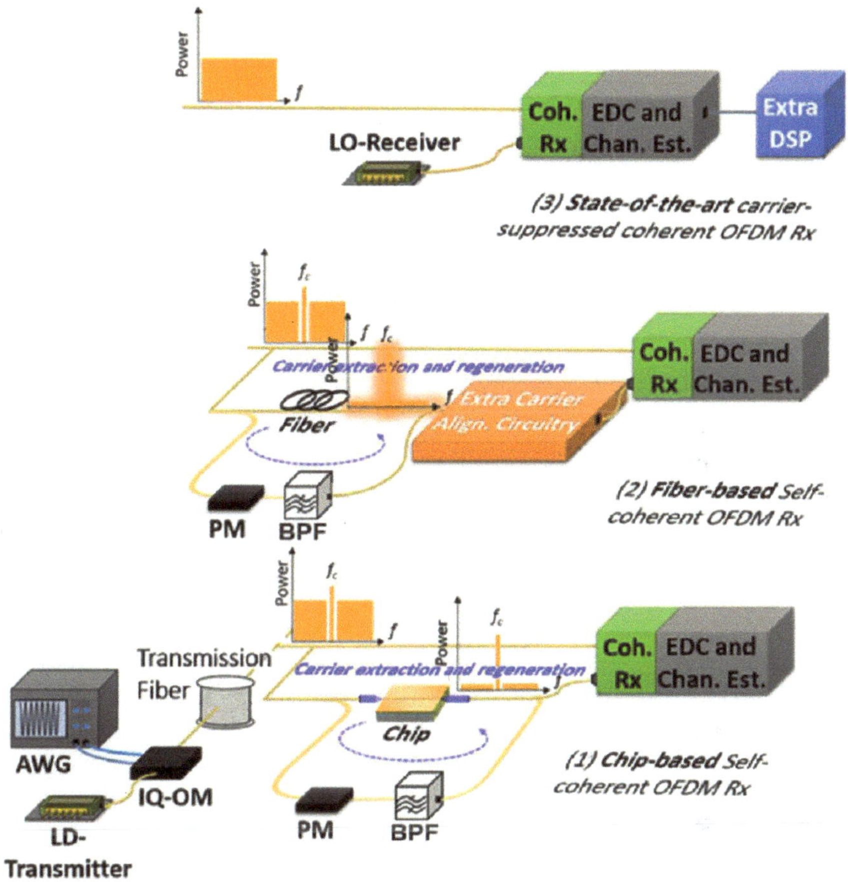

Fig. 4 (A) comparison of various coherent OFDM technologies. (1) An on-chip SBS implementation for carrier recovery at the receiver for Self-Co-OFDM signals. An arbitrary waveform generator (AWG) generates the signal that is modulated on the I and Q channels of an Optical modulator (OM) which has a laser diode (LD) transmitter as its optical output. The signal is transmitted over an optical fiber and then sent to the receiver circuit that consists of a power splitter. The first arm is fed to the signal port of the coherent receiver (Coh. Rx). The other arm is split again, and the upper arm is sent through the SBS chip, and the lower arm is fed to a combination of a phase modulator (PM) and a bandpass filter (BPF) that frequency shifts the signal by the Brillouin shift. The frequency shifted signal is counter-propagated with the signal in the SBS chip which amplifies the carrier as shown in the figure and is then fed to the LO port. (2) The fiber-based SBS amplification implementation of carrier recovery for Self-Co-OFDM. Here, the fiber is prone to phase and polarization fluctuations. (3) State-of-the-art Co-OFDM system where an additional LO is needed at the receiver. On-chip SBS filtering (1) at the receiver for a Self-Co-OFDM system has the lowest electronic signal processing requirements. *Adapted from Giacoumidis, E., Choudhary, A., Magi, E., Marpaung, D., Vu, K., Ma, P., Choi, D.Y., Madden, S., Corcoran, B., Pelusi, M., Eggleton, B.J., 2018. Chip-based Brillouin processing for carrier recovery in self-coherent optical communications. Optica 5 (10), 1191–1199.*

Since coherent optical communication systems require a coherent receiver with a local oscillator, this can significantly increase the cost of the entire system making it untenable for metro applications. Kramers Kronig (KK) detection (Mecozzi et al., 2016) is a recently demonstrated technique that enables the use of direct detection with a single photodiode. The authors showed that for strong carrier powers, the phase of the signal can be extracted by taking a Hilbert transform of the logarithm of the photocurrent. However, if the carrier power is high at the transmitter, this will result in very high nonlinearities during transmission causing significant impairments. To overcome this challenge narrowband SBS amplification was used (Zhang et al., 2018) at the receiver to increase the pump power thus enabling KK detection. Here the carrier was suppressed, and a weak 'carrier' was inserted during transmission with a weak SBS pump tone. The virtual carrier concept allows independent control of the carrier to signal ratio without modifying the biasing condition. At the receiver, the SBS pump pilot tone was amplified using injection locking which put a narrowband gain on the weak virtual carrier. This enabled KK detection at the receiver and resulted in the improvement of the OSNR sensitivity by 2.7 dB compared to a standard virtual carrier technique after propagation through 80 km of fiber. However, the current implementations are quite complex and would need to be simplified in order to be implemented in future communication networks.

Modern communication systems deploy wavelength division multiplexed systems, where multiple optical carriers are used on which the data is modulated. This increases the overall capacity of the communication system. As more spectrally efficient systems are being deployed, the requirement for lasers is becoming more stringent as the phase noise of the lasers limits the modulation formats that can be used. This means that external cavity lasers (ECL), with linewidths <100 kHz are required for 16 QAM and beyond. Such laser systems are bulky, expensive and complex to fabricated. For example, if a 100-channel WDM grid is deployed for a shorthaul link with 64 QAM, then 100 ECL lasers would be required, thus massively increasing the footprint of the transmitter. To overcome this challenge, frequency combs (Hillerkuss et al., 2011; Temprana et al., 2015) have been explored as a viable alternative to using a stack of lasers. A frequency comb is a laser source that produces phase locked laser lines of uniform separation, essentially performing the same functionality that is performed by a stack of ECL lasers. This drastically reduces the cost and size of the transmitter.

However, most optical frequency combs are generated using nonlinear optical processes, thus resulting in the generation of significant noise components around the carrier thus adversely affecting the carrier to noise power ratio (CNPR) of the transmission wavelength. This resulted in the limitation of modulation formats demonstrated using frequency combs to 16 QAM or below. Recently a technique to overcome this challenge was demonstrated using SBS-based frequency comb purification (Pelusi et al., 2016). In this technique, the frequency comb split and was passed through an optical fiber which behaves as the SBS medium. The other part of the signal was sent through a frequency shifter module which consisted of a modulator and an optical filter. The modulator was fed by a 10.8 GHz (SBS shift in fiber) RF tone, and an optical bandpass filter selected only the upper sideband. This frequency-shifted signal is then counter-propagated in the optical fiber. This counter-propagation enables SBS amplification of the frequency comb lines. This enhances the carrier to noise power ratio of the frequency comb. Then 64 QAM data was modulated on the carriers and detected at a coherent receiver. 40 lines were detected under the FEC limit, with 96 GB/s data on each carrier, thus enabling a total data rate of ~4 Tb/s (Pelusi et al., 2017). This concept is illustrated in Fig. 5. Chip-based purification was also demonstrated using a chalcogenide waveguide as the SBS medium (Choudhary et al., 2017c). The constellation diagrams with and without on-chip SBS are shown in Fig. 5C and it can be seen that SBS can remove the severe phase distortions from the constellation very successfully. This technique is interesting because the noisy frequency comb purifies itself through the SBS process.

One possible drawback of using combs compared to a stack of lasers is the reduced redundancy in case of failure, i.e., if the comb fails, then the entire network needs to be moved to another stand-by comb. In the case of individual lasers, a single failure of the laser does not affect the other wavelengths. Furthermore, the power consumption of the frequency comb can be quite high owing to the current power-intensive technique of comb generation.

In this section, we discussed the implementation of SBS in early non-coherent systems followed by some architectures for coherent optical communication systems which have been reducing in complexity and are thus promising for future networks. In the next section, we will discuss the future directions of SBS in communication systems and the steps that need to be taken for practical deployments.

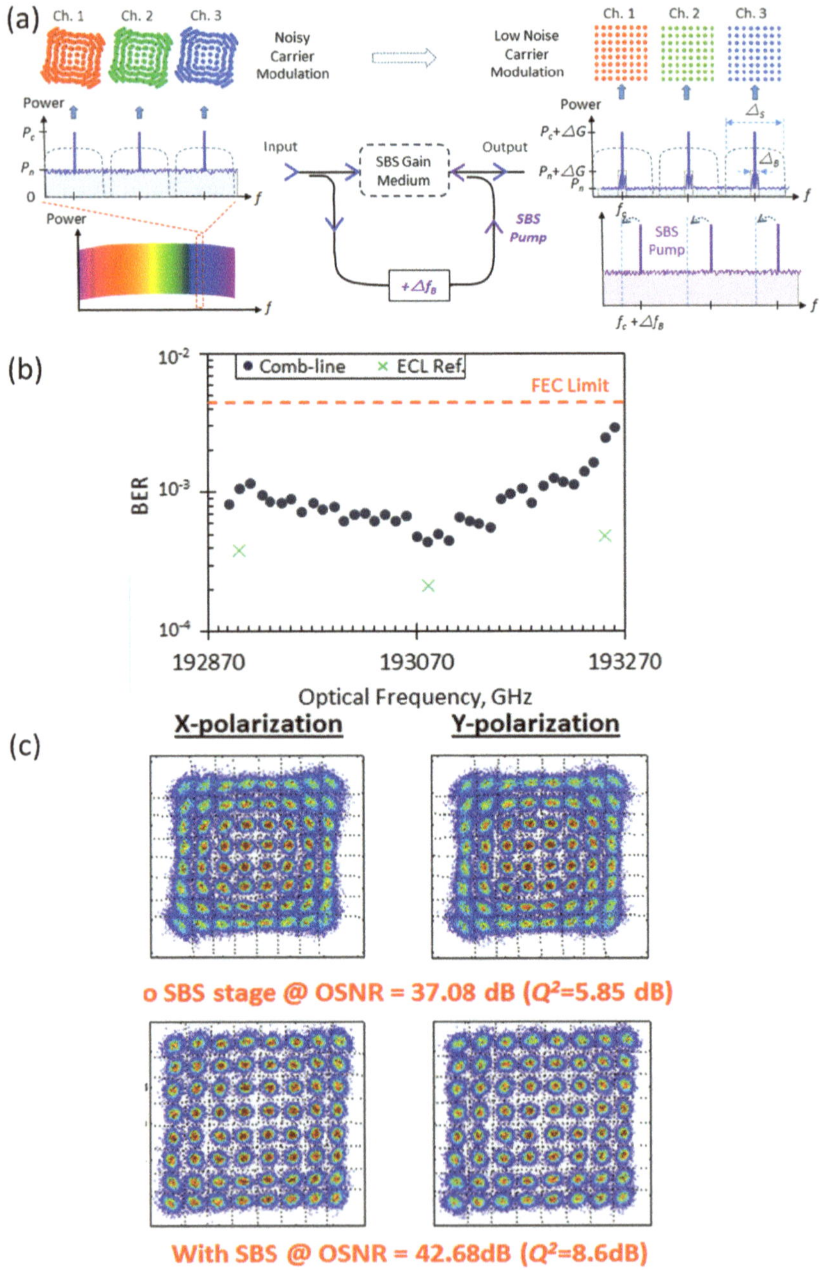

Fig. 5 Comb purification using SBS. (A) The concept: the noisy optical frequency comb is split and then sent through an SBS medium while the other arm is frequency-shifted by the SBS shift using optical modulation and then counter-propagated with the signal

4. Future directions for SBS in optical communication networks

The way we use optical systems to support global communications networks changes constantly, in big and small ways. One key push, as networks aim to support increasing data rates into the future, is for interfaces to optical systems (e.g., transmitters, receivers, etc.) to become smaller, more energy efficient and cheaper (at least in terms of $/bit). SBS has the potential to support these goals, and we see two particularly promising avenues to impact: In lowering the impact of laser phase noise on systems to help reduce the required digital signal processing in receivers, and in supporting the generation of low-noise multi-channel sources via optical frequency combs.

In modern optical communications systems, coherent detection of signals is becoming the dominant method over a widening range of contexts. Originally finding application in long-haul, high-capacity systems, coherent optical communication systems have pushed into metropolitan/regional networks and are looking to push further into short reach systems and data centre networks.

One of the key enablers of coherent detection is fast digital signal processing algorithms, able to correct for a number of impairments in signal generation, transmission and reception. An important step here is optical carrier recovery, where the decorrelated frequencies and phase states of the independent lasers at the transmitter and receiver need to be compensated for to extract phase and amplitude information.

However, in the push toward enabling coherent systems in new contexts, the extensive use of high-speed DSP is seen as a barrier to adoption. There has then been growing interest in DSP-lite or DSP-free coherent optical communications, in a strange echo of the initial explorations of

in the SBS medium. This effectively amplifies the comb lines power in a very narrowband region while the out-of-band noise is not amplified. This increases the carrier to noise power ratio of the optical frequency comb which improves the constellations received at the receiver. (B) The Bit Error rate (BER) as a function of optical carrier frequency. ~40 comb lines were simultaneously purified using SBS in an optical fiber and it was observed that all of the lines showed performance below the forward error correction (FEC) limit. For comparison, the BER is shown for an external cavity laser (ECL). (C) Constellation diagrams for the x- and y-polarizations without SBS and with on-chip SBS. It is clearly seen that the constellation diagrams show significant improvement when on-chip SBS is used (Choudhary et al., 2017c; Pelusi et al., 2017).

coherent optical communications in the 1980s. Optical carrier recovery is one aspect of DSP that can be reduced or removed by changes to the optical system, by either using ultra-stable lasers, or by recovering the optical carrier in the optical domain before detection.

SBS lasers (see also, chapter "Integrated Brillouin lasers and their applications" by Blumenthal et al.) have been noted for their relatively low linewidth for many years, and recent progress has moved toward providing such lasers in a compact form factor (Brodnik et al., 2021; Gundavarapu et al., 2019; Harrington et al., 2020), for compatibility with modern optical communication systems. An interesting cavity structure for SBS lasers on integrated platforms is micro-photonic resonators, particularly micro-ring, or micro-toroid cavities. Recent work has shown that such an integrated platform can potentially be co-integrated with an ultra-high quality factor reference cavity for stabilization (Brodnik et al., 2020). In that work, a chip based SBS laser is pumped with a semiconductor laser to provide low-linewidth lasing. This is further referenced to the resonances of an ultra-high-quality factor Fabry-Perot resonator ($Q > 1$ billion), which provides a spectral feature that is on the order of 100 kHz. Stabilizing to this cavity through feedback circuits enables a two-order of magnitude drop in integral linewidth, which is quoted to be around 40 Hz.

The use of such lasers as both a carrier for a 16-QAM signal and as a local oscillator in a coherent receiver enabled reception of the signal without carrier recovery DSP, while maintaining similar performance to a DSP enabled receiver (Brodnik et al., 2020). A follow-up article by the same authors highlighted the potential for extension to much high-order modulation formats (Brodnik et al., 2021), analyzing the fractional frequency stability and linewidth in term of system phase noise in the system to suggest that modulation formats up to 512 QAM could be supported by the scheme without resorting to DSP-based carrier recovery.

An alternative to using an ultra-stable laser at both ends of an optical communication system is to transmit the optical carrier along with the signal, and then to use that as a local oscillator at the receiver side. This 'self-homodyne' technique has been demonstrated in various contexts, including using SBS as a way to enable highly selective amplification of the optical carrier, to boost its power up for use as a local oscillator at the receiver side (Banchi et al., 2010; Giacoumidis et al., 2017, 2018; Lee et al., 1991; Smith et al., 1986). Recent work has focused on investigating the limits of this approach, and on adapting prior studies toward SBS-based optical carrier recovery systems that are more compatible with optical communication systems.

One key aspect of self-homodyne systems is how low the transmitted carrier can go in terms of power. To provide optical performance, most of the signal power needs to be in the data-carrying signal, as opposed to the optical carrier, so that the signal-to-noise ratio of the system can be maximized. In the context of SBS-based carrier recovery, this has been comprehensively studied and characterized (Pelusi et al., 2019). In such systems, SBS-based carrier recovery provides narrow-band amplification to boost the optical carrier above the system noise level. This study provided an indication of several interesting bounds to SBS-based carrier recovery. Using a generally accessible level of SBS gain (~30 dB), these studies were able to show that carrier-to-signal power ratios at the transmitter side of <20 dB were able to be used, essentially removing issues with achievable signal-to-noise ratios in self-coherent systems. The resulting self-homodyne system had a performance comparable to using an independent external cavity laser, indicating that the recovered carrier-to-noise ratio is high enough in a realistic receiver.

SBS-self homodyne receivers are interesting in that they can be made to be self-tracking—i.e., they can provide targeted narrow-band gain to the optical carrier, simply by using that carrier as a basis for exciting the SBS interaction. Typically, this requires the incoming carrier to be frequency shifted via single sideband modulation, to match the SBS shift (or Stokes shift) of the medium used for SBS gain. However, this SBS shift can change significantly with wavelength, which then removes the self-tracking capability as the incoming wavelength must be known to tune the frequency shifting stage to match the SBS shift. One way of avoiding this tuning step is to broaden the SBS gain through frequency dithering of the pump wave, which then provides some tolerance to uncertainty in the SBS shift (Zarifi et al., 2021). In that demonstration, a modest broadening of the SBS gain bandwidth to 150 MHz via pump dithering enabled high-fidelity SBS gain across the whole bandwidth of an erbium-doped fiber amplifier. This then showed that SBS gain can be made to be 'colorless' in terms of optical communication systems, which is a desirable aspect for coherent reception.

A remaining issue with SBS-based self-homodyne systems is the length of the SBS gain medium. A long SBS gain medium can introduce a temporal decorrelation between the signal and the recovered carrier, providing a mismatch between the phase of the recovered carrier and the signal. Decorrelation from propagation delay is particularly relevant in systems using optical fibers as gain media, as the decorrelating length can be several kilometers, relating a temporal decorrelation of tens of microseconds. Temporal decorrelation can be partially compensated by using a similar fiber length to delay the signal, however, a small fractional uncertainty in lengths can relate to a non-zero decorrelating delay. How significant this residual delay it is was characterized in Li et al. (2021), in which a residual delay of ~100 m out of 10 km was present in a SBS-based self-homodyne receiver. Even this imperfect delay matching was shown to significantly relax the tracking requirements for phase noise compensation, showing potential for enabling a 'DSP-lite' approach. Of note in this experiment is that chip-based SBS is likely to have de-correlation lengths of a few centimeters at most, and so chip-based carrier recovery is likely to provide very high fidelity (Giacoumidis et al., 2018).

While being able to relax DSP requirements is advantageous for expanding the impact of coherent communications, another goal for future ultra-high-capacity communication systems is massive parallelism to enable massive data-carrying capacity. One possible technology to support this is integrated optical frequency combs—known as micro-combs. As we highlighted in the previous section, SBS can be used to 'distil' frequency combs out of broadband noise (e.g., Pelusi et al., 2017). Recent work has also put SBS at the heart of generating frequency combs and shows similar advantages to using SBS for lasing (Harrington et al., 2020).

It is possible to use a compact, chip-based SBS laser to generate an optical frequency comb in a micro-ring resonator, with the microcomb taking on some of the low linewidth aspects of the SBS seed laser. Chip-based SBS lasers with linewidths on the order of 1 Hz have been demonstrated, based on SBS lasing in a micro-ring resonator (Gundavarapu et al., 2019). These lasers can then be used as a single wavelength pump for micro-comb generation in a second micro-resonator (Harrington et al., 2020). Ideally, the linewidth properties of the SBS pump laser should transfer onto the generated lines of the microcomb. When using such an exceptionally low-linewidth laser as a pump source, any phase noise added in the comb generation process is highlighted. In Harrington et al. (2020), a comb generated from a 1 Hz linewidth pump was shown to produce comb lines with an average fundamental linewidth of around 15 Hz. This is still exceptionally low linewidth and shows the potential to enable a very low linewidth multi-wavelength source. However, the integral linewidth (time averaged—useful for assessing impact on optical communication systems) of the comb lines is in the 1–10 kHz region, similar to that of lines produced when using a standard ECL as a pump. This highlights that there is potential for ultra-low linewidth, compact, multi-wavelength sources, but that further suppression of fundamental noise needs to be achieved.

Using SBS as a mechanism inside an optical cavity can also be useful for flexible comb generation. In Huang et al. (2019), a multiwavelength pump source was used in a cavity to produce a multiwavelength SBS laser, that through parametric mixing could produce a broadband optical frequency comb. By using a multiwavelength pump, the frequency spacing of the comb lines could be tuned, providing a flexible platform for comb generation. In this demonstration, the authors produced frequency combs with comb line spacing varied between 37.5 and 300 GHz. Fundamentally, this should provide a path toward low-linewidth pump generation *and* frequency comb generation within the same optical cavity, potentially significantly simplifying comb generation—which is key to providing optical frequency combs that are of interest to optical communications systems.

What has been clear over the course of the 40–50-year interaction of SBS and optical fiber communications is that there is a consistent potential for SBS to have a significant positive impact on optical communication systems, as a basis for novel photonic technologies. Demonstrations of promising device or sub-system technologies have emerged as the needs of optical communications systems have changed over time. These have provided unique and measurable advantages over other device technologies of the same era, continually showing potential for impact.

Clearly, though, this potential has yet to put SBS-based technologies front-and-centre in optical communication systems. Why might that be? We don't claim to have a definitive answer to this question, but there are some issues that we think might have played a role.

There is no really good existing one-size-fits-all solution to providing SBS as a functional block for optical communication systems. Various sub-system types and architectures have been trialed, and each is somewhat tailored to the target application at hand. This maybe makes it a little difficult to generate a device solution that can be used in multiple contexts. SBS is also an interaction that is sensitive to a lot of different factors. The Stokes shift is highly dependent on waveguide parameters, material parameters, temperature, strain, and is polarization sensitive. The narrowband nature of SBS gain is both a useful tool, and something that makes getting gain where you want it difficult. However, each of these issues has promising solutions—gain can be controlled through pump dithering, polarization sensitivity can be dealt with in a similar way by pumping in multiple polarization states. Working on integrated platforms gives a degree of control of waveguide and material parameters that is quite difficult to achieve over several kilometers of optical fiber.

A key remaining challenge is then to reduce the size, weight and power (SWaP) requirements of SBS-based devices, while keeping an eye on potential barriers to driving down device cost. The clear opportunity for this lies in producing full SBS sub-systems on an integrated chip platform. However, at this point, this is not as easy as pulling up a design and putting that onto a multi-project wafer run at your favorite photonics foundry. We highlight this challenge in Fig. 6. SBS systems require a number of different components to work. There is a gain medium with the required acousto-optic mode interactions, routing waveguides, active broadband gain elements, photodetectors, and often frequency shifting elements that rely on electro-optic modulators (Choudhary et al., 2017a). These functions are often produced on different material platforms, and so hybrid integration of multiple materials into a single and reliable photonic circuit architecture seem to be a requirement for future SBS chip device. Although the ambition of this task should not be understated, there are promising green shoots that point to this being a feasible approach. As an example, the generation of significant SBS gain in silicon nitride provides a CMOS-compatible platform for SBS. Such platforms have also been used for the hybrid integration through 'pick and place' type processes for active optical elements that can provide the basis for gain, modulation and photodetection. This is also not the only approach to achieving such hybrid integration of the necessary functions. Platforms such as silicon circuits loaded with soft-glass, thin-film lithium niobate, and the explosion of progress in 2D photonic materials are likely to provide multiple alternate routes to a hybrid integrated platform.

In summary, the interaction of SBS with optical communications is a tale of the good, the bad and the ugly. However, over time what each of those terms has meant has constantly changed, as physicists and engineers push the boundaries on what we can achieve with this mercurial nonlinear optical effect. The potential benefits of photonic systems based on SBS to optical communications systems are often significantly different to what might have been thought of only a decade ago. It remains to see if the good can outweigh the bad and the ugly to provide devices that have a positive impact globally on our increasingly connected society.

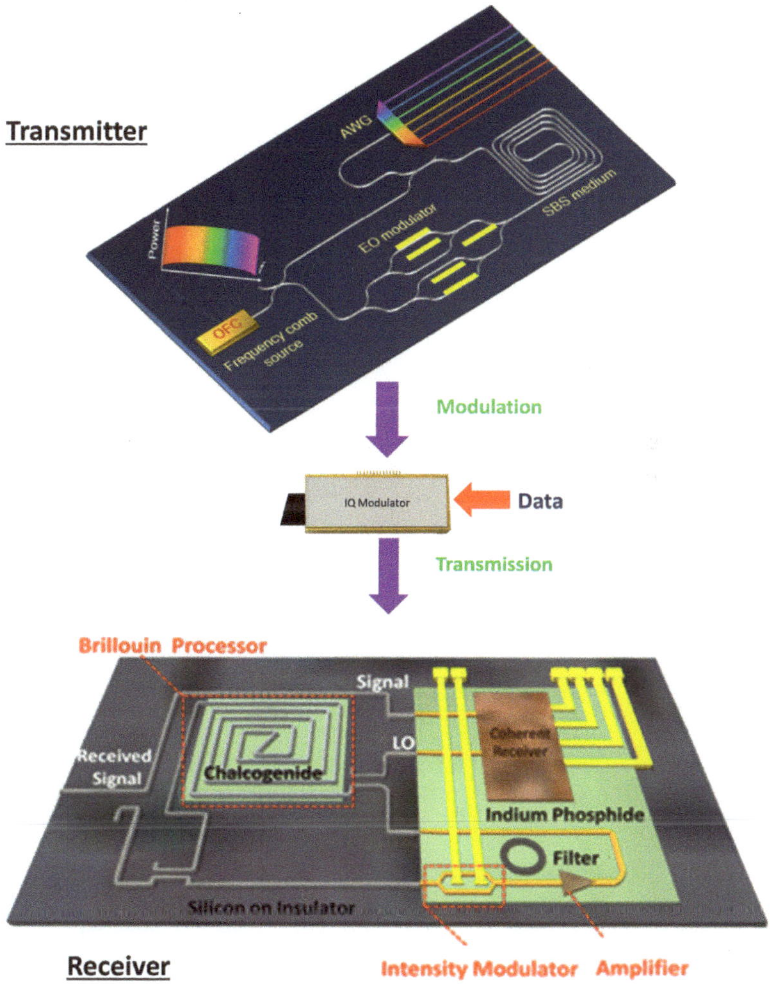

Fig. 6 An example schematic for a hybrid integrated photonic chip solution for SBS-based devices for optical communication systems. Amplification, electro-optic modulation, low-loss routing and a waveguide able to support stimulated Brillouin scattering are all functional blocks that need to be incorporated and might each be optimized using different material platforms. *Adapted from Giacoumidis, E., Choudhary, A., Magi, E., Marpaung, D., Vu, K., Ma, P., Choi, D.Y., Madden, S., Corcoran, B., Pelusi, M., Eggleton, B.J., 2018. Chip-based Brillouin processing for carrier recovery in self-coherent optical communications. Optica 5 (10), 1191–1199.*

References

Agrawal, G.P., 2002. Fiber-Optic Communication Systems, third ed. John Wiley & Sons.
Almaiman, A., Cao, Y., Ziyadi, M., Mohajerin-Ariaei, A., Liao, P., Bao, C., Alishahi, F., Fallahpour, A., Shamee, B., Ahmed, N., Willner, A.J., 2016. Experimental demonstration of phase-sensitive regeneration of a binary phase-shift keying channel without a phase-locked loop using Brillouin amplification. Opt. Lett. 41 (23), 5434–5437.
Aoki, Y., Tajima, K., Mito, I., 1988. Input power limits of single-mode optical fibers due to stimulated Brillouin scattering in optical communication systems. J. Light. Technol. 6 (5), 710–719.
Banchi, L., Presi, M., Proietti, R., Ciaramella, E., 2010. System feasibility of using stimulated Brillouin scattering in self coherent detection schemes. Opt. Exp. 18 (12), 12702–12707.
Brodnik, G.M., Harrington, M.W., Bose, D., Netherton, A.M., Zhang, W., Stern, L., Morton, P.A., Bowers, J.E., Papp, S.B., Blumenthal, D.J., 2020. Chip-scale, optical-frequency-stabilized PLL for DSP-free, low-power coherent QAM in the DCI. In: Optical Fiber Communication Conference (Optical Society of America), pp. M3A–6.
Brodnik, G.M., Harrington, M.W., Dallyn, J.H., Bose, D., Zhang, W., Stern, L., Morton, P.A., Behunin, R.O., Papp, S.B., Blumenthal, D.J., 2021. Optically synchronized fibre links using spectrally pure chip-scale lasers. Nat. Photonics 15 (8), 588–593.
Butler, D.L., Wey, J.S., Chbat, M.W., Burdge, G.L., Goldhar, J., 1995. Optical clock recovery from a data stream of an arbitrary bit rate by use of stimulated Brillouin scattering. Opt. Lett. 20 (6), 560–562.
Cabrera-Granado, E., Calderón, O.G., Melle, S., Gauthier, D.J., 2008. Observation of large 10-Gb/s SBS slow light delay with low distortion using an optimized gain profile. Opt. Exp. 16 (20), 16032–16042.
Cao, Y., Ziyadi, M., Almaiman, A., Mohajerin-Ariaei, A., Liao, P., Bao, C., Alishahi, F., Fallahpour, A., Shamee, B., Willner, A.J., Akasaka, Y., 2017. Pilot-tone-based self-homodyne detection using optical nonlinear wave mixing. Opt. Lett. 42 (9), 1840–1843.
Choudhary, A., Aryanfar, I., Shahnia, S., Morrison, B., Vu, K., Madden, S., Luther-Davies, B., Marpaung, D., Eggleton, B.J., 2016. Tailoring of the Brillouin gain for on-chip widely tunable and reconfigurable broadband microwave photonic filters. Opt. Lett. 41 (3), 436–439.
Choudhary, A., Giacoumidis, E., Pelusi, M., Magi, E., Marpaung, D., Inoue, T., Vu, K., Choi, D.Y., Ma, P., Madden, S., Corcoran, B., 2017a. On-chip Brillouin processing for coherent optical communications. In: 2017 22nd Microoptics Conference (MOC), Tokyo, Japan (IEEE), pp. 48–49. p. MOC17.
Choudhary, A., Morrison, B., Aryanfar, I., Shahnia, S., Pagani, M., Liu, Y., Vu, K., Madden, S., Marpaung, D., Eggleton, B.J., 2017b. Advanced integrated microwave signal processing with giant on-chip Brillouin gain. J. Light. Technol. 35 (4), 846–854.
Choudhary, A., Pelusi, M., Marpaung, D., Inoue, T., Vu, K., Ma, P., Choi, D.Y., Madden, S., Namiki, S., Eggleton, B.J., 2017c. On-chip Brillouin purification for frequency comb-based coherent optical communications. Opt. Lett. 42 (24), 5074–5077.
Chraplyvy, A.R., 1990. Limitations on lightwave communications imposed by optical-fiber nonlinearities. J. Light. Technol. 8 (10), 1548–1557.
Dahan, D., Mahlab, U., Levy, D., 2010. Stimulated Brillouin scattering based in-band OSNR monitoring technique for 40 Gbps and 100 Gbps optical transparent networks. Opt. Exp. 18 (15), 15769–15783.
Domingo, J.S., Pelayo, J., Villuendas, F., Heras, C.D., Pellejer, E., 2005. Very high resolution optical spectrometry by stimulated Brillouin scattering. IEEE Photon. Technol. Lett. 17 (4), 855–857.

Frisken, S., Baxter, G., Abakoumov, D., Zhou, H., Clarke, I., Poole, S., 2011. Flexible and grid-less wavelength selective switch using LCOS technology. In: Opt. Fiber Commun. Conference and Exposition/National Fiber Optic Engineers Conference, OSA Technical Digest (CD), p. OTuM3.

Fu, A., Pelusi, M.D., Eggleton, B.J., 2010, March. Simultaneous monitoring of in-band optical noise for WDM signals using stimulated Brillouin scattering. In: Optical Fiber Communication Conference (Optical Society of America), p. OWB2.

Gatti, D., Galzerano, G., Laporta, P., Longhi, S., Janner, D., Guglierame, A., Belmonte, M., 2008. Demonstration of differential phase-shift keying demodulation at 10 Gbit/s optimal fiber Bragg grating filters. Opt. Lett. 33 (13), 1512–1514.

Giacoumidis, E., Magi, E., Choudhary, A., Marpaung, D., Corcoran, B., Pelusi, M., Eggleton, B.J., 2017. Enhanced self-coherent optical OFDM using stimulated Brillouin scattering. In: Optical Fiber Communication Conference, Los Angeles (Optical Society of America), p. Th4I-2.

Giacoumidis, E., Choudhary, A., Magi, E., Marpaung, D., Vu, K., Ma, P., Choi, D.Y., Madden, S., Corcoran, B., Pelusi, M., Eggleton, B.J., 2018. Chip-based Brillouin processing for carrier recovery in self-coherent optical communications. Optica 5 (10), 1191–1199.

Gundavarapu, S., Brodnik, G.M., Puckett, M., Huffman, T., Bose, D., Behunin, R., Wu, J., Qiu, T., Pinho, C., Chauhan, N., Nohava, J., 2019. Sub-hertz fundamental linewidth photonic integrated Brillouin laser. Nat. Photonics 13 (1), 60–67.

Harrington, M.W., Brodnik, G.M., Briles, T.C., Stone, J.R., Streater, R.H., Papp, S.B., Blumenthal, D.J., 2020. Kerr soliton microcomb pumped by an integrated sbs laser for ultra-low linewidth WDM sources. In: 2020 Optical Fiber Communications Conference and Exhibition (OFC) (IEEE), p. T4G.6.

Hillerkuss, D., Schmogrow, R., Schellinger, T., Jordan, M., Winter, M., Huber, G., Vallaitis, T., Bonk, R., Kleinow, P., Frey, F., Roeger, M., 2011. 26 Tbits—1 line-rate super-channel transmission utilizing all-optical fast Fourier transform processing. Nat. Photonics 5 (6), 364–371.

Huang, C., Zheng, B., Zhang, N., Shu, C., 2016. Brillouin controlled phase matching in optical parametric processing of coherent signals. IEEE Photon. Technol. Lett. 28 (21), 2347–2350.

Huang, C., Zhang, N., Shu, C., 2017. SBS-enhanced FWM for polarization division multiplexed signals in coherent communication systems. Opt. Lett. 42 (21), 4271–4274.

Huang, Y., Li, Q., Han, J., Jia, Z., Yu, Y., Yang, Y., Xiao, J., Wu, J., Zhang, D., Huang, Y., Qin, W., 2019. Temporal soliton and optical frequency comb generation in a Brillouin laser cavity. Optica 6 (12), 1491–1497.

Ippen, E.P., Stolen, R.H., 1972. Stimulated Brillouin scattering in optical fibers. Appl. Phys. Lett. 21 (11), 539–541.

Iredale, T.B., Pelusi, M., Eggleton, B.J., 2007, June. Highly-sensitive all-optical in-band OSNR monitor using stimulated Brillouin scattering. In: COIN-ACOFT 2007-Joint International Conference on the Optical Internet and the 32nd Australian Conference on Optical Fibre Technology (IEEE), pp. 1–3.

Johnson, C., Demarest, K., Allen, C., Hui, R., Peddanarappagari, K.V., Zhu, B., 1999. Multiwavelength all-optical clock recovery. IEEE Photon. Technol. Lett. 11 (7), 895–897.

Keiser, G., 2000. Optical Fiber Communications. vol. 2 McGraw-Hill, New York.

Kumar, S., Deen, M.J., 2014. Fiber Optic Communications: Fundamentals and Applications. John Wiley & Sons.

Lee, Y.H., Wu, J., Kao, M.S., Tsao, H.W., 1991. Wavelength division and subcarrier system based on Brillouin amplification. In: Proc. SPIE (International Society for Optics and Photonics) Adv. Fiber Commun. Technol. vol. 1579, pp. 155–166.

Li, C., Merklein, M., Liu, Y., Choudhary, A., Eggleton, B.J., Corcoran, B., 2021. Effective linewidth reduction in self-homodyne coherent reception by stimulated Brillouin scattering-based optical carrier recovery. Opt. Exp. 29 (16), 25697–25708.

Liu, X., Hu, H., Chandrasekhar, S., Jopson, R.M., Gnauck, A.H., Dinu, M., Xie, C., Winzer, P.J., 2014. Generation of 1.024-Tb/s Nyquist-WDM phase-conjugated twin vector waves by a polarization-insensitive optical parametric amplifier for fiber-nonlinearity-tolerant transmission. Opt. Exp. 22 (6), 6478–6485.

Marazzi, L., Parolari, P., Brunero, M., Ferrario, M., Gatti, D., Longhi, S., Martinelli, M., 2011. Brillouin-based DPSK demodulation. In: CLEO: Science and Innovations. Optical Society of America, p. CTuX3.

Mecozzi, A., Antonelli, C., Shtaif, M., 2016. Kramers–Kronig coherent receiver. Optica 3 (11), 1220–1227.

Montes, C., Rubenchik, A.M., 1992. Stimulated Brillouin scattering from trains of solitons in optical fibers: information degradation. J. Opt. Soc. Am. B 9 (10), 1857–1875.

Pelusi, M., Choudhary, A., Inoue, T., Marpaung, D., Eggleton, B., Tan, H.N., Solis-Trapala, K., Namiki, S., 2016. Low noise, regeneration of optical frequency comb-lines for 64QAM enabled by SBS gain. In: 2016 21st OptoElectronics and Communications Conference (OECC) held jointly with 2016 International Conference on Photonics in Switching (PS), Niigata, Japan (IEEE), p. PD1-3.

Pelusi, M., Choudhary, A., Inoue, T., Marpaung, D., Eggleton, B.J., Solis-Trapala, K., Tan, H.N., Namiki, S., 2017. Low noise frequency comb carriers for 64-QAM via a Brillouin comb amplifier. Opt. Exp. 25 (15), 17847–17863.

Pelusi, M., Inoue, T., Namiki, S., 2019. Enhanced carrier to noise ratio by Brillouin amplification for optical communications. J. Light. Technol. 38 (2), 319–331.

Porzi, C., Chin, S., Trita, A., Fresi, F., Berrettini, G., Mezosi, G., Ghelfi, P., Giuliani, G., Potì, L., Sorel, M., Thévenaz, L., 2013. Application of brillouin-based continuously tunable optical delay line to contention resolution between asynchronous optical packets. J. Light. Technol. 31 (17), 2888–2896.

Shannon, C.E., 1948. A mathematical theory of communication. Bell Syst. Tech. J. 27 (3), 379–423.

Smith, R.G., 1972. Optical power handling capacity of low loss optical fibers as determined by stimulated Raman and Brillouin scattering. Appl. Optics 11 (11), 2489–2494.

Smith, D.W., Atkins, C.G., Cotter, D., Wyatt, R., 1986. Application of Brillouin amplification in coherent optical transmission. In: Opt. Fiber Commun. Conference (Optical Society of America), p. WE3.

Smith, S.P., Zarinetchi, F., Ezekiel, S., 1991. Narrow-linewidth stimulated Brillouin fiber laser and applications. Opt. Lett. 16 (6), 393–395.

Temprana, E., Myslivets, E., Kuo, B.P., Liu, L., Ataie, V., Alic, N., Radic, S., 2015. Overcoming Kerr-induced capacity limit in optical fiber transmission. Science 348 (6242), 1445–1448.

Tkach, R.W., Chraplyvy, A.R., Derosier, R.M., 1989. Performance of a WDM network based on stimulated Brillouin scattering. IEEE Photon. Technol. Lett. 1 (5), 111–113.

Van Deventer, M.O., Boot, A.J., 1994. Polarization properties of stimulated Brillouin scattering in single-mode fibers. J. Light. Technol. 12 (4), 585–590.

Wei, W., Yi, L., Jaouën, Y., Morvan, M., Hu, W., 2015. Ultra-selective flexible add and drop multiplexer using rectangular optical filters based on stimulated Brillouin scattering. Opt. Exp. 23 (15), 19010–19021.

Yi, L., Zhang, T., Li, Z., Zhou, J., Dong, Y., Hu, W., 2013. Secure optical communication using stimulated Brillouin scattering in optical fiber. Opt. Commun. 290, 146–151.

Zadok, A., Eyal, A., Tur, M., 2011. Stimulated Brillouin scattering slow light in optical fibers. Appl. Optics 50 (25), E38–E49.

Zarifi, A., Merklein, M., Liu, Y., Choudhary, A., Eggleton, B.J., Corcoran, B., 2021. Wide-range optical carrier recovery via broadened Brillouin filters. Opt. Lett. 46 (2), 166–169.

Zhang, Y., Yi, L., Zhang, T., Li, Z., Hu, W., 2013. Mutual optical format conversion between on-off keying and binary phase-shift keying based on stimulated brillouin scattering. Chin. Opt. Lett. 11 (10), 100601.

Zhang, Q., Zheng, B., Shu, C., 2018. Kramers–Kronig detection with Brillouin-amplified virtual carrier. Opt. Lett. 43 (6), 1367–1370.

Zhou, X., Shalaby, H.H., Chao, L., Cheng, T.H., Ye, P., 2000. A performance analysis of all-optical clock extraction circuit based on stimulated Brillouin scattering. J. Light. Technol. 18 (10), 1453.

CHAPTER TWELVE

Slow light, dynamic gratings, and light storage

Birgit Stiller[a,*], Herbert Winful[b], Robert Boyd[c,d], and Moritz Merklein[e]
[a]Max Planck Institute for the Science of Light, Erlangen, Germany
[b]University of Michigan, Ann Arbor, MI, United States
[c]University of Ottawa, Ottawa, ON, Canada
[d]University of Rochester, Rochester, NY, United States
[e]School of Physics, The University of Sydney, Sydney, NSW, Australia
[*]Corresponding author: e-mail address: birgit.stiller@mpl.mpg.de

Contents

1. Introduction

In addition to sensing, lasing, and microwave photonics, stimulated Brillouin scattering (SBS) is also promising for applications in delaying, storing, and processing optical information. In this chapter, we explain and compare several approaches to use SBS for delaying and storing light. We differentiate the following approaches: slow light based on SBS, Brillouin dynamic gratings (BDGs), quasi-light storage, and light storage. We introduce each of the concepts, elaborate on advantages and limitations, and

Semiconductors and Semimetals, Volume 110
ISSN 0080-8784
https://doi.org/10.1016/bs.semsem.2022.06.001

review the advances achieved in the past 30 years. Parts of this chapter are based on a recent review paper on Brillouin-based delay techniques (Merklein et al., 2018).

The typical application field of signal processing and information delay is optical data transmission; despite the high-speed data transmission, there is a critical need to delay or buffer optical signals. Consecutively this permits signal processing or switching and rerouting, which takes place in the nodes of optical networks. The buffering is ideally an all-optical process as it avoids electro-optical conversion and maintains a large bandwidth. However, changing the speed of light is not only of technological relevance but also of fundamental interest, for example, to enhance nonlinear effects (Corcoran et al., 2009; Li et al., 2011a; Merklein et al., 2015a; Monat et al., 2010).

The third-order nonlinear effect of SBS has been the basis for several concepts of all-optical data delay. An overview on four different Brillouin-based delay schemes is given in Fig. 1 (Merklein et al., 2018). The underlying principle of how the optical information delay is achieved, despite all being based on SBS, is different in all four schemes:

- The signal delay can originate from a change in the group index n_g (slow light).
- The delay can be caused by the time of flight between the fiber input and a movable reflector in the fiber (BDG approach).
- The signal delay can be achieved by a convolution of the signal with a frequency comb via SBS that copies the signal (quasi-light storage).
- The signal is delayed by resonantly transferring the optical signal to an acoustic wave that travels very slowly in the waveguide (light storage).

Fig. 1A shows the operation principle of SBS slow (and fast) light. In more general terms, slow light schemes can be categorized into two distinct groups: material slow-light and structural slow-light (Boyd, 2012; Monat et al., 2010). Examples of material slow light are electromagnetically induced transparency (EIT) in atoms (Boyd and Gauthier, 2009) or SBS-induced slow light (Okawachi et al., 2005; Song et al., 2005b), while Bragg gratings (Mok et al., 2006) and photonic crystals (Baba, 2008; Krauss, 2007, 2008) fall in the category of structural slow light. SBS slow light can be regarded as a form of material slow light in most cases as it relies on the material-specific acoustic resonance. It has a lot of compelling properties to delay optical signals with the most important points being: room temperature operation, implementation only relies on telecommunication components and can be integrated on a chip, the delay is tunable via the optical

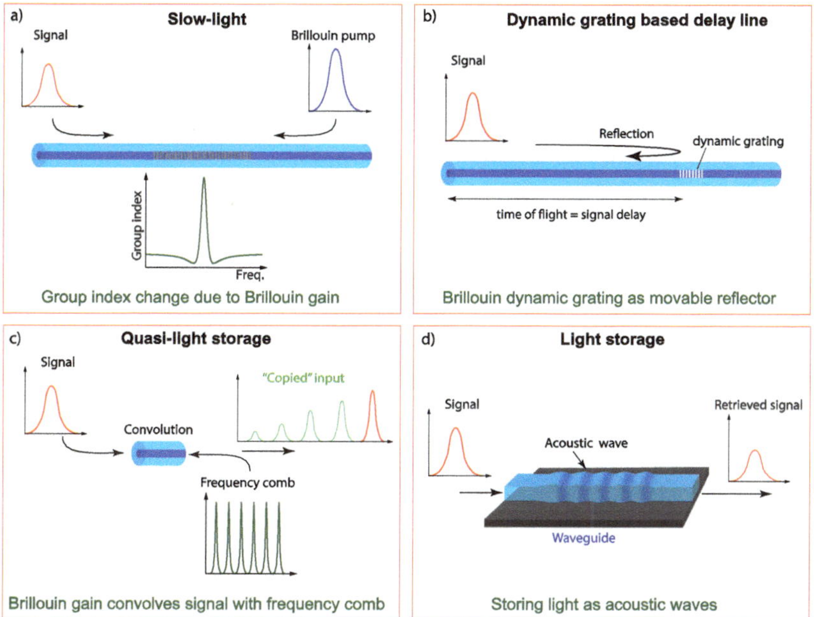

Fig. 1 (A) Signal delay due to a change in the group index caused by SBS slow-light. (B) Tunable delay by reflecting a signal from a movable dynamic grating. (C) Convolution of a signal with a frequency comb generates copies of the original signal. (D) Transferring optical data to an acoustic wave and retrieving it afterwards causes a delay due to the five orders of magnitude difference between the speed of sound and the speed of light. *From Merklein, M., Stiller, B., Eggleton, B.J., 2018. Brillouin-based light storage and delay techniques. J. Opt. (United Kingdom) 20 (8), 083003. https://doi.org/10.1088/2040-8986/aad081.*

pump power and works at all wavelength of the transparency window of the guiding structure (which distinguishes it from most other material slow-light implementations that rely on the resonance frequency of the atoms that are used) (Merklein et al., 2018). This flexibility allows SBS slow light to be implemented in chip-scale waveguides (Pant et al., 2012) and resonators (Kim et al., 2015). As the achievable delay is proportional to the SBS gain, tailored high Brillouin gain waveguides improve not only the form factor but also the overall performance. The principle in very short is as follows: An optical signal is launched into an optical fiber or waveguide and is counterpropagated by an optical pulse. If the signal frequency detuning coincides with the Brillouin frequency shift (BFS) Ω_B, the signal pulse experiences a delay due to the induced change in the group index. It will be explained in more detailed in Section 2 of this chapter.

Even with all beneficial properties that SBS slow light provides, this field of research faces one major challenge that so far could not be overcome, that is, the limitation in fractional delay imposed by the time-bandwidth product. The fractional delay is defined as the time delay normalized by the pulse duration, for example, a fractional delay of one means that a pulse is delayed by one pulse width. The time-bandwidth product sets a limit to the achievable fractional delay, meaning with an increase in bandwidth of the signal, the delay time decreases. Therefore, different other delay schemes based on SBS, known as Brillouin dynamic grating (BDG)-based delay lines (Chin and Thévenaz, 2012; Song et al., 2009a), quasi-light storage (Preussler et al., 2009) and light storage (Dong et al., 2015; Merklein et al., 2017; Zhu et al., 2007b), were put forward to overcome these limitations while maintaining all the other previously listed advantages of an SBS-based delay scheme. The schematic of a tunable delay line based on BDGs is shown in Fig. 1B. The BDG-based scheme achieves a tunable delay by writing a dynamic grating via SBS at different positions in an optical fiber and reflecting a signal from this grating structure. In a sense it is the fiber analog to a movable reflector in bulk optics and the delay is given by the time of flight of the signal from the fiber input to the BDG reflector and back. The position of the BDG can be dynamically changed by altering the pumps that create it. In this way, a tunable delay is achieved where the delay time is proportional to the twice the time of flight from the fiber input to the BDG.

SBS quasi-light storage relies on the frequency-time coherence and uses SBS to multiply an optical signal with a frequency comb, generating a large number of copies of the original optical signal. The basic operation principle of quasi-light storage is illustrated in Fig. 1C and relies on the amplification that the narrow bandwidth Brillouin scattering effect provides. An optical data pulse is counter-propagated by an optical pump, which is a frequency comb. By combining the spectrum of the input pulse with the counter-propagating frequency comb via SBS, a pulse train in the time domain is generated that consists of copies of the original data pulse. A fast optical switch is needed to select the copy of the data pulse that has the desired delay relative to the original signal.

In the Brillouin light storage scheme, the optical data are delayed by being transferred to an acoustic wave that propagates 5 orders of magnitude slower than light (Fig. 1D). The transfer of the data is achieved by a counter-propagating short write pulse that is frequency offset by the BFS Ω_B. The lower frequency write pulse completely depletes the optical data pulse while creating an acoustic wave in the optical fiber. A time delayed read pulse restores the optical data pulse in a readout process while the acoustic wave gets depleted. SBS light storage was first shown in a highly nonlinear fiber in 2007 (Zhu et al., 2007b). However, the high power requirement of more than 100 W was a limiting factor and the bandwidth of the signal that can be stored appeared to be limited due to the narrow bandwidth of the SBS response. Implementations in a silica resonator greatly reduced the required power and significantly increased the storage time (Dong et al., 2015) because of the high optical quality factor, however, limited the bandwidth even further. In 2015 Brillouin light storage was shown in an integrated circuit (Merklein et al., 2015b). These engineered high-gain waveguides not only allowed to greatly reduce the power requirements but also allowed broadening of the SBS response to more than a GHz (Merklein et al., 2017). The power and flexibility of Brillouin-based light storage in high-gain waveguides were further illustrated by showing the coherence of the process and the multi-wavelength capacity (Merklein et al., 2017).

A timeline of the seminal demonstrations of the SBS-based delay schemes is shown in Fig. 2. It also shows the implementation of the schemes in different platforms following the initial fiber-based demonstrations, such as on-chip implementation and resonators. The introduction of new platforms not only reduced the footprint but also increased the performance of the schemes in terms of power consumption, delay, and bandwidth. In this sense, the time axis also represents an overall performance increase within the respective schemes.

In the following sections we will dive in more detail into the different Brillouin-based schemes.

2. SBS slow light

Slow light has attracted the interest of many research groups working in different fields of optical sciences and engineering (Boyd and Gauthier, 2009; Krauss, 2008). In a wider context, there are approaches based on photonic crystals (Baba, 2008; Krauss, 2007, 2008), Bragg gratings (Mok et al., 2006), atomic clouds (Boyd and Gauthier, 2009), Bose–Einstein condensates (Hau et al., 1999), and optomechanical systems (Safavi-Naeini et al., 2011; Weis et al., 2010). All these systems rely on the fact that an optical resonant structure is accompanied with a change in the group index n_g. This index change is a direct consequence of the Kramers–Kronig relation that links the real and imaginary part of the resonant refractive index response (Lenz et al., 2001; Mok and Eggleton, 2005). Whereas all these

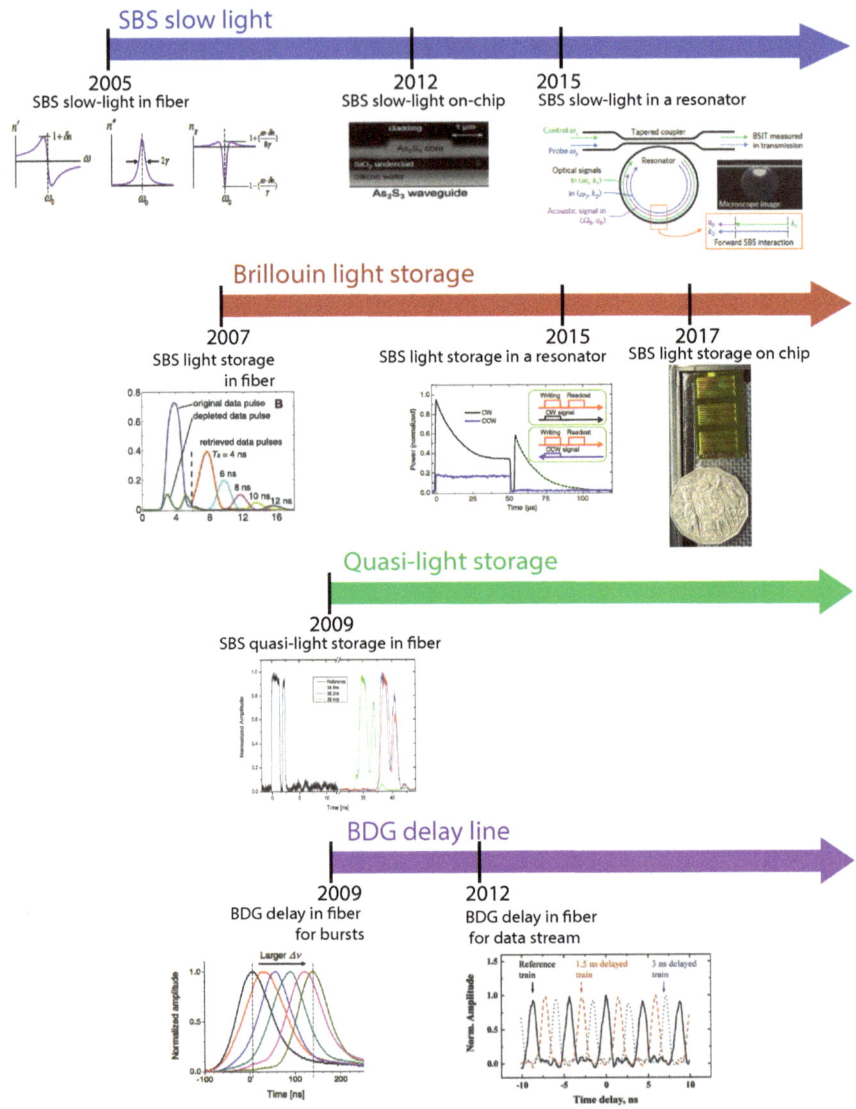

Fig. 2 Timeline of the development of SBS-based delay schemes. The first demonstration of SBS slow light in fiber initiated the field in 2005 (Okawachi et al., 2005; Song et al., 2005b). SBS-based light storage and quasi-light storage were introduced in 2007 (Zhu et al., 2007b) and 2009 (Preussler et al., 2009), respectively, to overcome the shortcomings in the achievable fractional delay in SBS-based slow light. Also in 2009 BDG-based tunable delay was shown (Song et al., 2009a). The timeline furthermore illustrates the introduction of new platforms to the individual schemes which increased the performance and capacity of SBS-based delay schemes (Chin and Thévenaz, 2012; Dong et al., 2015; Kim et al., 2015; Merklein et al., 2017; Pant et al., 2012). *From Merklein, M., Stiller, B., Eggleton, B.J., 2018. Brillouin-based light storage and delay techniques. J. Opt. (United Kingdom) 20 (8), 083003. https://doi.org/10.1088/2040-8986/aad081.*

different slow-light schemes are reducing the group velocity v_g of a light signal, they can be categorized into two distinct groups: material slow light (e.g., EIT in atoms or SBS-induced slow light) and structural slow light (e.g., Bragg gratings and photonic crystals; Boyd, 2012; Monat et al., 2010).

Structural slow-light schemes can be used to enhance nonlinear effects (Corcoran et al., 2009; Li et al., 2011a; Merklein et al., 2015a), which does not occur in the case of material slow light (Boyd, 2012; Monat et al., 2010). This enhancement effect of nonlinear effects can be assigned to an enhanced signal path length and a build-up of the optical fields in the slow-light region of the slow-light structure. An overview of slow light-enhanced nonlinear effects is given in Monat et al. (2010). Here, we will focus specifically on SBS-based slow-light schemes.

Slow light describes a rapid change of the group index n_g in the vicinity of a resonance and as SBS is a narrow-band (MHz range) optical resonance, it can be used to slow down light pulses.[a] In the following we outline the formalism for describing SBS-based slow light, which results in a change of the group index in the vicinity of the Stokes resonance.

The description starts from the coupled-mode equations (Agrawal, 2001; Boyd, 2003) and we focus on the equation for the Stokes wave as we want to investigate the effect of the gain resonance on the group index. The coupled mode equation of the Stokes wave has the following form (a detailed description of the coupled mode equations can be found in Agrawal (2001):

$$-\frac{\partial A_S}{\partial z} + \frac{n}{c}\frac{\partial A_S}{\partial t} = \frac{g_B}{2A_{\text{eff}}^{\text{ao}}} A_P Q^*. \quad (1)$$

Here A_S and A_P are the slowly varying envelope of the Stokes and pump fields, respectively. To calculate the acoustic mode envelope Q^* we assume the steady state for the acoustic mode. The following derivation follows the description given in Agrawal (2001). Here we do not assume that the acoustic and optical modes are perfectly phase-matched but instead introduce a complex detuning term $i(\Omega_B - \Omega)$. Inserting the steady-state solution for Q^* (see Agrawal, 2001) in Eq. (1) yields:

[a] Note that in fact SBS consists of two resonances, the SBS gain resonance and the loss resonance. In the loss resonance case the signal is upshifted in frequency by the BFS Ω_B and hence energy is transferred from the signal to the pump and hence SBS can also be used for fast light generation. However, in this paper we only focus on SBS slow light.

$$-\frac{\partial A_S}{\partial z}+\frac{n}{c}\frac{\partial A_S}{\partial t}=\frac{g_B}{2A_{\text{eff}}^{\text{ao}}}A_S|A_P|^2\left(\frac{1}{1+i\delta}\right). \tag{2}$$

Here the fractional detuning $\delta = 2(\Omega_B - \Omega)/\Gamma_B$ was introduced. Furthermore we can identify $|A_P|^2$ with the power of the pump wave P_p. This allows us to summarize the right-hand side of Eq. (1) using a complex gain coefficient $g_c(\delta)$:

$$-\frac{\partial A_S}{\partial z}+\frac{n}{c}\frac{\partial A_S}{\partial t}=g_c(\delta)A_S. \tag{3}$$

The complex gain coefficient $g_c(\delta)$ is given by:

$$g_c(\delta)=\frac{g_B P_P}{A_{\text{eff}}^{\text{ao}}}\left(\frac{1}{1+i\delta}\right). \tag{4}$$

The real part of the complex gain $g_c(\delta)$ gives the canonical Brillouin gain coefficient $g_B(\Omega)$. The imaginary part, however, is responsible for a Brillouin-induced change in the refractive index Δn_B, which is given by (Agrawal, 2001):

$$\Delta n_B=\frac{c}{2\omega_S}\text{Im}(g_c)=\frac{c}{2\omega_S}\frac{g_B\,P_P}{A_{\text{eff}}^{\text{ao}}}\left(\frac{\delta}{1+\delta^2}\right), \tag{5}$$

with the Stokes frequency ω_S. This change in the index follows directly from the Kramers–Kronig relations. Even though the absolute changes in refractive index are usually small ($\approx 10^{-8}$), the group index still can change significantly as the refractive index change occurs over a narrow frequency bandwidth $\Delta\omega$. This index change is about two orders of magnitude larger in chalcogenide waveguides due to the larger Brillouin gain coefficient g_B and the smaller effective acousto-optic mode area $A_{\text{eff}}^{\text{ao}}$.

The SBS-induced index change in an optical fiber (standard SMF) and a chalcogenide waveguide in the vicinity of the Stokes resonance is shown in Fig. 3A and C for different pump powers at a wavelength of 1.55 μm (the parameters of the photonic waveguide can be found in Pant et al. (2011)).

The induced index change can be translated into a change of the group index n_g in the presence of SBS, which leads to the following expression:

$$n_g=n+\frac{c\,g_B}{\Gamma_B\,A_{\text{eff}}^{\text{ao}}}\frac{1-\delta^2}{(1+\delta^2)^2}\cdot P_P. \tag{6}$$

The SBS-induced group index change in an SMF and a chalcogenide waveguide is shown in Fig. 3B and D, respectively. One can see that the achieved

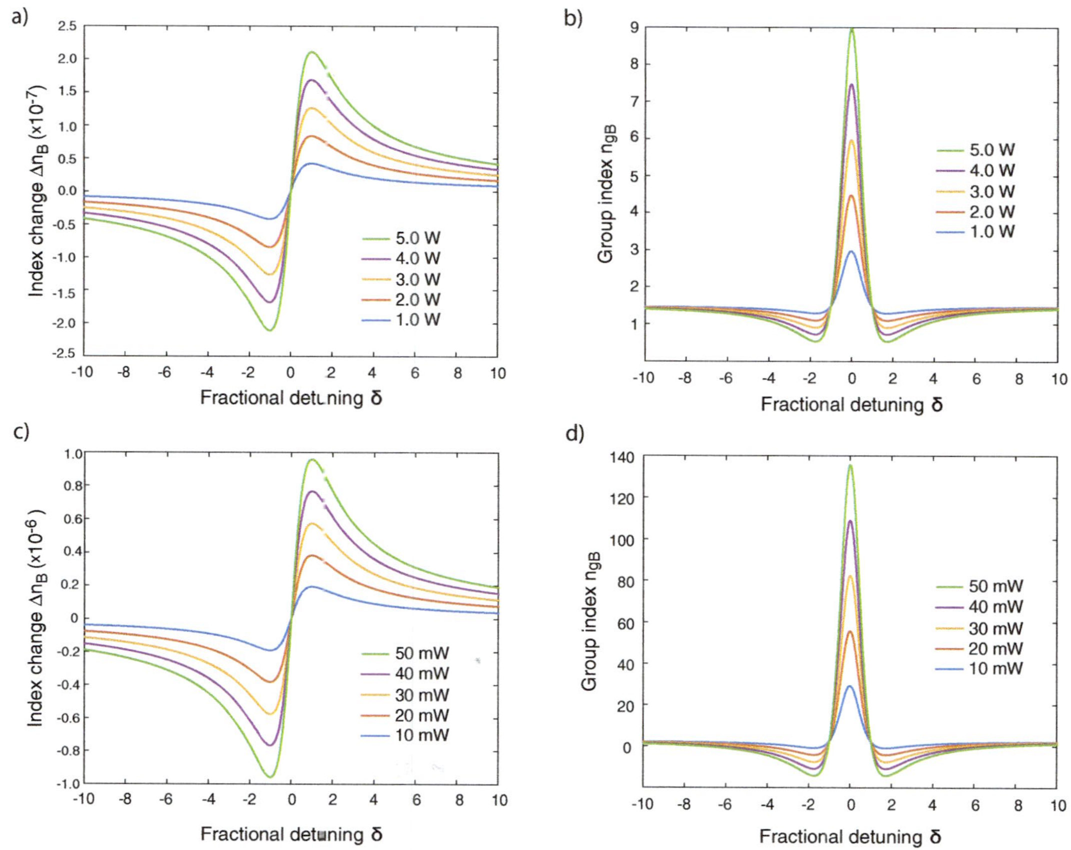

Fig. 3 (A) Refractive index change induced by the Stokes resonance in a standard single mode fiber for different pump powers. (B) Corresponding group index change induced by SBS. (C) Refractive index change induced in a high Brillouin gain chalcogenide waveguide and (D) corresponding group index change. *Taken from Merklein, M., Stiller, B., Eggleton, B.J., 2018. Brillouin-based light storage and delay techniques. J. Opt. (United Kingdom) 20 (8), 083003. https://doi.org/10.1088/2040-8986/aad081.*

group index change in chalcogenide waveguides is much larger than in fiber, even for a much lower input power. This large effect is a consequence of the orders of magnitude higher gain. However, the fundamental challenges SBS slow light is facing remain, that is, limited fractional delay and pulse distortions.

2.1 Experimental implementations

SBS slow light in optical fiber was first experimentally demonstrated in 2005 in two seminal papers published in close succession (Okawachi et al., 2005; Song et al., 2005b). In the work by Song et al. (2005b), the dependence of the delay on the Brillouin gain/pump power was shown (see Fig. 4A). SBS fast light was also shown for the first time in this paper. At the same time, the work from Okawachi et al. (2005) showed the delay of a 15-ns pulse using SBS slow-light in optical fiber (see Fig. 4B).

The concept of SBS slow light was later implemented in on-chip waveguides (Pant et al., 2012). In this demonstration, large changes in the group index Δn_g were achieved due to the orders of magnitude larger Brillouin gain compared to standard silica fibers (see Fig. 5A). However, even though large changes in the group index could be shown, the fractional delay (the number of pulse widths the pulse can be delayed) was still limited to only around one pulse width (Thévenaz, 2008).

Slow light could also be observed in a resonator using a scheme called Brillouin induced transparency (Kim et al., 2015) (see Fig. 5). In this demonstration, a very large absolute delay of 110 μs was achieved; however, the fractional delay was still limited to around two.

In the following years, many attempts were pursued to increase the fractional delay, extend the bandwidth, and to minimize distortions in SBS

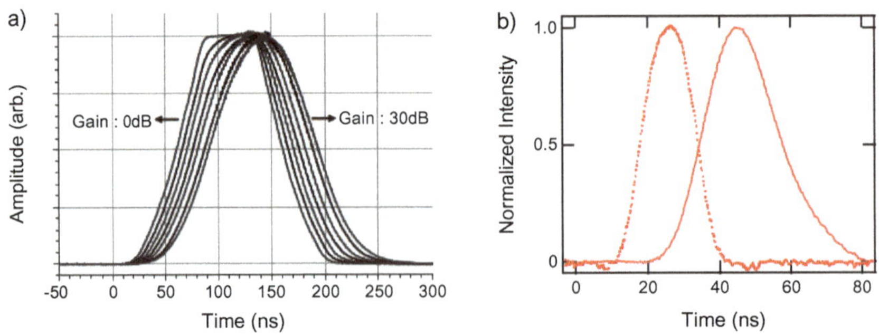

Fig. 4 (A) Tunable delay of a 100-ns pulse for different Brillouin gain values (Song et al., 2005b). (B) Pulse delay of a 15-ns Stokes pulse (Okawachi et al., 2005).

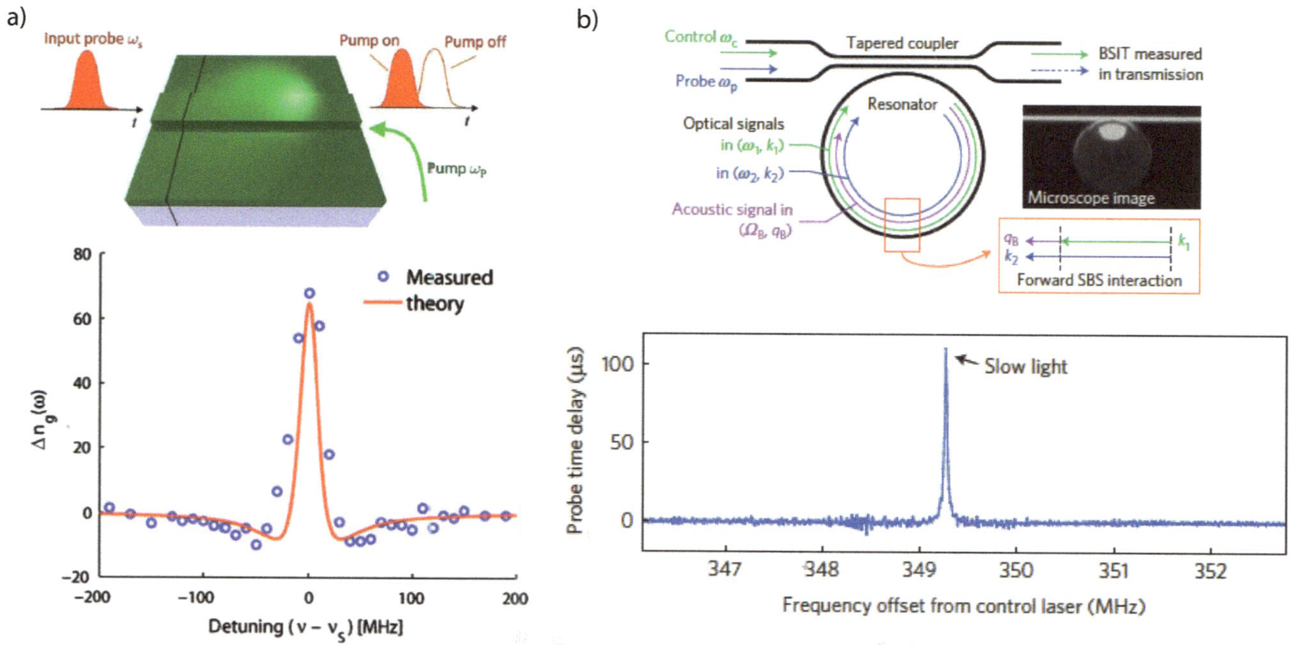

Fig. 5 (A) Change of the group index of an on-chip SBS slow-light implementation (Pant et al., 2012). (B) Delay generated via Brillouin induced transparency in a microresonator (Kim et al., 2015). *Panel A: Reprinted with permission from Pant, R., Byrnes, A., Poulton, C. G., Li, E., Choi, D.-Y., Madden, S., Luther-Davies, B., Eggleton, B. J., 2012. Photonic-chip-based tunable slow and fast light via stimulated Brillouin scattering. Opt. Lett. 37 (5), 969. https://doi.org/10.1364/OL.37.000969. http://ol.osa.org/abstract.cfm?URI=ol-37-5-969, OSA.*

slow-light schemes (González Herráez et al., 2006; Pant et al., 2008; Schneider et al., 2006a, b, 2007; Yi et al., 2007; Zadok et al., 2006; Zhu et al., 2007a). One great advantage of SBS slow light in optical fibers or integrated waveguides is that the resonance that generates the slow-light behavior can be altered and tailored by the profile of the pump laser as opposed to a resonance given by the material or the physical structure. The first demonstration of altering the SBS gain profile relied on two optical pumps in close vicinity to each other (Song et al., 2005a). Using a pump that included two frequencies provided a great increase in control of the dispersion of the generated SBS resonance and even allowed switching between SBS slow and fast light (Song et al., 2005a).

Following these seminal demonstrations, different forms of tailoring the pump profile emerged that can be categorized in the following way: using multiple discrete frequency pumps or continuously broadening a single pump profile. In the first case, best slow-light performance was achieved using three separate tones (Minardo et al., 2006; Shi et al., 2007). Continuously broadening the pump spectrum allowed slowing down of light signals with much larger bandwidth (exceeding 10 GHz), compared to the narrowband intrinsic Brillouin linewidth (González Herráez et al., 2006; Pant et al., 2008; Schneider et al., 2006a; Yi et al., 2007; Zadok et al., 2006; Zhu et al., 2007a).

However, broadening the SBS gain to delay broadband signals did not come without an additional cost; the required pump power to maintain the same normalized delay increases proportionally to the broadening. If one wants to maintain the same absolute delay, the required pump power increases quadratically, as the achievable delay is not only proportional to the peak power but also proportional to the bandwidth of the resonance. Hence to achieve the same fractional delay for a 10-GHz broadband signal compared to a signal narrower than the intrinsic Brillouin linewidth, 350 times higher pump power is required (Thévenaz, 2008).

Further efforts were undertaken to increase the fractional delay in SBS slow-light schemes which have been limited to a few pulse widths even with the great achievements in broadening the spectral response. Cascading several slow-light segments allowed increasing the fractional delay (Song et al., 2005c), however with the cost of system complexity and only a small improvement in overall delay.

With all the aforementioned challenges and proposed solutions, it was shown by Khurgin (2006) that a delay scheme based on optical amplification, such as SBS slow light, still is only able to delay a series of maximum five bits. The number five follows from realistic power budget considerations in a fiber-based slow-light setup and is a consequence of the fact that the required gain in such a delay scheme increases in a nonlinear way with the number of delayed bits (Khurgin, 2006).

Besides power consideration another major obstacle for SBS slow light are distortions of the optical pulse (González-Herráez et al., 2012; Zhang et al., 2014). An analytical expression was found by González-Herráez et al. (2012), showing that in a linear slow-light system pulse distortions are inevitable. On the one hand, there is a broadening of the pulse amplitude due to the SBS acting as a low pass filter for the pulse bandwidth. A reduction of the bandwidth of the pulse corresponds to a broadening of the pulse in the time domain. On the other hand, there are phase distortions due to the strong dispersion caused by the SBS resonance. As the amplitude response is purely real and the phase response is purely imaginary both effects cannot cancel each other (González-Herráez et al., 2012). Compensating the phase distortion using second-order chromatic dispersion in a fiber is also not very promising as the phase distortion only includes odd terms (González-Herráez et al., 2012). One way, however, to compensate for distortions are nonlinear slow-light systems (Chin et al., 2009; Mok et al., 2006), which have their own challenges as their transfer function depends on the power and shape of the input pulse.

3. Brillouin dynamic gratings

BDGs are moving periodic structures created in dielectric media through the photoelastic effect induced by two counterpropagating light waves detuned by the acoustic frequency. The grating travels at the sound velocity which is five orders of magnitude slower than that of light and hence the grating can be considered quasistationary with respect to the inducing light waves. The grating scatters the forward wave backward with a frequency downshift equal to the acoustic frequency, a shift understandable as the Doppler shift due to reflection from a receding mirror. The term Brillouin dynamic grating (BDG) was coined by Song who first demonstrated the existence of such a grating in SBS (Song et al., 2008). Since then numerous applications have been found for these gratings (Chin and Thévenaz, 2012; Chin et al., 2010a; Dong et al., 2010; Song, 2011; Song and Yoon, 2010; Song et al., 2008, 2009a; Zou and Chen, 2013). They include reconfigurable delay lines, high spatial resolution sensors, optical signal processing, and optical spectrometry. In this section we review the physics of BDG formation and their application to reconfigurable delay lines.

3.1 Physics of BDG formation and readout

As shown in Fig. 6, a forward pump wave at frequency ν_1 interacts with a backward Stokes wave at frequency ν_2 through an acoustic wave of frequency $\nu_B = \nu_1 - \nu_2$ generated through electrostriction which then modulates the refractive index through the photoelastic effect. Phase matching requires that the wave vectors of the three waves satisfy $k_B = k_1 - k_2$, which yields the relation

$$\nu_B = \frac{2nV}{\lambda} \tag{7}$$

where V is the acoustic velocity, n is the refractive index, and λ is the free-space wavelength of the pump wave. In a birefringent fiber there are two eigenmodes of the light wave characterized by orthogonal polarizations and different refractive indices n_x (slow) and n_y (fast). The acoustic wave is a longitudinal wave that can couple to light of any polarization. For a given acoustic frequency and velocity, it is possible to achieve phase matching with different optical wavelength and refractive index combinations. For two light waves polarized along the fast and slow axes, the phase matching requirement yields

$$n_x\nu_x = n_y\nu_y, \tag{8}$$

where ν_x and ν_y are the optical frequencies that satisfy the same phase matching condition for the same BFS. A grating created with pump and

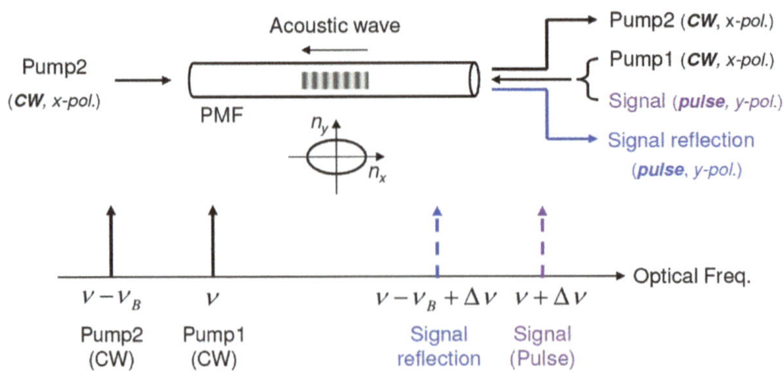

Fig. 6 Fiber Bragg grating formation and readout from Song et al. (2009b).

Stokes polarized along the x-axis and detuned by ν_B can nevertheless reflect a y-polarized wave detuned by $\nu_B + \nu_y$. The fiber birefringence Δn determines the frequency difference $\Delta\nu$ between the writing and probe beams:

$$\Delta\nu = \frac{\Delta b}{n}\nu = \frac{\Delta\nu_B}{\nu_B}\nu \tag{9}$$

Fig. 6 illustrates the relationship between the pump, Stokes, probe, and reflected signal frequencies.

3.2 Reconfigurable delay lines

A BDG can perform any of the functions that can be realized with static fiber Bragg gratings. The one advantage they have over FBGs is that they are reconfigurable. A BDG localized in an optical fiber acts therefore like a lumped mirror and can be used as an optical delay line. In that application, counterpropagating pump beams polarized along one axis of a polarization maintaining fiber (PMF) create a BDG. In order to achieve spatial localization, there are a number of approaches that have been demonstrated. The first approach, one that works even with CW beams, is to apply a strain gradient to the fiber. The strain distribution induces a change in the BFS, which varies along the fiber. A BDG is formed at the location in the fiber where the BFS equals the frequency difference between forward and backward pump waves. The advantage of this method, when used with CW pumps, is that the grating is continuously pumped and does not decay while it is being probed. The probe signal is polarized along the orthogonal axis of the PMF. The signal delay is just the round trip time to the BDG and back. To reconfigure the delay line, one changes the frequency difference between the pump waves which causes the grating to be formed at a different position where the BFS equals that difference. In one experiment, a pulse of width 82 ns was delayed by 132 ns, a fractional delay of 1.6 (Song et al., 2009a).

Another way to localize BDGs is to use short pump pulses such that a grating is formed only where the counterpropagating pulses overlap in the fiber (Chin and Thévenaz, 2012). The location of the grating is then tuned by varying the delay between the forward and backward pump pulses so that they meet at different points within the fiber. Fig. 7A illustrates the principle of this localization scheme and Fig. 7B shows experimental delay measurements in which 600-ps pulses were delayed up to 1.15 μs (Chin and Thévenaz, 2012).

Gratings formed this way are truly transient as they only formed during the spatiotemporal overlap and decay within the acoustic lifetime. The signal pulse must arrive within the acoustic lifetime of about 10 ns (silica). This also

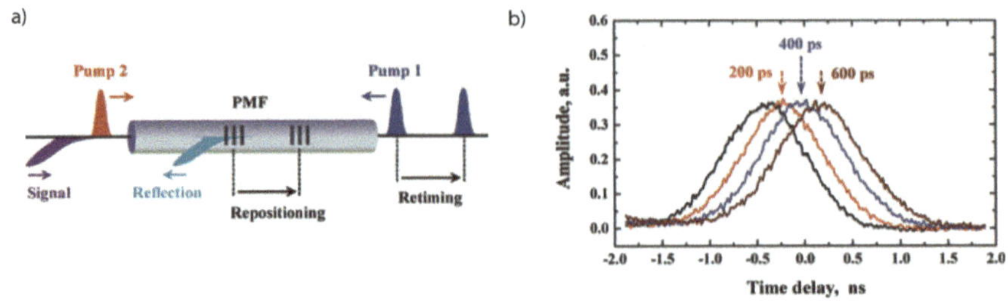

Fig. 7 (A) Principle of BDG-based reconfigurable delay line in polarization maintaining fiber (Chin and Thévenaz, 2012). (B) Measured delayed pulses from BDGs formed at different locations in the fiber (Chin and Thévenaz, 2012).

means that one cannot delay signals consisting of data streams. One way to overcome this limitation is to use pulse trains as pumps with a period shorter than the acoustic lifetime (Chin and Thévenaz, 2012). This way the grating is refreshed by successive pulses in the train. The fiber should be short enough that only one set of pump pulses meets in the fiber. With use of this approach, 840-ps pulses were delayed by up to 3.3 ns, a fractional delay of 4. The fiber used in this case was only 40 cm, which limits the achievable delay. With longer fibers the counterpropagating pump pulses would overlap at several places along the fiber, thus creating a series of gratings which would lead to multiple reflections with different delays causing signal distortions. To avoid this effect, the use of advanced modulation concepts was proposed (Antman et al., 2012a, b, 2013, 2014; Zadok et al., 2012). An example of such a pump modulation format is the Golomb code. With these modulated pumps, a grating is only formed at the location of the correlation peak (Antman et al., 2012a). This makes it possible to use long fibers and achieve large fractional delays. In one experiment, 1-ns pulses were delayed by 770 ns (Antman et al., 2013). Another correlation approach to BDG generation is based on chaotic lasers. A grating is formed in the fiber only at the correlation peak of the counterpropagating chaotic laser pump and Stokes beams (Zhu et al., 2021).

4. SBS-based quasi-light storage

Another concept to delay optical signals relies on utilizing time-frequency coherence to generate multiple copies of an optical input signal (Jamshidi et al., 2012; Preussler et al., 2009, 2011b; Schneider et al., 2010). One of the copies is then selected using an optical switch. As the signal is copied rather than stored, the concept was termed quasi-light storage (QLS).

QLS achieves multiple copies of the optical signal by multiplication in the frequency domain with a frequency comb. That process corresponds to a convolution in the time domain of the pulse train (time domain representation of the frequency comb) with the signal pulse. The result is a train of copies of the signal pulse in the time domain as shown in Fig. 8.

As illustrated in Fig. 8, if the optical signal pulse is sampled with a frequency comb that consists of ideal delta functions one receives a pulse train

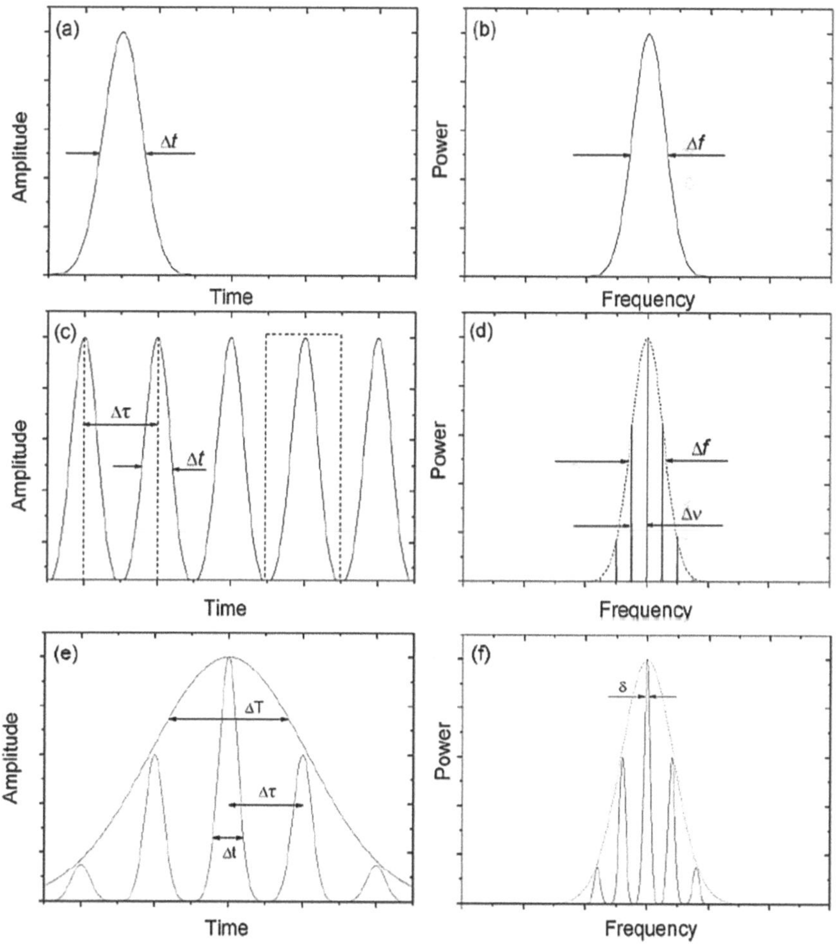

Fig. 8 Time domain and the frequency domain representation of the principle of QLS (Preussler et al., 2009). (A and B) Optical pulse in the time domain and its corresponding frequency representation. (C and D) Pulse train and sampled frequency spectrum of signal using frequency comb of delta peaks. (E and F) show a pulse train and sampled frequency spectrum with frequency comb lines with finite bandwidth.

with constant amplitude in the time domain. So the challenge is to achieve narrowband sampling in the frequency domain of the signal to achieve the longest possible delays.

These requirements make SBS a perfect match to perform the multiplication of the optical signal and the frequency comb due to the narrowband nature of the SBS gain resonance. The signal counter-propagates a frequency comb in an optical fiber that is offset in frequency by the BFS and acts as the Brillouin pump. The narrower the comb lines the larger the achievable delay and hence operating in the high pump power regime, which is known to reduce the Brillouin gain bandwidth to around 10 MHz, allows for the longest storage times. Using this technique initial experiments showed delays of around 100 ns (Preussler et al., 2009).

Following these proof-of-principle demonstrations, it was shown that further reducing the effective Brillouin linewidth can further increase the achievable delay time. To reduce the effective Brillouin linewidth Preussler et al. overlaid the Brillouin gain response with two Brillouin loss resonances, left and right of the maximum gain peak. Using this technique the effective Brillouin gain bandwidth could be reduced to 3.4 MHz with a corresponding delay time of over 140 ns (Preussler et al., 2011a).

The experimental results of that demonstration are shown in Fig. 9 for an optical bit pattern. Fig. 9 shows multiple copies of the initial bit pattern for more than 140 ns. A fast optical switch could then be used to select one of the delayed bit patterns after the desired delay time.

As outlined in the above section quasi-light storage is generating a train of copies of the optical signal pulses using optical sampling and an optical switch can be used to obtain a single delayed copy of the signal. The achievable delay using QLS is quite large and hence overcomes some of the

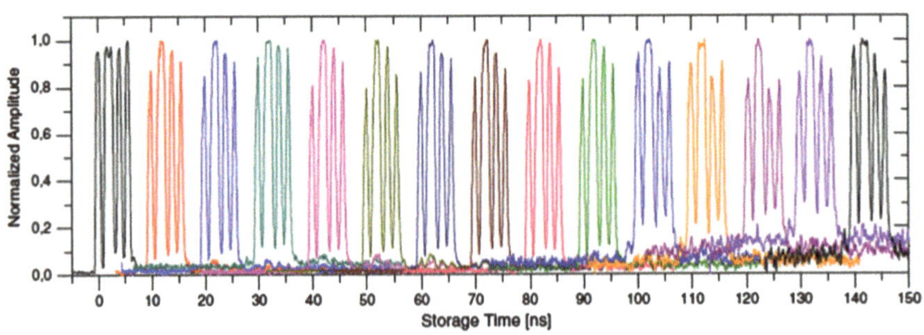

Fig. 9 Multiple copies of an optical bit pattern generated via QLS with reduced effective SBS gain bandwidth using a combination of Brillouin gain and loss resonance (Preussler et al., 2011b).

shortcomings of SBS slow light in regards to fractional delay times for single optical pulses or bursts of optical data. Methods to extend QLS to continuous data streams might be possible but are not obvious at this point.

5. SBS-based light storage

This approach uses SBS to store light in the form of acoustic waves. The SBS process acts as a coherent coupling mechanism between optical pulses that carry information and acoustic waves that are a much slower moving carrier of the information. The information is completely transferred from optical to acoustic waves, which travel several orders of magnitude slower than the speed of light. Note that the main difference to SBS-based slow light is that the information does not stay in the optical domain where the data pulse experiences a different group velocity, as in the case of SBS-based slow light, but changes the carrier of the information into the acoustic domain.

The basic principle is shown in Fig. 10A and applies to all platforms, such as integrated photonic waveguides or optical fibers. The optical pulse is stored as an acoustic excitation via a so-called write pulse that counter-propagates the data pulse and is frequency down-shifted by the BFS. An optical read pulse subsequently transfers the information back from the acoustic to the optical domain.

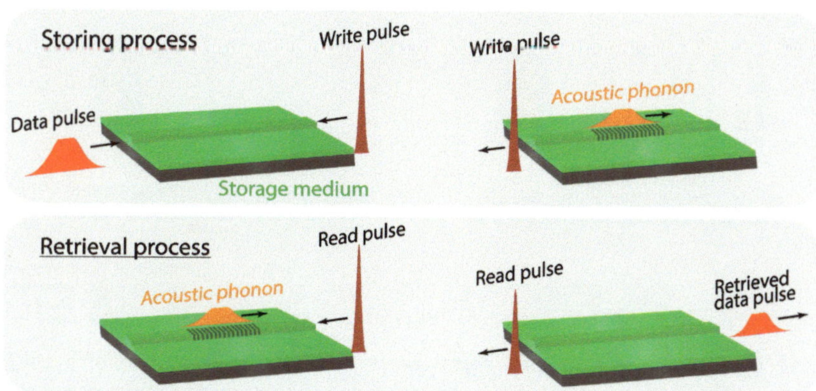

Fig. 10 Basic principle of Brillouin-based light storage (Merklein et al., 2015b). An optical data pulse is depleted by a strong counter-propagating write pulse, storing the data pulse as an acoustic wave. In the retrieval process, a read pulse depletes the acoustic wave, converting the data pulse back into the optical domain.

The first proof-of-principle implementation was realized in highly nonlinear fiber (Zhu et al., 2007b) and the main results are summarized in Fig. 11. With this approach, Zhu et al. were able to store 2 ns long pulses for up to 12 ns, which is a fractional delay of 6 and hence exceeded most SBS slow-light demonstrations (see also Fig. 18). The required power is exceeding 100 W and the storage time is limited by the short acoustic lifetime of the hyper-sound phonons—in the order of several nanoseconds.

A very interesting theoretical idea has been subsequently proposed using chirped optical pulses (Winful et al., 2013), depicted in Fig. 12. This not only leads to much higher read-out efficiencies but also enables the storage of a larger bandwidth and more flexibility in terms of pulse shape retrieval

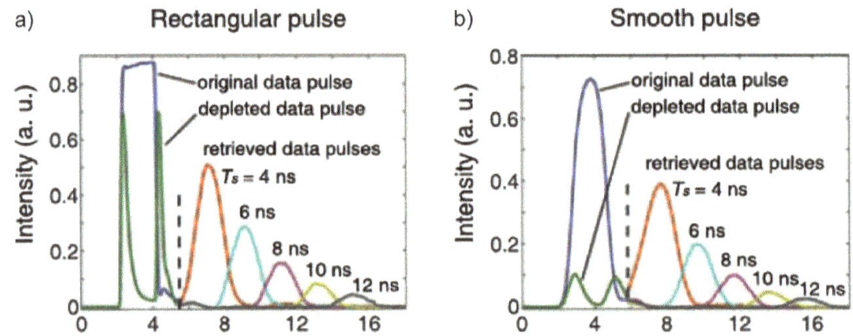

Fig. 11 (A) Stored rectangular pulse in an optical fiber (Zhu et al., 2007b). Note that the pulses right of the *dashed line* were multiplied by a factor of 2. (B) Storage and retrieval of smooth pulses with the retrieved pulses multiplied by a factor of 2 (Zhu et al., 2007b).

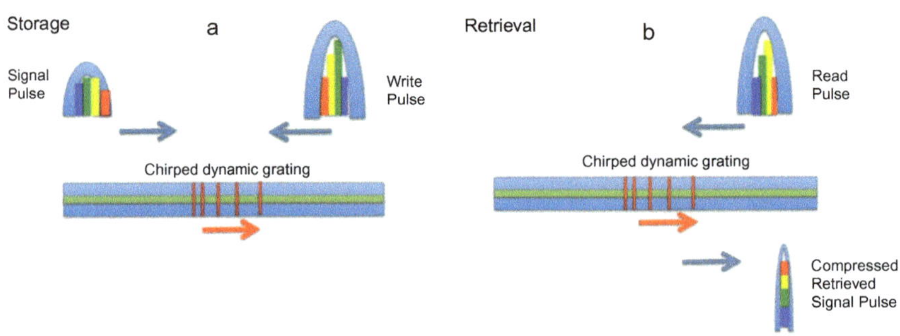

Fig. 12 (A) Storage of a chirped signal pulse through the formation of a chirped acoustic grating with a similarly chirped write pulse. (B) Retrieval of compressed signal by interrogating grating with oppositely chirped read pulse. *From Winful, H. G., Kabakova, I. V., Eggleton, B. J., 2013. Model for distributed feedback Brillouin lasers. Opt. Express 21 (13), 16191–16199. https://doi.org/10.1364/OE.21.016191. http://www.opticsexpress.org/abstract.cfm?URI=oe-21-13-16191.*

and manipulation. The influence of the pulse area of the write and read pulses has also been studied theoretically within this work (Dong et al., 2015). To date, the experimental realization of the strongly chirped pulses memory has not yet been achieved.

Waveguides with a strong local optoacoustic coupling are a potential solution to overcome the high power requirement for the control pulses required in highly nonlinear fiber. Options are materials such as soft glasses or small microstructures such as photonic crystal fibers. Chalcogenide waveguides (As_2S_3) with micrometer dimension on a photonic integrated chip have been demonstrated to have a particularly high Brillouin gain and can therefore provide the base for efficient optoacoustic conversion via SBS. The waveguides are designed such that the optical and acoustic waves are confined within the waveguide structure and hence have a strong optoacoustic overlap.

The first on-chip demonstration of the Brillouin-based memory has been published in Merklein et al. (2017), where not only was an optical power reduction achieved, but also an extension of the bandwidth to the GHz-regime and the phase coherence of the concept were experimentally proven. Fig. 13 shows the main results of the on-chip Brillouin-based memory.

As in the fiber approach, an optical data pulse is transferred to the acoustic domain by a counter-propagating write pulse (Fig. 13A). While increasing the power of the write pulse, the data pulse gets depleted and the information is held in the acoustic wave. The write power was reduced by a factor of 20, compared to highly nonlinear fiber, to around 5 W peak power. The read pulse, which is located at the same frequency as the write pulse, is then sent into the optical chip, which depletes the acoustic wave and transfers the information back to the optical domain. The original pulse can be retrieved after several nanoseconds (Fig. 13B) as the acoustic lifetime in As_2S_3 is similar as in silica and highly nonlinear fiber. Also here the delay is tunable by adjusting the time difference between the optical write and read pulses. In this experimental demonstration, a 2.5-ns optical pulse was stored. The pulse shape of the retrieved pulses after different storage times is compared in Fig. 13C and one can see, the pulse shape is conserved and small features of the original pulse can be successfully retrieved. The maintenance of the pulse shape and the ability to resolve small features in the pulse shape is an indication of the high bandwidth, that can be achieved in high Brillouin gain waveguides and that even shorter optical pulses can be stored; this was experimentally demonstrated in Jaksch et al. (2017).

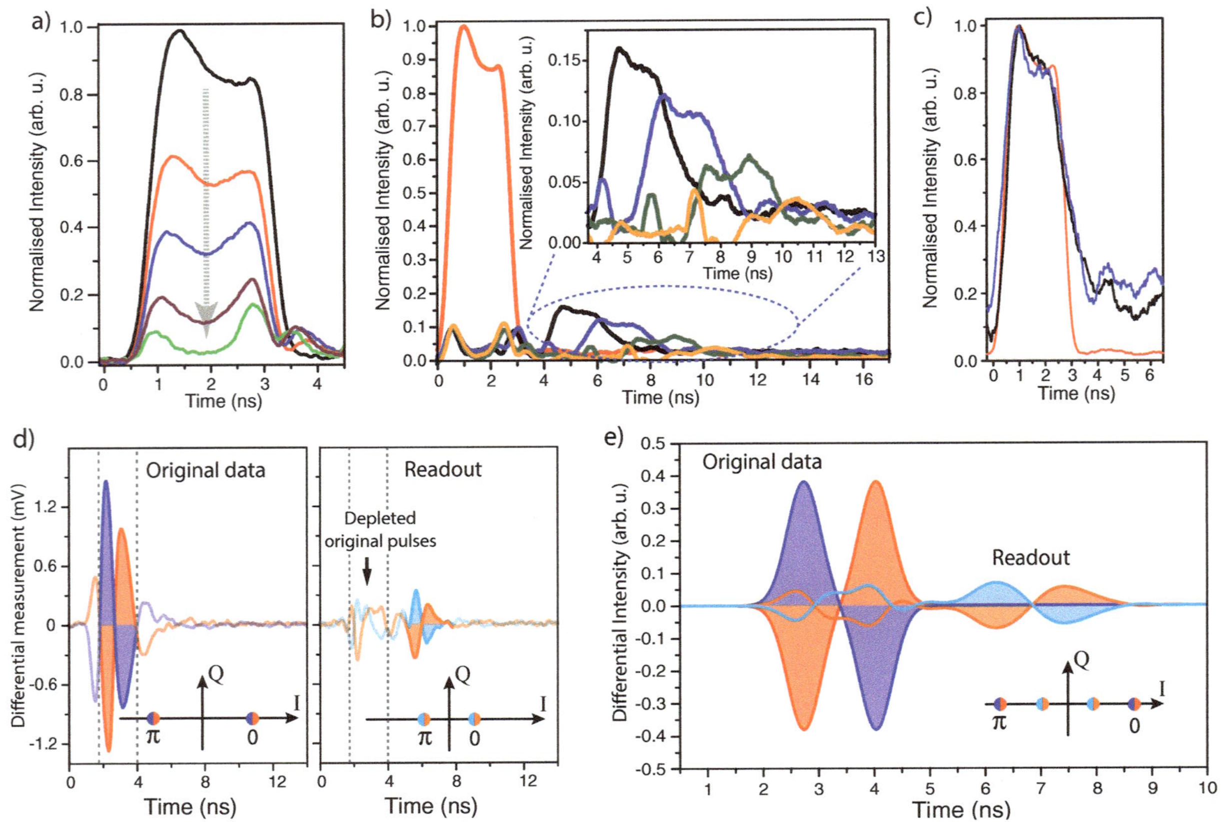

Fig. 13 See figure legend on opposite page.

The optical pulse can carry amplitude and phase information, which can be retrieved coherently. In Merklein et al. (2017) and Jaksch et al. (2017), the linearity of retrieving several amplitude levels has been experimentally demonstrated, which leads to an extended capacity of the Brillouin-based memory. Modern optical communication systems do encode not only information on the signal amplitude but also the optical phase. As SBS is a phase-sensitive process, one can distinguish the initial phase of the original pulses respective to each other. In the experiment, two data pulses were encoded with a π-phase shift (Fig. 13D and E) and detected via homodyne detection, beating the pulses with a local oscillator at the same wavelength. When switching on the write and read pulses, the optical pulses are depleted, the phase information is transferred to the acoustic domain and is retrieved again several nanoseconds later.

The previously described measurements show coherent storage of a phase- and amplitude-encoded signal at one wavelength. However, information systems often use frequency multiplexing, wherefore storage on several wavelength channels is useful. The concept of Brillouin-based memory has been experimentally proven to be able to operate at several wavelength channels, closely separated by only 25 GHz (Stiller et al., 2019) and the principle is shown in Fig. 14. The crosstalk between these channels is so low that it can be neglected, which is due to the strict phase-matching condition of backward Brillouin scattering that allows the interaction of only specific optical wavelengths with the written acoustic phonon. This is particularly

Fig. 13 (A) Storing process: The optical data pulse is depleted by the counter-propagating write pulse, transferring the carried information to the acoustic phonon. The curves show the transmitted data pulse for different write pulse powers: *black*: original data, *red*: 1.1 W, *blue*: 2.0 W, *purple*: 3.0 W, *green*: 4.7 W. Depletion of more than 90% can be achieved. (B) Retrieved data pulses after different storage times. The inset shows a zoomed-in version of the retrieved data pulses. (C) Original data pulse (*red*) superimposed with the retrieved data pulses after 3.5 ns (*black*) and 5.5 ns (*blue*), respectively; the details of the original shape can be distinguished in the retrieved data pulses. (D) Two phase levels of two 500 ps optical pulses, either 0 and π or π and 0 (depicted by the *red* and *blue trace*, respectively) are retrieved via homodyne detection after a storage time of 3.5 ns (*orange* and *light blue trace*). The insets show the two phases in phase space for the original (*red/blue*) and retrieved (*orange/light blue*) pulses. (E) Simulated phase response of the system (the color schemes of the two encoded phases and the phase space diagram are according to the measurements depicted in (D). *Adapted from Merklein, M., Stiller, B., Vu, K., Madden, S. J., Eggleton, B. J., 2017. A chip-integrated coherent photonic-phononic memory. Nat. Commun. 8 (1), 574. https://doi.org/10.1038/s41467-017-00717-y. http://www.nature.com/articles/s41467-017-00717-y, licensed under Creative Commons Attribution 4.0 International License.*

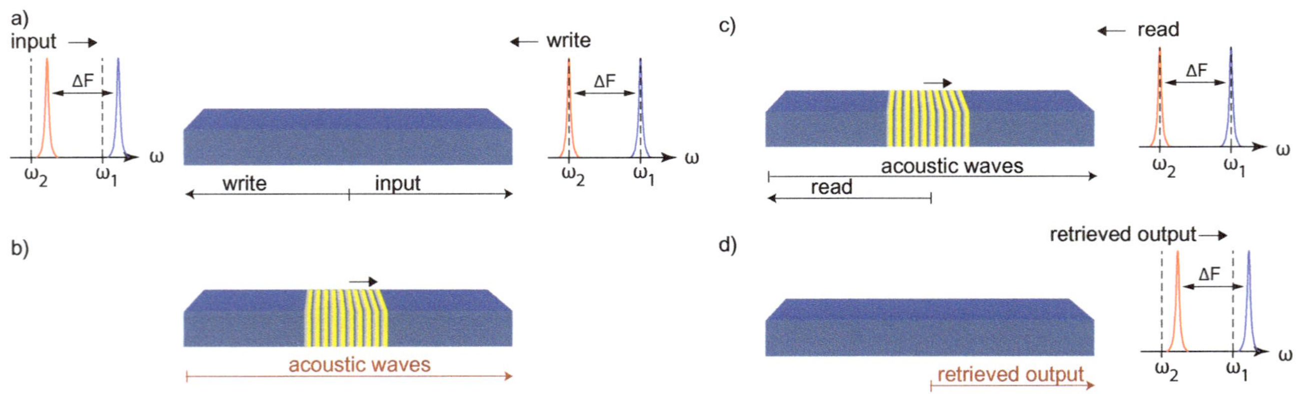

Fig. 14 Principle of using multiple frequencies in Brillouin-based memory. Two different frequency channels are used for the data and control pulses. (A) The data enters the waveguide from the left side, spaced by frequency difference ΔF. The write pulses for the respective frequency channels counter-propagate the data pulses and convert the optical information from both channels to acoustic phonons. (B) Two acoustic waves carrying the information propagate at the same position and have similar acoustic frequency. (C) The read pulses for the respective frequency channel enter the waveguide from the right side and convert the information back to the acoustic domain. (D) The data are again separated in frequency as the original data and leaves the waveguide. *Taken from Stiller, B., Merklein, M., Vu, K., Ma, P., Madden, S. J., Poulton, C. G., Eggleton, B. J., 2019. Cross talk-free coherent multi-wavelength Brillouin interaction. APL Photonics 4 (4). https://doi.org/10.1063/1.5087180.*

interesting as the acoustic frequency of optical channels separated by only 25 GHz is very similar (only 1 MHz difference). The reason for the separation of the different processes is that the acoustic phonons are spatially extended and the read pulse accumulates a phase mismatch that suppresses interaction after about several hundreds of micrometers of interaction length. This leads to a very strict frequency bound of the Brillouin-based memory.

The same phase-matching condition is responsible for the nonreciprocity of the concept. It has been experimentally demonstrated that Brillouin-based memory can act as a nonreciprocal device over the whole transmission window of the optical waveguide. An isolation of 9 dB can be achieved (Merklein et al., 2020). In Fig. 15 C and D two optical pulses traveling in opposite directions are measured. One is counter-propagating the write and read pulses, one is co-propagating to them. While one data pulse is stored and retrieved (Fig. 15C), the other one is not affected and conserves amplitude, phase, and pulse shape (Fig. 15D).

Also cascaded storage (Stiller et al., 2018) and storage of short optical pulses down to 100 ps (Jaksch et al., 2017) have been experimentally shown. The latter allows for a storage time of 100 times the pulse width of the original optical pulses. This exceeds SBS slow light approaches by two orders of magnitude. For describing Brillouin-based memory with short pulses the conventional coupled mode equations have to be adapted such as proposed in Piotrowski et al. (2021) because some approximations used for the conventional formalism do not hold anymore. An experimental measurement of the storage and retrieval of 200-ps-long pulses including the linearity of the amplitude states is shown in Fig. 15A and B.

The Brillouin-based memory waveguide (or optical fiber) approach has various advantages. It allows for storage of optical data with a GHz-bandwidth encoded at any wavelength within the transparency window of the waveguide material independent of any resonator modes (Stiller et al., 2019). Being independent of a resonator also means that the approach does not rely on a structural resonance but the process can be arbitrarily positioned at any location in the waveguide, controlled by external pulses only (Stiller et al., 2018). Using As_2S_3 as photonic waveguide material moreover provides an exceptional high optoacoustic gain which leads to a great improvement in terms of optical power requirements. Together with the coherence of the process, the multi-wavelength ability and the conservation of the pulse shape, the Brillouin-based light storage scheme can be a powerful concept for optical interconnects, not only for long-distance communications but also on a smaller scale, on a photonic chip.

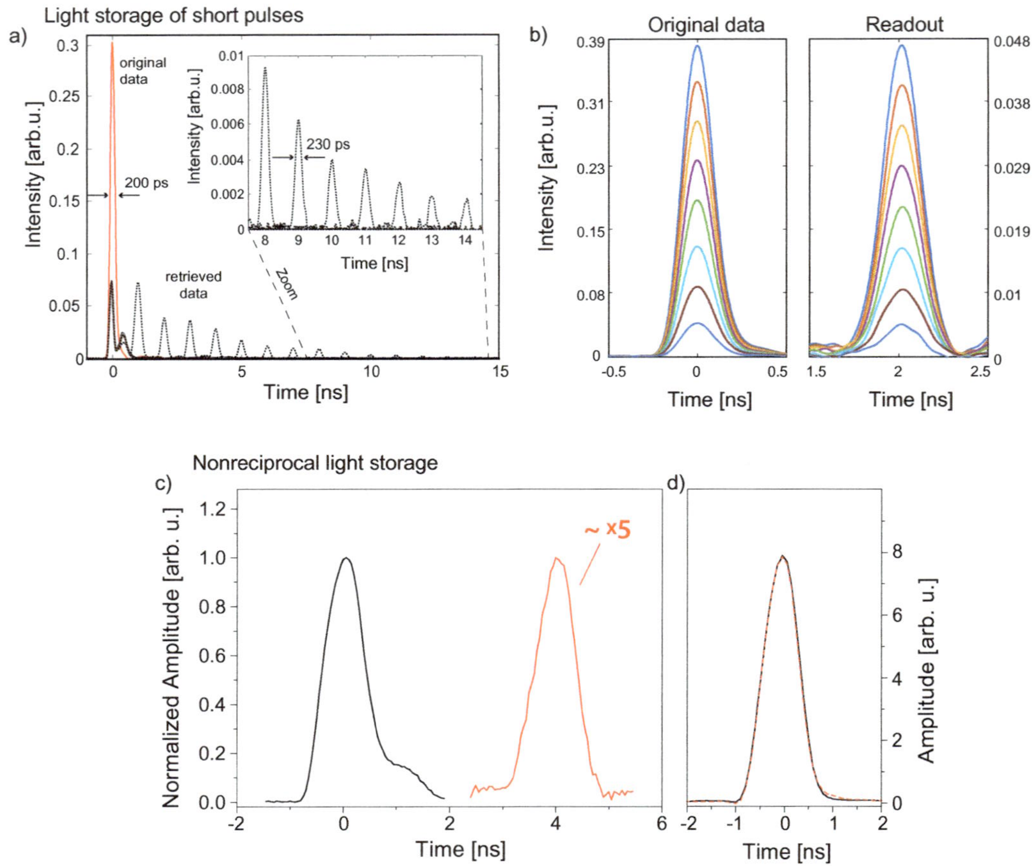

Fig. 15 (A) Original and delayed data for 200-ps-long optical pulses. A fractional delay of 70 pulse widths is achieved. (B) Amplitude storage of a 200-ps-long optical pulse. The amplitude of the read-out pulse scales linearly with the amplitude of the original pulse (Jaksch et al., 2017). (C) and (D) Nonreciprocal storage. (C) In one direction the data is stored and delayed by 4 ns whereby (D) the data propagating in the other direction at the same time are not affected (Merklein et al., 2020).

One limitation, however, is the acoustic lifetime of only several nanoseconds in amorphous media such as silica and chalcogenide glass. An attempt to overcome this limit in waveguides is to reinforce the decaying acoustic phonons by optical refresh pulses (Stiller et al., 2020). The principle is shown in Fig. 16. Here, the same process as the writing process is used (a Stokes process) to transfer energy from optical refresh pulses to the acoustic waves (Fig. 16A)IV and V. Not the optical data pulse initiates the energy transfer to the acoustic domain but the existing acoustic phonon initiates the Stokes process and receives energy from the optical refresh pulses. Fig. 16B and C shows the storage and retrieval of amplitude and phase after 40 ns which is four times the intrinsic acoustic lifetime. This concept has also been implemented in optomechanical resonators by changing periodically the coupling from the resonator to the thermal bath (Lake et al., 2021).

Another way of increasing the storage time is to use high-Q micro-resonators. Also, the power requirements to couple the two waves are greatly reduced. The acoustic lifetime achieved in these resonators is much longer and hence when used to store light pulses, much longer storage times can be achieved. Approaches based on the breathing modes of optomechanical resonators will be described in the next chapter. Here, we only mention the Brillouin-based resonator approach (Dong et al., 2015; Kim et al., 2015) where the light field does not couple to a radial breathing mode but an acoustic whispering gallery mode that travels around the circumference of the resonator. The pump, Stokes and acoustic wave have to fulfill the well-known Brillouin phase matching conditions. This approach combines some of the advantages of both schemes—the SBS approach and the micro-resonator approach.

The Brillouin-based resonator approach allows for nonreciprocal light storage (Dong et al., 2015) (see Fig. 17) as it only couples optical signal waves that copropagate with the write and read pulses. (Note that more recently optomechanically induced nonreciprocity in micro-resonators was also shown, where only copropagating laser fields couple to a radial mechanical mode (Shen et al., 2016).) The phase-matching in the Brillouin light storage scheme assures that the frequencies of the pulses in the storage process are preserved, which solves one of the main issues of previous resonator-based approaches for light storage. However, the frequencies are still bound to the resonances given by the resonator, and therefore cannot be continuously tuned or adjusted. Also, the bandwidth of the stored pulses is still very limited. This limitation arises due to the narrowband nature of the resonances in the resonator configuration. Furthermore, the pulse shape was not preserved in these demonstrations (see Fig. 17).

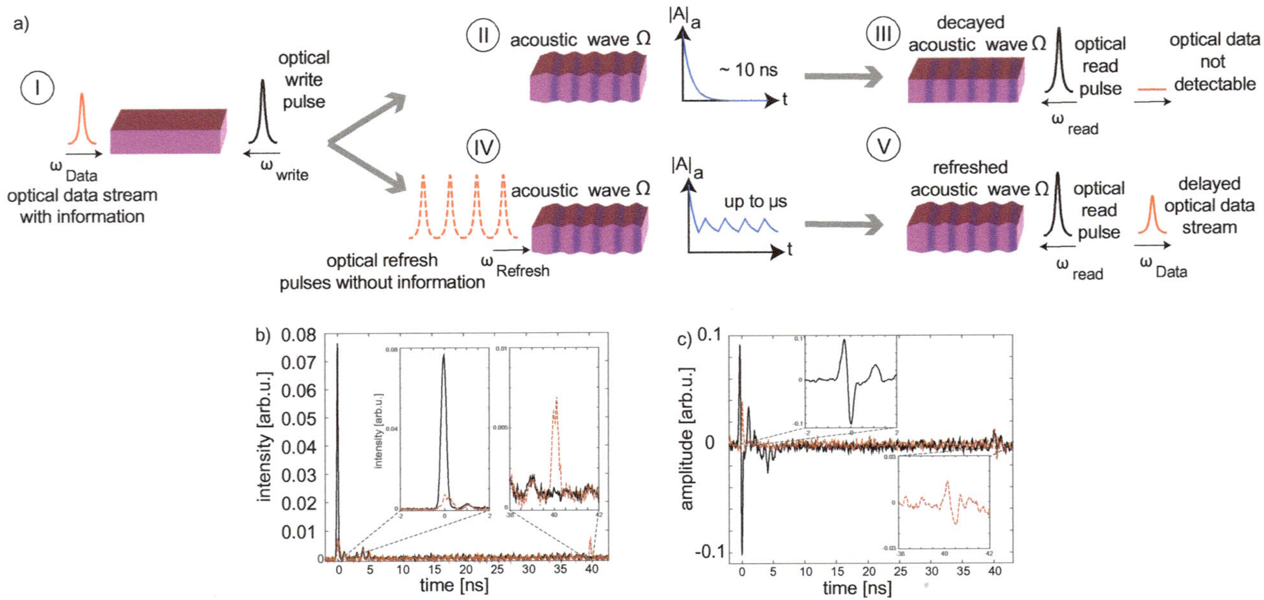

Fig. 16 (A) Principle of coherently refreshing acoustic phonons. (I–III) describes the normal Brillouin-based memory process where it can be seen that the acoustic decay limits the storage time which means that when the read pulse arrives to late, the original information is not detectable anymore. (IV) Refresh pulses transfer energy to the acoustic wave and amplifies them such that a read pulse injected after the acoustic lifetime can still retrieve the previously stored optical information. (B) Storage and retrieval of the amplitude of an optical pulse after 40 ns. (C) Storage and retrieval of the phase of two optical pulses after 40 ns. The two optical pulses were encoded with a π phase shift and are detected with homodyne detection. *Taken from Stiller, B., Merklein, M., Wolff, C., Vu, K., Ma, P., Madden, S. J., Eggleton, B. J., 2020. Coherently refreshing hypersonic phonons for light storage. Optica 7 (5), 492–497. https://doi.org/10.1364/optica.386535.*

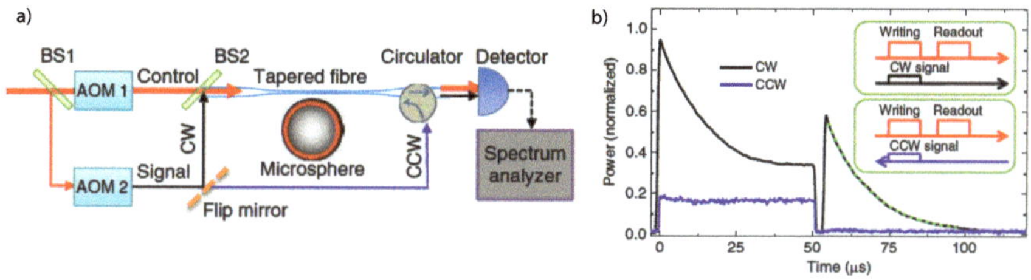

Fig. 17 (A) Experimental setup for Brillouin-based light storage in a resonator (Dong et al., 2015). (B) Nonreciprocal storage of light pulses using Brillouin optomechanical interactions in a resonator (Dong et al., 2015).

Table 1 Comparison of the different platforms—optical fiber, micro-resonator and softglass on-chip waveguides—used to implement SBS-based light storage schemes.

	Fiber (Zhu et al., 2007b)	**Resonator (Dong et al., 2015)**	**Waveguide (Merklein et al., 2017)**
Power req.	>100 W	<0.5 mW	<5 W
Abs. delay	12 ns	3 μs (lifetime 14 μs)	14 ns (Jaksch et al., 2017)
Frac. delay	6	0.06	70 (Jaksch et al., 2017)
Bandwidth	200 MHz	Tens of kHz	>1 GHz
Wavelength multipl.	Possible	Limited	Exp. demonstrated

From Merklein, M., Stiller, B., Eggleton, B.J., 2018. Brillouin-based light storage and delay techniques. J. Opt. (United Kingdom) 20 (8), 083003. https://doi.org/10.1088/2040-8986/aad081.

Brillouin-based light storage has been shown in different platforms with an overview given in Table 1. The resonator approach has clear advantages in terms of power requirement and absolute delay. However, it has a very narrow bandwidth and limitations concerning wavelength tuning and wavelength multiplexing. Engineered photonic–phononic waveguides show that Brillouin-based light storage is compatible with high bandwidth optical signals (exceeding a GHz) and allows for continuous wavelength tuning and wavelength multiplexing.

6. Storage in optomechanical resonators

In addition to the Brillouin-based approaches that are mostly realized in waveguides, we would like to give a short outlook on optical delay approaches that use optomechanical resonators which efficiently couple optical cavity modes to mechanical degrees of freedom. Due to the high Q-factors of these resonators, the power requirements for coupling the optical to mechanical or acoustic waves are greatly reduced. Moreover, the lifetime of acoustic modes can be orders of magnitudes larger than the high frequency traveling acoustic waves in an optical fiber and hence much longer storage times can be achieved.

It was demonstrated by Fiore et al. that the previously described concept of storing light as acoustic waves can be transferred from optical fiber to silica resonators (Fiore et al., 2011). The resonators used in this demonstration have a diameter of around 30 μm and offer acoustic and optical Q-factors of 3000 and 0.5×10^5, respectively. With the much larger acoustic Q-factors compared to optical fiber, storage times in the microsecond range could be demonstrated. Furthermore, it was shown that this storage process is also fully coherent (Fiore et al., 2013).

As in this case the optical modes couple to a transverse breathing mode, the mechanical breathing mode can be excited at a certain optical frequency and afterward read out by any other optical cavity mode. Therefore, it allows transferring a quantum state from a certain frequency to another, which has potential applications in quantum signal processing (Hill et al., 2012). Despite state conversion being a desired feature for quantum networks, this nonfrequency preserving nature can be a problem in light storage for classical communication applications, as storage of multiplexed data streams at different wavelength channels will lead to cross-talk.

Another challenge the resonator-based memory faces is that the retrieved data pulse depends not only on the wavelength but also on the profile and intensity of the read pulse. This might be a hindrance when using information encoded in the amplitude, as particular care has to be taken concerning the writing and reading conditions.

Optomechanical storage has also been proposed in optomechanical crystal arrays (Chang et al., 2011) and in the quantum domain for single phonons (Galland et al., 2014; Wallucks et al., 2020). Some concepts like reinforcing acoustic phonons (Stiller et al., 2020) have also been proven useful for optomechanical approaches (Lake et al., 2021), in this particular case to achieve much longer storage times by dynamically changing the mechanical dissipation rate.

7. Discussion, summary, and outlook

In this chapter, we have given an overview on the different Brillouin-based delay techniques. Over the years the performance of the schemes improved, either by implementing new platforms with higher Brillouin gain

Table 2 Overview of the four different SBS-based delay schemes described in this chapter.

SBS-based slow-light	BDG delay line	Quasi-light storage	SBS light storage
✓ Implemented in fiber, chip and resonator	✓ Implemented in fiber	✓ Implemented in fiber	✓ Implemented in fiber, chip and resonator
✓ Room temperature operation	✓ Room temperature operation	✓ Room temperature operation	✓ Room temperature operation
✓ Delay controlled via pump	✓ Delay controlled via spatial pump overlap	✓ Delay controlled via optical switch	✓ Delay controlled via timing read pulse
✗ Limited fractional delay (<10)	✓ Large fractional delay (>10)	✓ Large fractional delay (>10)	✓ Large fractional delay (>10)
	✗ Delay limited by length of PMF	✗ Delay limited by phonon lifetime	✗ Delay limited by phonon lifetime
	✗ Low signal output		

Source: Taken from Merklein, M., Stiller, B., Eggleton, B.J., 2018. Brillouin-based light storage and delay techniques. J. Opt. (United Kingdom) 20 (8), 083003. https://doi.org/10.1088/2040-8986/aad081.

or by techniques to improve the scheme using, for example, pump modulation. A summary of the advantages and existing challenges of the four different Brillouin delay schemes is given in Table 2 (Merklein et al., 2018).

Common advantages of all schemes are room temperature operation and implementation with usual telecommunication components. It is also possible to integrate all concepts into a photonic circuit, which is a major advantage compared to schemes based an atom clouds or Bose–Einstein condensates in terms of compatibility with current networks.

7.1 Delay time

The important parameter of the delay time can be controlled in all Brillouin-based schemes. It is continuously adjustable either by the pump power (SBS slow light), the spatial overlap of the two pumps that create the BDG (in the BDG scheme) or the arrival time of the read pulse (light storage). In the quasi-light storage scheme, the delay time can be coarsely set by selecting the individual copy of the original signal while finer tuning of the delay time can be achieved by altering the repetition rate of the frequency comb that is convoluted with the signal.

In SBS slow-light schemes the delay time can be easily controlled by the pump power; however, the achievable fractional delay is limited to around 1 for tolerable distortions due to constraints imposed by the time-bandwidth product. BDG-based delay lines, quasi-light storage and light storage overcome this barrier. Quasi-light storage and Brillouin light storage allow for large fractional delays, but are limited by the phonon lifetime. In the quasi-light storage case, this limitation could be partially overcome by tailoring the Brillouin spectrum, overlying gain and loss resonances. One way to improve the delay time in the Brillouin light storage case is to reinforce the decaying acoustic waves which has been demonstrated in Stiller et al. (2020). However, more settings can be imagined by altering the phonon confinement and dissipation.

BDG-based delay schemes achieve very large fractional delays for bursts of pulses and fractional delays of 10 for continuous data streams could be achieved. A solution for distortions due to reflections from sidelobes or coding noise is for example the use of Golomb codes. Even though the BDG approach allows for very large delays and is compatible with data streams it faces the challenge of fairly long lengths of polarization maintaining fiber and hence does not allow integration on a chip. More severe is however the low reflectivity of the dynamic grating of only −64 dB (advanced modulation schemes). There seems not yet to be a straightforward solution how to increase the reflectivity of BDGs that have a bandwidth compatible with optical data streams.

An overview of the fractional delays achieved using Brillouin-based delay techniques is shown in Fig. 18.

7.2 Application to optical data streams

SBS slow-light might not be applicable to delay continuous optical data streams but has found useful applications in microwave photonics (Santagiustina et al., 2012). A basic scheme of a microwave photonic link that harnesses slow light is shown in Fig. 19. The bandwidth requirements are more relaxed in analog microwave signal processing schemes compared to high bit rate optical data streams, and hence the former greatly benefit from the advantages of SBS slow-light without being strongly impacted by the penalty in bandwidth and fractional delay. Therefore SBS slow light can be harnessed in microwave photonics to induce a phase shift that is required for many microwave applications, such as tuning of microwave filters (Sancho et al., 2010), phase shifting, and true time delay schemes required for phased array antennas (Aryanfar et al., 2017; Chin et al., 2010b; Choudhary et al., 2017; Li et al., 2011b; Pagani et al., 2014). How far quasi-light storage and Brillouin light storage can play a role in the field of microwave photonics is still to be shown.

Quasi-light storage and light storage show great potential to delay optical data signals. The feasibility was shown by storing multiple bits encoded in an on-off modulated signal. It has been demonstrated that Brillouin light

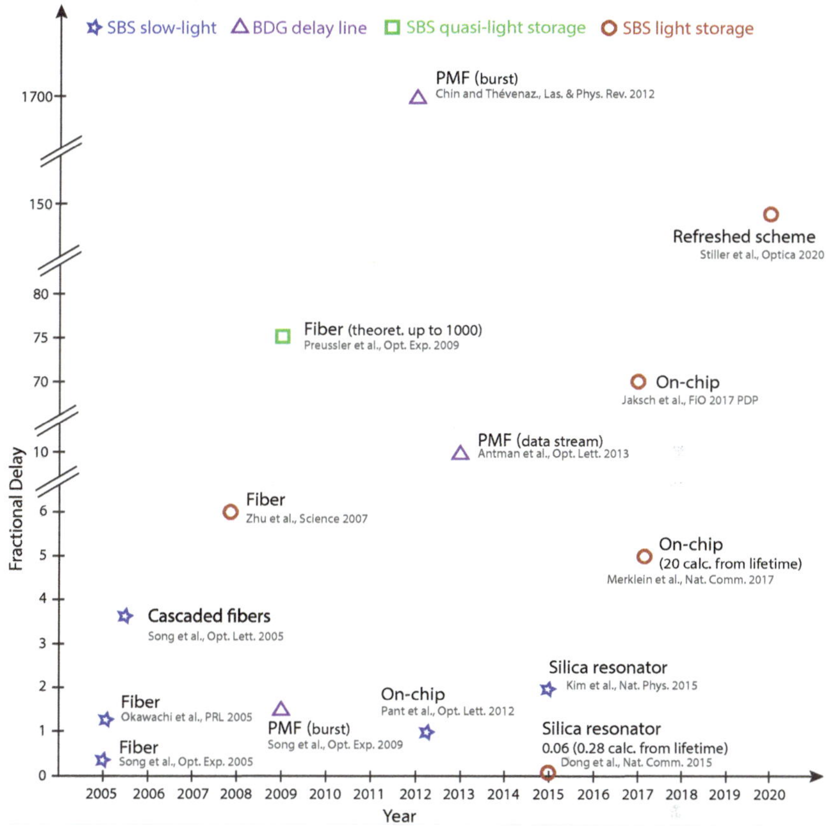

Fig. 18 Overview of the achieved fractional delays in different Brillouin-based delay and storage techniques. *Updated from Merklein, M., Stiller, B., Eggleton, B.J., 2018. Brillouin-based light storage and delay techniques. J. Opt. (United Kingdom) 20 (8), 083003. https://doi.org/10.1088/2040-8986/aad081.*

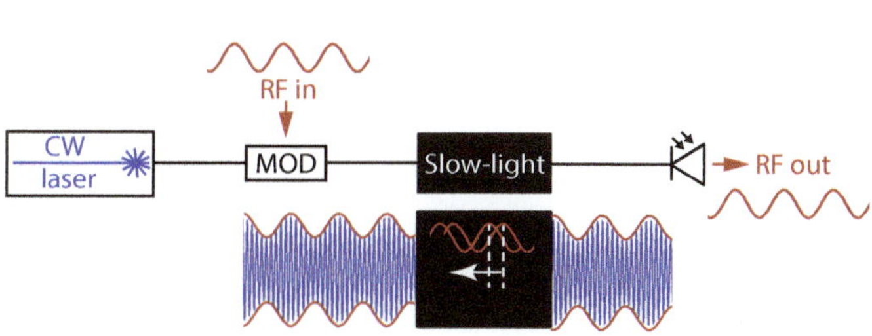

Fig. 19 Scheme of a microwave photonic link that uses slow or fast light to shift the phase of an RF signal. *From Santagiustina, M., Eisenstein, G., Thevenaz, L., Capmany, J., Mork, J., Reithmaier, J. P., De Rossi, A., Sales, S., Yvind, K., Combrie, S., Bourderionnet, J., 2012. Slow light devices and their applications to microwaves and photonics. IEEE Photonics Soc. Newsletter 26 (February), 6–13. http://infoscience.epfl.ch/record/174930/files/IEEEPhotSocNewsletterFeb2012.pdf.*

storage can be used to store multiple pulses with different amplitude and phase levels. Also, the compatibility with wavelength multiplexing schemes was already demonstrated. However, both schemes have been so far only applied to bursts of optical data. The feasibility for storing and delaying continuous data streams still has to be shown. Here, the fact that quasi-light storage generates a great number of copies of the signal could be a hindrance. Also, the requirements for the fast optical switch, in that case, might cause a major bottleneck. In the Brillouin light storage case, the delaying and storing of continuous bit streams will create challenges for the write and the read processes as any residual data pulses, that are either not efficiently stored or retrieved, will lead to an increase in the bit error rate.

Acknowledgments

The authors would like to thank Prof. Alexander Gaeta for valuable input and feedback to the manuscript.

References

Agrawal, G.G.P., 2001. Nonlinear fiber optics. In: Kelley, P.L., Kaminow, I.P., Agrawal, G.G.P. (Eds.), Optics and Photonics. Academic Press, p. 458. http://books.google.com/books?hl=en&lr=&id=1HZ6ROsqz2gC&oi=fnd&pg=PP2&dq=Applications+of+Nonlinear+Fiber+Optics&ots=4-AlXStz-R&sig=yoOKs2bIvVinwJUQvascfv1v8iU%5Cnhttp://books.google.com/books?id=wjHP0oAVcScC&pgis=1%5Cnhttp://www.springerlink.com/index/q02.

Antman, Y., Levanon, N., Zadok, A., 2012a. Low-noise delays from dynamic Brillouin gratings based on perfect Golomb coding of pump waves. Opt. Lett. 37 (24), 5259–5261. http://www.ncbi.nlm.nih.gov/pubmed/23258071.

Antman, Y., Primerov, N., Sancho, J., Thevenaz, L., Zadok, A., 2012b. Localized and stationary dynamic gratings via stimulated Brillouin scattering with phase modulated pumps. Opt. Express 20 (7), 7807–7821. http://www.ncbi.nlm.nih.gov/pubmed/22453458.

Antman, Y., Yaron, L., Langer, T., Tur, M., Levanon, N., Zadok, A., 2013. Experimental demonstration of localized Brillouin gratings with low off-peak reflectivity established by perfect Golomb codes. Opt. Lett. 38 (22), 4701–4704. http://www.ncbi.nlm.nih.gov/pubmed/24322110.

Antman, Y., Yaron, L., Langer, T., Tur, M., Zadok, A., 2014. Variable delay of Gbit/s data using coded Brillouin dynamic gratings. In: SPIE OPTO. International Society for Optics and Photonics, February, p. 89980W. http://proceedings.spiedigitallibrary.org/proceeding.aspx?doi=10.1117/12.2047534.

Aryanfar, I., Marpaung, D., Choudhary, A., Liu, Y., Vu, K., Choi, D.-Y., Ma, P., Madden, S., Eggleton, B.J., 2017. Chip-based Brillouin radio frequency photonic phase shifter and wideband time delay. Opt. Lett. 42 (7), 1313. https://doi.org/10.1364/OL.42.001313. https://www.osapublishing.org/abstract.cfm?URI=ol-42-7-1313.

Baba, T., 2008. Slow light in photonic crystals. Nat. Photonics 2 (8), 465–473. https://doi.org/10.1038/nphoton.2008.146.

Boyd, R.W., 2003. Nonlinear Optics. Academic Press, ISBN: 0121216829.

Boyd, R.W., 2012. Material slow light and structural slow light: similarities and differences for nonlinear optics. J. Opt. Soc. Am. B 29 (9), 2644. https://doi.org/10.1364/JOSAB.29.002644. https://www.osapublishing.org/abstract.cfm?URI=josab-29-9-2644.

Boyd, R.W., Gauthier, D.J., 2009. Controlling the velocity of light pulses. Science 326 (5956), 1074–1077. https://doi.org/10.1126/science.1170885. http://www.ncbi.nlm.nih.gov/pubmed/19965419. http://www.sciencemag.org/cgi/doi/10.1126/science.1170885.

Chang, D.E., Safavi-Naeini, A.H., Hafezi, M., Painter, O., 2011. Slowing and stopping light using an optomechanical crystal array. New J. Phys. 13. https://doi.org/10.1088/1367-2630/13/2/023003.

Chin, S., Thévenaz, L., 2012. Tunable photonic delay lines in optical fibers. Laser Photonics Rev. 6 (6), 724–738. https://doi.org/10.1002/lpor.201100038.

Chin, S., Gonzalez-Herraez, M., Thevenaz, L., 2009. Complete compensation of pulse broadening in an amplifier-based slow light system using a nonlinear regeneration element. Opt. Express 17 (24), 21910–21917. https://doi.org/10.1364/OE.17.021910.

Chin, S., Primerov, N., Thevenaz, L., 2010a. Photonic delay line for broadband optical signals, based on dynamic grating reflectors in fibers. In: European Conference on Optical Communication, ECOC, 1–2, pp. 19–21. vol.

Chin, S., Thévenaz, L., Sancho, J., Sales, S., Capmany, J., Berger, P., Bourderionnet, J., Dolfi, D., 2010b. Broadband true time delay for microwave signal processing, using slow light based on stimulated Brillouin scattering in optical fibers. Opt. Express 18 (21), 22599–22613. http://www.ncbi.nlm.nih.gov/pubmed/20941157.

Choudhary, A., Liu, Y., Morrison, B., Vu, K., Choi, D.-Y., Ma, P., Madden, S., Marpaung, D., Eggleton, B.J., 2017. High-resolution, on-chip RF photonic signal processor using Brillouin gain shaping and RF interference. Sci. Rep. 7 (1), 5932. https://doi.org/10.1038/s41598-017-06270-4. http://www.nature.com/articles/s41598-017-06270-4.

Corcoran, B., Monat, C., Grillet, C., Moss, D.J., Eggleton, B.J., White, T.P., O'Faolain, L., Krauss, T.F., 2009. Green light emission in silicon through slow-light enhanced third-harmonic generation in photonic-crystal waveguides. Nat. Photonics 3 (April), 206–210. https://doi.org/10.1038/nphoton.2009.28. http://www.nature.com/doifinder/10.1038/nphoton.2009.28.

Dong, Y., Chen, L., Bao, X., 2010. Characterization of the Brillouin grating spectra in a polarization-maintaining fiber. Opt. Express 18 (18), 18960–18967. https://doi.org/10.1364/OE.18.018960.

Dong, C.-H., Shen, Z., Zou, C.-L., Zhang, Y.-L., Fu, W., Guo, G.-C., 2015. Brillouin-scattering-induced transparency and non-reciprocal light storage. Nat. Commun. 6, 6193. https://doi.org/10.1038/ncomms7193.

Fiore, V., Yang, Y., Kuzyk, M.C., Barbour, R., Tian, L., Wang, H., 2011. Storing optical information as a mechanical excitation in a silica optomechanical resonator. Phys. Rev. Lett. 107 (13), 1–5. https://doi.org/10.1103/PhysRevLett.107.133601.

Fiore, V., Dong, C., Kuzyk, M.C., Wang, H., 2013. Optomechanical light storage in a silica microresonator. Phys. Rev. A 87 (2), 023812. https://doi.org/10.1103/PhysRevA.87.023812.

Galland, C., Sangouard, N., Piro, N., Gisin, N., Kippenberg, T.J., 2014. Heralded single-phonon preparation, storage, and readout in cavity optomechanics. Phys. Rev. Lett. 112 (14), 1–6. https://doi.org/10.1103/PhysRevLett.112.143602.

González Herráez, M., Song, K.Y., Thévenaz, L., 2006. Arbitrary-bandwidth Brillouin slow light in optical fibers. Opt. Express 14 (4), 1395–1400. https://doi.org/10.1364/OE.14.001395. http://www.ncbi.nlm.nih.gov/pubmed/19503463.

González-Herráez, M., Martin-Lopez, S., Thévenaz, L., 2012. Analytical expression of pulse broadening in an arbitrary linear slow light medium. Opt. Lett. 37 (15),

3171–3173. https://doi.org/10.1364/OL.37.003171. http://www.ncbi.nlm.nih.gov/pubmed/22859122.
Hau, L.V., Harris, S.E., Dutton, Z., Behroozi, C.H., 1999. Light speed reduction to 17 metres per second in an ultracold atomic gas. Nature 397 (6720), 594–598. https://doi.org/10.1038/17561. http://www.nature.com/nature/journal/v397/n6720/full/397594a0.html%5Cn. http://www.nature.com/nature/journal/v397/n6720/pdf/397594a0.pdf.
Hill, J.T., Safavi-Naeini, A.H., Chan, J., Painter, O., 2012. Coherent optical wavelength conversion via cavity optomechanics. Nat. Commun. 3, 1196. https://doi.org/10.1038/ncomms2201. http://www.ncbi.nlm.nih.gov/pubmed/23149741.
Jaksch, K., Merklein, M., Vu, K., Ma, P., Madden, S.J., Eggleton, B.J., Stiller, B., 2017. Brillouin-based light storage of 200ps-long pulses for 70 pulse widths. In: Frontiers in Optics 2017, OSA, Washington, D.C. https://www.osapublishing.org/abstract.cfm?URI=FiO-2017-FTh4A.5.
Jamshidi, K., Preuler, S., Wiatrek, A., Schneider, T., 2012. A review to the all-optical quasi-light storage. IEEE J. Sel. Top. Quantum Electron. 18 (2), 884–890. https://doi.org/10.1109/JSTQE.2011.2134836.
Khurgin, J.B., 2006. Performance limits on delay lines based on optical amplifiers. Opt. Lett. 31 (7), 948–950.
Kim, J., Kuzyk, M.C., Han, K., Wang, H., Bahl, G., 2015. Non-reciprocal Brillouin scattering induced transparency. Nat. Phys. 11 (3), 275–280. https://doi.org/10.1038/nphys3236.
Krauss, T.F., 2007. Slow light in photonic crystal waveguides. J. Phys. D Appl. Phys. 40 (9), 2666–2670. https://doi.org/10.1088/0022-3727/40/9/S07. http://stacks.iop.org/0022-3727/40/i=9/a=S07?key=crossref.24bb17b68f0b42bf18eba40b5deffc2d.
Krauss, T.F., 2008. Why do we need slow light? Nat. Photonics 2 (8), 448–450. https://doi.org/10.1038/nphoton.2008.139. http://www.nature.com/nphoton/journal/v2/n8/abs/nphoton.2008.139.html.
Lake, D.P., Mitchell, M., Sukachev, D.D., Barclay, P.E., 2021. Processing light with an optically tunable mechanical memory. Nat. Commun. 12 (1), 663. https://doi.org/10.1038/s41467-021-20899-w.
Lenz, G., Eggleton, B., Madsen, C., Slusher, R., 2001. Optical delay lines based on optical filters. IEEE J. Quantum Electron. 37 (4), 525–532. https://doi.org/10.1109/3.914401. http://ieeexplore.ieee.org/lpdocs/epic03/wrapper.htm?arnumber=914401.
Li, J., O'Faolain, L., Rey, I.H., Krauss, T.F., 2011a. Four-wave mixing in photonic crystal waveguides: slow light enhancement and limitations. Opt. Express 19 (5), 4458–4463. http://www.ncbi.nlm.nih.gov/pubmed/21369277.
Li, W., Zhu, N.H., Wang, L.X., Wang, J.S., Liu, J.G., Liu, Y., Qi, X.Q., Xie, L., Chen, W., Wang, X., Han, W., 2011b. True-time delay line with separate carrier tuning using dual-parallel MZM and stimulated Brillouin scattering-induced slow light. Opt. Express 19 (13), 12312. https://doi.org/10.1364/OE.19.012312. https://www.osapublishing.org/oe/abstract.cfm?uri=oe-19-13-12312.
Merklein, M., Kabakova, I.V., Büttner, T.F.S., Choi, D.-Y., Luther-Davies, B., Madden, S.J., Eggleton, B.J., 2015a. Enhancing and inhibiting stimulated Brillouin scattering in photonic integrated circuits. Nat. Commun. 6, 6396. https://doi.org/10.1038/ncomms7396.
Merklein, M., Stiller, B., Vu, K., Madden, S.J., Eggleton, B., 2015b. Storing light as sound in a photonic integrated circuit. In: Frontiers in Optics 2015, OSA, Washington, D.C. https://www.osapublishing.org/abstract.cfm?URI=FiO-2015-FW6C.6.
Merklein, M., Stiller, B., Vu, K., Madden, S.J., Eggleton, B.J., 2017. A chip-integrated coherent photonic-phononic memory. Nat. Commun. 8 (1), 574. https://doi.org/10.1038/s41467-017-00717-y. http://www.nature.com/articles/s41467-017-00717-y.
Merklein, M., Stiller, B., Eggleton, B.J., 2018. Brillouin-based light storage and delay techniques. J. Opt. (United Kingdom) 20 (8), 083003. https://doi.org/10.1088/2040-8986/aad081.

Merklein, M., Stiller, B., Vu, K., Ma, P., Madden, S.J., Eggleton, B.J., 2020. On-chip broadband nonreciprocal light storage. Nanophotonics, 75–82. https://doi.org/10.1515/nanoph-2020-0371.

Minardo, A., Bernini, R., Zeni, L., 2006. Low distortion Brillouin slow light in optical fibers using AM modulation. Opt. Express 14 (13), 5866–5876. https://doi.org/10.1364/OE.14.005866.

Mok, J.T., Eggleton, B.J., 2005. Photonics: expect more delays. Nature 433 (7028), 811–812. https://doi.org/10.1038/433811a. http://www.ncbi.nlm.nih.gov/pubmed/24957657. http://www.ncbi.nlm.nih.gov/pubmed/15729323.

Mok, J.T., de Sterke, C.M., Littler, I.C.M., Eggleton, B.J., 2006. Dispersionless slow light using gap solitons. Nat. Phys. 2 (11), 775–780. https://doi.org/10.1038/nphys438. http://www.nature.com/doifinder/10.1038/nphys438.

Monat, C., de Sterke, M., Eggleton, B.J., 2010. Slow light enhanced nonlinear optics in periodic structures. J. Opt. 12 (10), 104003. https://doi.org/10.1088/2040-8978/12/10/104003. http://stacks.iop.org/2040-8986/12/i=10/a=104003?key=crossref.8866be39d295139d5be22e30af6d64da.

Okawachi, Y., Bigelow, M., Sharping, J., Zhu, Z., Schweinsberg, A., Gauthier, D., Boyd, R., Gaeta, A., 2005. Tunable all-optical delays via Brillouin slow light in an optical fiber. Phys. Rev. Lett. 94 (15), 153902. https://doi.org/10.1103/PhysRevLett.94.153902.

Pagani, M., Marpaung, D., Choi, D.-Y., Madden, S.J., Luther-Davies, B., Eggleton, B.J., 2014. Tunable wideband microwave photonic phase shifter using on-chip stimulated Brillouin scattering. Opt. Express 22 (23), 28810–28818. https://doi.org/10.1364/OE.22.028810.

Pant, R., Stenner, M.D., Neifeld, M.A., Gauthier, D.J., 2008. Optimal pump profile designs for broadband SBS slow-light systems. Opt. Express 16 (4), 2764–2777. https://doi.org/10.1364/OE.16.002764.

Pant, R., Poulton, C.G., Choi, D.-Y., Mcfarlane, H., Hile, S., Li, E., Thevenaz, L., Luther-Davies, B., Madden, S.J., Eggleton, B.J., 2011. On-chip stimulated Brillouin scattering. Opt. Express 19 (9), 8285–8290. https://doi.org/10.1364/OE.19.008285. http://www.opticsexpress.org/abstract.cfm?URI=oe-19-9-8285.

Pant, R., Byrnes, A., Poulton, C.G., Li, E., Choi, D.-Y., Madden, S., Luther-Davies, B., Eggleton, B.J., 2012. Photonic-chip-based tunable slow and fast light via stimulated Brillouin scattering. Opt. Lett. 37 (5), 969. https://doi.org/10.1364/OL.37.000969. http://ol.osa.org/abstract.cfm?URI=ol-37-5-969.

Piotrowski, J., Schmidt, M.K., Stiller, B., Poulton, C.G., Steel, M.J., 2021. Picosecond acoustic dynamics in stimulated Brillouin scattering. Opt. Lett. 46 (12), 2972–2975. https://doi.org/10.1364/ol.424701.

Preussler, S., Jamshidi, K., Wiatrek, A., Henker, R., Bunge, C.-A., Schneider, T., 2009. Quasi-light-storage based on time-frequency coherence. Opt. Express 17 (18), 15790–15798. https://doi.org/10.1364/OE.17.015790.

Preussler, S., Wiatrek, A., Jamshidi, K., Schneider, T., 2011a. Brillouin scattering gain bandwidth reduction down to 3.4 MHz. Opt. Express 19 (9), 8565–8570. https://doi.org/10.1364/OE.19.008565.

Preussler, S., Wiatrek, A., Jamshidi, K., Schneider, T., 2011b. Quasi-light-storage enhancement by reducing the Brillouin gain bandwidth. Appl. Opt. 50 (22), 4252–4256. https://doi.org/10.1364/AO.50.004252.

Safavi-Naeini, A.H., Mayer Alegre, T.P., Chan, J., Eichenfield, M., Winger, M., Lin, Q., Hill, J.T., Chang, D.E., Painter, O., 2011. Electromagnetically induced transparency and slow light with optomechanics. Nature 472 (7341), 69–73. https://doi.org/10.1038/nature09933. http://www.ncbi.nlm.nih.gov/pubmed/21412237.

Sancho, J., Chin, S., Sagues, M., Loayssa, A., Lloret, J., Gasulla, I., Sales, S., Thevenaz, L., Capmany, J., 2010. Dynamic microwave photonic filter using separate carrier tuning based on stimulated Brillouin scattering in fibers. IEEE Photonics Technol. Lett. 22 (23), 1753–1755. https://doi.org/10.1109/LPT.2010.2082514.

Santagiustina, M., Eisenstein, G., Thevenaz, L., Capmany, J., Mork, J., Reithmaier, J.P., De Rossi, A., Sales, S., Yvind, K., Combrie, S., Bourderionnet, J., 2012. Slow light devices and their applications to microwaves and photonics. IEEE Photonics Soc. Newsletter 26 (February), 6–13. http://infoscience.epfl.ch/record/174930/files/IEEE PhotSoc Newsletter Feb2012.pdf.

Schneider, T., Junker, M., Lauterbach, K.-U., 2006a. Potential ultra wide slow-light bandwidth enhancement. Opt. Express 14 (23), 11082–11087. https://doi.org/10.1364/OE.14.011082. http://www.ncbi.nlm.nih.gov/pubmed/19529523.

Schneider, T., Junker, M., Lauterbach, K.U., Henker, R., 2006b. Distortion reduction in cascaded slow light delays. Electron. Lett. 42 (19), 1110. https://doi.org/10.1049/el:20062235. http://digital-library.theiet.org/content/journals/10.1049/el_20062235.

Schneider, T., Junker, M., Lauterbach, K.-U., 2007. Time delay enhancement in stimulated-Brillouin-scattering-based slow-light systems. Opt. Lett. 32 (October), 220–222. https://doi.org/10.1364/OL.32.000220.

Schneider, T., Jamshidi, K., Preussler, S., 2010. Quasi-light storage: a method for the tunable storage of optical packets with a potential delay-bandwidth product of several thousand bits. J. Lightwave Technol. 28 (17), 2586–2592. https://doi.org/10.1109/JLT.2010.2056355.

Shen, Z., Zhang, Y.-L., Chen, Y., Zou, C.-L., Xiao, Y.-F., Zou, X.-B., Sun, F.-W., Guo, G.-C., Dong, C.-H., 2016. Experimental realization of optomechanically induced non-reciprocity. Nat. Photonics 10 (10), 657–661. https://doi.org/10.1038/nphoton.2016.161. http://arxiv.org/abs/1604.02297.

Shi, Z., Pant, R., Zhu, Z., Stenner, M.D., Neifeld, M.A., Gauthier, D.J., Boyd, R.W., 2007. Design of a tunable time-delay element using multiple gain lines for increased fractional delay with high data fidelity. Opt. Lett. 32 (14), 1986–1988. https://doi.org/10.1364/OL.32.001986.

Song, K.Y., 2011. Operation of Brillouin dynamic grating in single-mode optical fibers. Opt. Lett. 36 (23), 4686–4688. https://doi.org/10.1364/OL.36.004686. https://www.osapublishing.org/abstract.cfm?URI=ol-36-23-4686%5Cnhttp://www.ncbi.nlm.nih.gov/pubmed/22139284.

Song, K.Y., Yoon, H.J., 2010. Observation of narrowband intrinsic spectra of Brillouin dynamic gratings. Opt. Lett. 35 (17), 2958–2960. https://doi.org/10.1364/OL.35.002958.

Song, K.Y., González Herráez, M., Thévenaz, L., 2005a. Gain-assisted pulse advancement using single and double Brillouin gain peaks in optical fibers. Opt. Express 13 (24), 9758–9765. https://doi.org/10.1364/OPEX.13.009758. http://www.ncbi.nlm.nih.gov/pubmed/19503183.

Song, K.Y., Herráez, M., Thévenaz, L., 2005b. Observation of pulse delaying and advancement in optical fibers using stimulated Brillouin scattering. Opt. Express 13 (1), 82–88. https://doi.org/10.1364/OPEX.13.000082.

Song, K.Y., Herráez, M.G., Thévenaz, L., 2005c. Long optically controlled delays in optical fibers. Opt. Lett. 30 (14), 1782–1784. https://doi.org/10.1364/OL.30.001782.

Song, K.Y., Zou, W., He, Z., Hotate, K., 2008. All-optical dynamic grating generation based on Brillouin scattering in polarization-maintaining fiber. Opt. Lett. 33 (9), 926. https://doi.org/10.1364/OL.33.000926. https://www.osapublishing.org/abstract.cfm?URI=ol-33-9-926.

Song, K.Y., Lee, K., Lee, S.B., 2009a. Tunable optical delays based on Brillouin dynamic grating in optical fibers. Opt. Express 17 (12), 10344–10349. https://doi.org/10.1364/OE.17.010344.

Song, K.Y., Zou, W., He, Z., Hotate, K., 2009b. Optical time-domain measurement of Brillouin dynamic grating spectrum in a polarization-maintaining fiber. Opt. Lett. 34 (9), 1381–1383. https://doi.org/10.1364/OL.34.001381. http://www.ncbi.nlm.nih.gov/pubmed/19412279.

Stiller, B., Merklein, M., Wolff, C., Vu, K., Ma, P., Poulton, C.G., Madden, S.J., Eggleton, B.J., 2018. On-chip multi-stage optical delay based on cascaded Brillouin light storage. Opt. Lett. 43, 4321–4324.

Stiller, B., Merklein, M., Vu, K., Ma, P., Madden, S.J., Poulton, C.G., Eggleton, B.J., 2019. Cross talk-free coherent multi-wavelength Brillouin interaction. APL Photonics 4 (4), 040802. https://doi.org/10.1063/1.5087180.

Stiller, B., Merklein, M., Wolff, C., Vu, K., Ma, P., Madden, S.J., Eggleton, B.J., 2020. Coherently refreshing hypersonic phonons for light storage. Optica 7 (5), 492–497. https://doi.org/10.1364/optica.386535.

Thévenaz, L., 2008. Slow and fast light in optical fibres. Nat. Photonics 2 (8), 474–481. https://doi.org/10.1038/nphoton.2008.147. http://www.nature.com/nphoton/journal/v2/n8/abs/nphoton.2008.147.html.

Wallucks, A., Marinković, I., Hensen, B., Stockill, R., Gröblacher, S., 2020. A quantum memory at telecom wavelengths. Nat. Phys. 16 (7), 772–777. https://doi.org/10.1038/s41567-020-0891-z.

Weis, S., Riviere, R., Deleglise, S., Gavartin, E., Arcizet, O., Schliesser, A., Kippenberg, T.J., 2010. Optomechanically induced transparency. Science 330 (6010), 1520–1523. https://doi.org/10.1126/science.1195596. http://www.sciencemag.org/cgi/doi/10.1126/science.1195596.

Winful, H.G., Kabakova, I.V., Eggleton, B.J., 2013. Model for distributed feedback Brillouin lasers. Opt. Express 21 (13), 16191–16199. https://doi.org/10.1364/OE.21.016191. http://www.opticsexpress.org/abstract.cfm?URI=oe-21-13-16191.

Yi, L., Jaouen, Y., Hu, W., Su, Y., Bigo, S., 2007. Improved slow-light performance of 10 Gb/s NRZ, PSBT and DPSK signals in fiber broadband SBS. Opt. Express 15 (25), 16972–16979. http://www.ncbi.nlm.nih.gov/pubmed/19550988.

Zadok, A., Eyal, A., Tur, M., 2006. Extended delay of broadband signals in stimulated Brillouin scattering slow light using synthesized pump chirp. Opt. Express 14 (19), 8498–8505. https://doi.org/10.1364/OE.14.008498.

Zadok, A., Antman, Y., Primerov, N., Denisov, A., Sancho, J., Thevenaz, L., 2012. Random-access distributed fiber sensing. Laser Photonics Rev. 6 (5), 1–5. https://doi.org/10.1002/lpor.201200013.

Zhang, L., Soto, M.A., Thévenaz, L., 2014. Minimizing distortion and enlarging group delay in Brillouin slow light systems by gain profile optimization. In: Asia Communications and Photonics Conference, vol. 1. ATh2C. 4.

Zhu, Z., Dawes, A.M.C., Gauthier, D.J., Zhang, L., Willner, A.E., 2007a. Broadband SBS slow light in an optical fiber. J. Lightwave Technol. 25 (1), 201–206. https://doi.org/10.1109/JLT.2006.887188. http://ieeexplore.ieee.org/document/4137625/.

Zhu, Z., Gauthier, D.J., Boyd, R.W., 2007b. Stored light in an optical fiber via stimulated Brillouin scattering. Science 318 (5857), 1748–1750. https://doi.org/10.1126/science.1149066. http://www.ncbi.nlm.nih.gov/pubmed/18079395.

Zhu, Y., Wu, X., Wang, K., Shuai, W., Zhang, J., Qiao, L., Wang, T., Gao, S., Zhang, M., 2021. Experimental observation of chaotic Brillouin dynamic grating. Opt. Lett. 46 (23), 5814–5817. https://doi.org/10.1364/ol.441783.

Zou, W., Chen, J., 2013. All-optical generation of Brillouin dynamic grating based on multiple acoustic modes in a single-mode dispersion-shifted fiber. Opt. Express 21 (12), 14771–14779. https://doi.org/10.1364/OE.21.014771. http://www.ncbi.nlm.nih.gov/pubmed/23787664.

CHAPTER THIRTEEN

Nonreciprocity in Brillouin scattering

Nils T. Otterstrom[a], Eric A. Kittlaus[b], Donggyu B. Sohn[c], and Gaurav Bahl[c,*]
[a]Photonic and Phononic Microsystems, Sandia National Laboratories, Albuquerque, NM, United States
[b]Jet Propulsion Laboratory, California Institute of Technology, Pasadena, CA, United States
[c]Mechanical Science and Engineering, University of Illinois at Urbana-Champaign, Champaign, IL, United States
*Corresponding author: e-mail address: bahl@illinois.edu

Contents

1. Introduction

Nonreciprocal devices—especially isolators and circulators—play a crucial role in optical signal processing systems (Jalas et al., 2013; Sounas and Alù, 2017). As the description implies, these devices produce an optical response that is not reciprocal such that forward- and backward-propagating light experience distinct transformations. In the case of an isolator, for instance, light propagating in the forward direction is transmitted while light-propagating in the backward direction is extinguished. Fig. 1 shows a general optical medium (or device) that has three input ports, each with a forward- (a_i) and backward-propagating (b_i) fields. The relationship

Semiconductors and Semimetals, Volume 110
ISSN 0080-8784
https://doi.org/10.1016/bs.semsem.2022.05.006

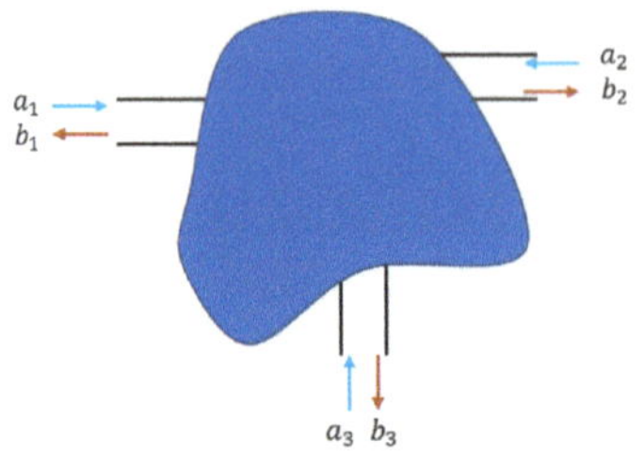

$$\begin{bmatrix} b_1 \\ b_2 \\ b_3 \end{bmatrix} = \begin{bmatrix} S_{11} & S_{12} & S_{13} \\ S_{21} & S_{22} & S_{23} \\ S_{31} & S_{32} & S_{33} \end{bmatrix} \begin{bmatrix} a_1 \\ a_2 \\ a_3 \end{bmatrix}$$

$$\bar{b} = S\bar{a} \qquad S^{\mathrm{T}} \neq S$$

Fig. 1 Scattering matrix formalism for nonreciprocal optical systems. General system with input fields a_i and output fields b_i. Reciprocity is broken when the scattering matrix is not symmetric.

between input and output modes is characterized by the scattering matrix S. For nonreciprocal systems, the scattering matrix is not symmetric (Jalas et al., 2013).

Traditional methods of producing optical nonreciprocity center on the use of magneto-optic Faraday rotation (Adam et al., 2002; Dötsch et al., 2005). In this effect, the precession of the magnetic dipoles within a magneto-optic material under an applied magnetic field produces a nonreciprocal phase delay between left- and right-circularly polarized light. This effect produces a form of nonreciprocal polarization rotation, which, combined with polarizers or polarizing beam splitters, is the most popular and commercially viable approach to produce nonreciprocal photonic elements such as isolators and circulators.

As optical systems and applications are increasingly moving toward chip-scale optics, it has been difficult to translate these magneto-optic based technologies to integrated systems and devices (Sounas and Alù, 2017). Magneto-optic materials are typically lossy and generally not compatible with standard CMOS processing in foundries. Moreover, this approach requires co-integration of magnetic field sources that can be difficult to shield, posing challenges for systems that are magnetic field sensitive (e.g., Zeeman shifts in atomic systems).

In view of the challenges faced by traditional magneto-optic approaches, alternative strategies such as spatiotemporal modulation and related nonlinear optical processes have emerged to achieve nonreciprocal light propagation without relying on magneto-optic-based materials (Sounas and Alù, 2017). As we will explore in this chapter, traveling-wave Brillouin optomechanical systems are particularly well-suited to this task (Kim et al., 2015; Kittlaus et al., 2018; Poulton et al., 2012; Sohn et al., 2018).

2. Nonreciprocity arising from Brillouin phase matching

The characteristic nonreciprocal dynamics of Brillouin interactions are fundamentally derived from the principle of phase matching. Specifically, in this three-wave process, the sum of the wavevectors of the annihilated fields must be equal to that of the generated fields (see Chapter "SBS-based fiber sensors" by Zadok et al.). This relationship gives rise to what can be viewed as two distinct cases of nonreciprocity: (1) the scattering regime, defined by when a particular elastic wave is supplied (in addition to one of the optical fields), and (2) the stimulated gain regime, in which the two optical fields are provided. In what follows, we examine each of these regimes in detail.

2.1 Scattering regime

We first examine the scattering regime, in which we consider the linear Brillouin scattering of light from pre-existing phonons in the medium. We start by considering optical and acoustic plane waves propagating within a bulk isotropic system with nonzero photoelastic coupling (Boyd, 2020). The system is biased by an acoustic field with frequency Ω_B and dispersion-restricted wavevector $\bar{q}(\Omega_B)$, which in the simple 2-D diagram in Fig. 2A makes an arbitrary angle θ with respect to the injected probe wave of frequency ω_1 and wavevector $\bar{k}(\omega_1)$.

To see how nonreciprocity arises in this case, we consider all possible anti-Stokes and Stokes processes that could potentially be permitted by forward and backward optical fields, which yields four possible sets of phase-matching conditions, given by

$$\begin{aligned} \bar{k}(\omega_2) &= \bar{k}(\omega_1) + \bar{q}(\Omega_B) \\ \omega_2 &= \omega_1 + \Omega_B, \end{aligned} \tag{1}$$

$$\begin{aligned} \bar{k}(\omega_0) &= \bar{k}(\omega_1) - \bar{q}(\Omega_B) \\ \omega_0 &= \omega_1 - \Omega_B, \end{aligned} \tag{2}$$

$$\begin{aligned} -\bar{k}(\omega_2) &= -\bar{k}(\omega_1) + \bar{q}(\Omega_B) \\ \omega_2 &= \omega_1 + \Omega_B, \end{aligned} \tag{3}$$

$$\begin{aligned} -\bar{k}(\omega_0) &= -\bar{k}(\omega_1) - \bar{q}(\Omega_B) \\ \omega_0 &= \omega_1 - \Omega_B. \end{aligned} \tag{4}$$

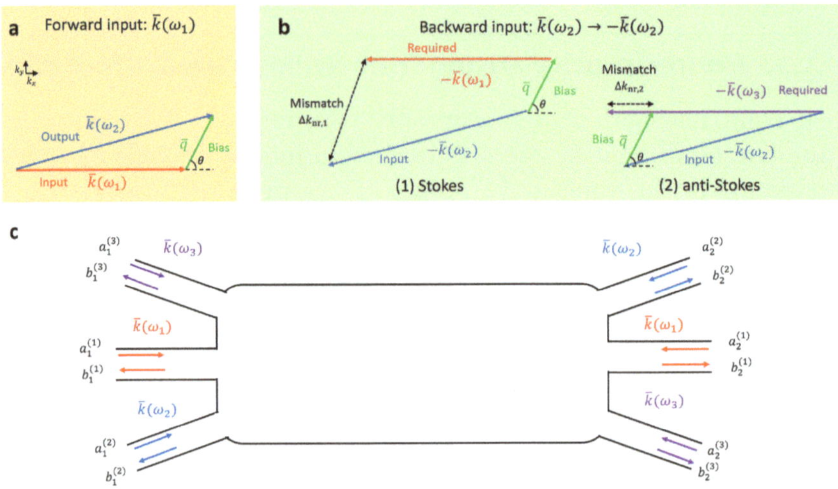

Fig. 2 Origin of nonreciprocity in linear Brillouin interactions. (A) A phase-matched forward anti-Stokes process for an input probe field ω_1 with an externally applied acoustic field, creating the output anti-Stokes field ω_2 with wavevector $\bar{k}(\omega_2)$. (B) When the output anti-Stokes field of case (A) is applied in the backward direction (i.e., $\bar{k}(\omega_2) \rightarrow -\bar{k}(\omega_2)$), neither the (1) Stokes nor (2) anti-Stokes processes in the backward direction are phase-matched. (C) In the limit of $\Delta k_{nr,1}L \gg 2.78$ and $\Delta k_{nr,2}L \gg 2.78$, this case results in the diagram (C) and scattering matrix given by Eq. (9).

Here, the first two sets of equations represent the phase matching requirements for the anti-Stokes and Stokes processes in the forward direction, and the last two sets of equations are the phase matching requirements for the anti-Stokes and Stokes processes in the backward direction.

Considering a medium with a single monotonic optical dispersion relation, it is clear that only one set of conditions can be true for any particular dispersion-allowed phonon frequency and wavevector.[a] For instance, given that the phonon field permits anti-Stokes scattering in the forward direction [i.e., Eq. (1) is satisfied], no backward scattering processes are permitted by phase matching, yielding a nonreciprocal response.

2.1.1 Nonreciprocal phase mismatch

To quantify the degree of nonreciprocity produced through these phase-matching constraints, we consider a phase-matched anti-Stokes process that satisfies

[a] Note that for the case of a 1-D waveguide systems (not the example considered here), maximally two sets of conditions [either Eqs. (1)–(2) or (3)–(4)] can be simultaneously true.

$$\begin{aligned}\overline{k}(\omega_2) &= \overline{k}(\omega_1) + \overline{q}(\Omega_B) \\ \omega_2 &= \omega_1 + \Omega_B,\end{aligned} \tag{5}$$

where again, the phonon field makes an arbitrary angle θ with respect to the injected probe wave of frequency ω_1 and wavevector $\overline{k}(\omega_1)$.

Following the definition of a nonreciprocal scattering matrix (as shown in Fig. 1), the direction and frequency of light will define a set of distinct input and output ports. As such, the relevant backward process corresponding to the anti-Stokes process above is a Stokes process wherein the output anti-Stokes light of wavevector $\overline{k}(\omega_2)$ is retro-reflected (i.e., $\overline{k}(\omega_2) \rightarrow -\overline{k}(\omega_2)$) and subsequently interacts with the acoustic phonons, as depicted in Fig. 2B(1).

As is clear geometrically from Fig. 2B(1), this backward process is not phase-matched with the original probe at ω_1, and the resulting phase mismatch is given by

$$\overline{\Delta k}_{nr,1} = -2\overline{q}(\Omega_B). \tag{6}$$

In the case that $|\overline{\Delta k}_{nr,1}|L = n\pi$ or $|\overline{\Delta k}_{nr,1}|L \gg 2.78$,[b] this scattering process is prohibited by the phase-matching constraint.

In certain applications, it is desirable that no backward scattering occur, even to a frequency that would constitute a distinct port. To fully consider these possibilities, we also examine the potential anti-Stokes scattering process to a frequency ω_3 and wavevector $-\overline{k}(\omega_3)$. As shown in Fig. 2B(2), it is clear that for a monotonic single optical dispersion relation, this case would also fail to satisfy the phase matching condition. The resulting phase mismatch is given by

$$\overline{\Delta k}_{\mathrm{nr},2} = -\overline{k}(\omega_3) - (\overline{q}(\Omega_B) - \overline{k}(\omega_2)). \tag{7}$$

If we neglect higher order dispersion, the magnitude of the phase-mismatch becomes

$$|\overline{\Delta k}_{\mathrm{nr},2}| = \frac{1}{v_g}(\omega_3 - \omega_1) = \frac{2\Omega_B}{v_g} \tag{8}$$

where v_g is the optical group velocity. Hence, the Stokes process arising from a backward-propagating pump will be suppressed by a wavevector

[b] The scattered power is proportional to $\mathrm{sinc}^2(|\overline{\Delta k}_{nr,1}|L/2)$, which goes to zero at $|\overline{\Delta k}_{\mathrm{nr},1}|L/2 = n\pi$ and has a half-width at half-max at $|\overline{\Delta k}_{nr,1}|L = 2.78$.

mismatch proportional to ratio of the Brillouin frequency to the group velocity of light, and in the case that $|\overline{\Delta k}_{nr,2}|L \gg 1$, no scattering will occur.

Altogether, we find that the supplied acoustic field allows only a single forward-scattering process according to the phase-matching condition. Characteristic of nonreciprocal systems, this case (under the conditions $|\overline{\Delta k}_{nr,1}|L \gg 2.78$ and $|\overline{\Delta k}_{nr,2}|L \gg 2.78$) yields a nonsymmetric scattering matrix given by

$$\begin{bmatrix} b_1^{(1)} \\ b_1^{(2)} \\ b_1^{(3)} \\ b_2^{(1)} \\ b_2^{(2)} \\ b_2^{(3)} \end{bmatrix} = \begin{bmatrix} 0 & 0 & 0 & 1 & 0 & 0 \\ 0 & 0 & 0 & 0 & 1 & 0 \\ 0 & 0 & 0 & 0 & 0 & 1 \\ \sqrt{1-\eta} & \sqrt{\eta} & 0 & 0 & 0 & 0 \\ \sqrt{\eta} & \sqrt{1-\eta} & 0 & 0 & 0 & 0 \\ 0 & 0 & 1 & 0 & 0 & 0 \end{bmatrix} \begin{bmatrix} a_1^{(1)} \\ a_1^{(2)} \\ a_1^{(3)} \\ a_2^{(1)} \\ a_2^{(2)} \\ a_2^{(3)} \end{bmatrix}, \tag{9}$$

where η is the scattering efficiency given by the ratio of scattered optical power relative to the input optical power, and the superscripts and subscripts indicate the sides and direction/frequency the ports, respectively, as illustrated in Fig. 2C.

Note that, while we considered an acoustic field that is phase-matched to a specific forward anti-Stokes process, it is typically the case that an elastic wave which produces phase-matched scattering for any one process will not phase-match to another Brillouin process. This direction-dependent phase matching forms the basis for Brillouin-based optical nonreciprocity.

2.1.2 Example: Intermodal Brillouin scattering

Next, let us see how this general framework can be applied to a 1-D waveguide system that produces intermodal Brillouin scattering. In this example, the same optical waveguide conveys light in two transverse spatial modes (with wavevectors $k_1(\omega)$ and $k_2(\omega)$, respectively). In this waveguide system, both forward and backward propagation of light are allowed. With careful design, we can find an acoustic wave that is phase-matched to the anti-Stokes process between symmetric (pump) and antisymmetric (anti-Stokes) spatial modes with a wavevector given by $q(\Omega_B) = k_2(\omega_2) - k_1(\omega_1)$ (Fig. 3).

To quantify the degree of nonreciprocity produced by phase matching, we will examine whether this same phonon can participate in a backward

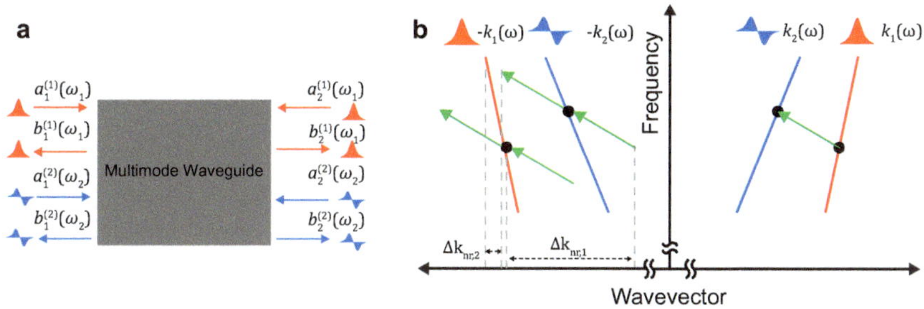

Fig. 3 Nonreciprocity in intermodal Brillouin scattering. (A) A four-port multimode waveguide system. (B) Dispersion diagram for intermodal Brillouin scattering highlighting the nonreciprocal phase mismatches $\Delta k_{\mathrm{nr},1}$ and $\Delta k_{\mathrm{nr},2}$, with $\Delta k_{\mathrm{nr},1} \gg \Delta k_{\mathrm{nr},2}$. A phonon field that satisfies phase matching for a forward anti-Stokes process (producing ω_2) does not phase match for either Stokes or anti-Stokes processes, with wavevector mismatches $\Delta k_{\mathrm{nr},1}$ and $\Delta k_{\mathrm{nr},2}$, respectively.

anti-Stokes or Stokes process. Considering the phase-matching condition, we see that a backward Stokes process yields a phase mismatch given by

$$\Delta k_{\mathrm{nr},1} = 2q(\Omega_\mathrm{B}), \tag{10}$$

whereas a backward anti-Stokes process produces a phase mismatch given by

$$\Delta k_{\mathrm{nr},2} = |-k(\omega_3) - (q(\Omega_\mathrm{B}) - k(\omega_2))| = \frac{2\Omega_\mathrm{B}}{v_\mathrm{g}}. \tag{11}$$

where $\omega_3 = \omega_2 + \Omega_\mathrm{B}$. Thus, for $\Delta k_{\mathrm{nr},1}L$, $\Delta k_{\mathrm{nr},2}L \gg 2.78$, the scattering is inherently nonreciprocal. We can express the scattering matrix for this nonreciprocal process as

$$\begin{bmatrix} b_1^{(1)}(\omega_1) \\ b_1^{(2)}(\omega_2) \\ b_2^{(1)}(\omega_1) \\ b_2^{(2)}(\omega_2) \end{bmatrix} = \begin{bmatrix} 0 & 0 & 1 & 0 \\ 0 & 0 & 0 & 1 \\ \sqrt{1-\eta} & \sqrt{\eta} & 0 & 0 \\ \sqrt{\eta} & \sqrt{1-\eta} & 0 & 0 \end{bmatrix} \begin{bmatrix} a_1^{(1)}(\omega_1) \\ a_1^{(2)}(\omega_2) \\ a_2^{(1)}(\omega_1) \\ a_2^{(2)}(\omega_2) \end{bmatrix}, \tag{12}$$

where the input (a) and output mode (b) amplitudes have subscripts and superscripts that indicate the port and mode, respectively, and where η is the scattering efficiency. The scattering matrix is nonsymmetric, revealing the nonreciprocal response of the multimode waveguide system. It is worth noting that the bandwidth of this nonreciprocal scattering response in principle has no restrictions if the slopes of dispersion relations of mode 1 and mode 2 are closely matched (i.e., when $v_{g,\,1} \approx v_{g,\,2}$) (Yu and Fan, 2009).

2.2 Stimulated gain regime

So far, we have seen how Brillouin light scattering can be biased by phonons within the material, leading to nonreciprocal effects. The available scattering pathways are restricted by wavevector and energy conservation conditions, that is, the phase matching relationship. In contrast to this "scattering regime," through the process of stimulated Brillouin scattering, a strong pump wave can also dictate the directionality of *stimulated* optical gain or loss experienced by a probe signal. Once again, the Brillouin phase matching condition determines whether a probe will experience gain (or loss) or pass through the medium unaffected.

The situation can be explored through a geometric representation (Fig. 4) similar to that of spontaneous scattering. Here, a strong, forward-propagating pump with wavevector $\overline{k}_p$ is injected along the x-axis of an isotropic dielectric medium. We consider a red-detuned probe with wavevector $\overline{k}_{pr}$ that propagates through the same volume as the pump.

For stimulated Brillouin scattering to occur, energy conservation requires that the frequency of the mediating elastic wave is equal to the difference frequency between the pump and probe waves. This frequency choice fixes the wavenumber $|\overline{q}|$ of the Brillouin-active elastic wave through the mechanical dispersion relation of the medium. This can be geometrically visualized (Fig. 4) as a dashed circle signifying the magnitude of the

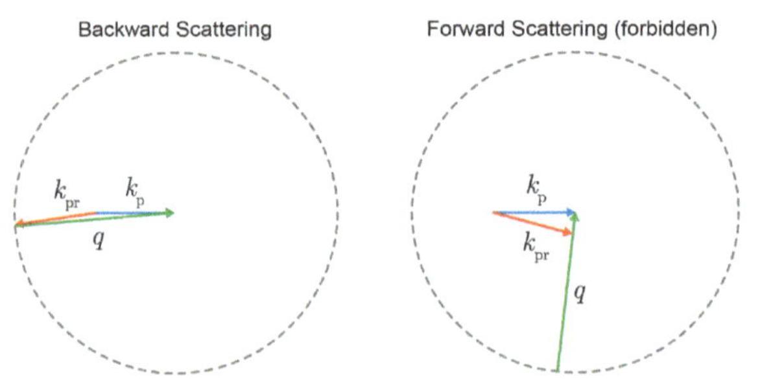

Fig. 4 Geometric representation of phase matching for stimulated Brillouin scattering. A strong pump with wavevector $\overline{k}_p$ and a red-detuned probe wave with $\overline{k}_{pr}$ pass through an isotropic medium. Their frequency difference sets the wavenumber $|q|$ of Brillouin-active phonons through the mechanical dispersion relation and energy conservation relation. Under these circumstances, phase-matched SBS can occur under certain directions, such as is the case for backward scattering (*left*). However, it cannot simultaneously be achieved for copropagating optical waves (*right*).

wavevector associated with phonons of the required frequency. Given a pump wave and a selected pump-probe detuning, we find that the Brillouin phase matching will be satisfied when $\overline{k}_{\mathrm{p}} - \overline{q} - \overline{k}_{\mathrm{pr}} \approx 0$, leading to SBS back-scatter from pump to probe waves. By contrast, forward-direct SBS cannot typically occur in bulk media, since bulk acoustic phonons carry very large momentum that can only be bridged by photons traveling in opposite directions. This leads to the well-known preferential occurrence of backward-directed SBS lasers and amplifiers, that counter-propagate relative to the pump.

As a result, the presence of the pump breaks propagation symmetry for probe waves, leading to nonreciprocity. Specifically, the pump wave sets the allowable configuration for energy transfer to occur based on the available mechanical dispersion relation. In this familiar context of backward SBS, gain or loss for the probe only occurs if it is counter-propagating with respect to the pump. There is no Brillouin interaction when the probe is reversed, leading to a nonreciprocal response with respect to propagation of the probe wave.

To further illustrate the nonreciprocal behavior of SBS, we examine the simplified case of a 1-D system (e.g., waveguide, ring resonator, or exactly co/counterpropagating light in a bulk medium) as shown in Fig. 5A. Here, the allowed configurations for SBS can be readily understood from the optical and mechanical dispersion relations in frequency–wavevector space, as shown in Fig. 5B–D. For the backward-SBS process the mechanical dispersion relation determines two solutions relative to the pump where the frequency and wavevector of the probe are correctly phase matched. For a red-detuned probe with backward wavevector k_{pr} relative to the pump wavevector k_{p}, (see Fig. 5B and D) we find the familiar case of stimulated Brillouin gain in which energy transfer occurs from the strong pump to the relatively weaker probe, along with the creation of forward propagating phonons with wavevector q. In this simplified geometric picture shown above, no other optical energy transfer process is possible for red detuning (e.g., in the forward direction) since there is no combination of phonon and destination optical state to constitute an alternate allowed photonic transition. Similarly, for a backward blue-detuned probe (Fig. 5C and D) we find that stimulated Brillouin "loss" is phase matched, through which energy is transferred from the weaker probe to the stronger pump, along with the creation of backward propagating phonons (wavevector $-q$). Once again, we find that a forward-directed blue-detuned probe does not experience any gain or loss.

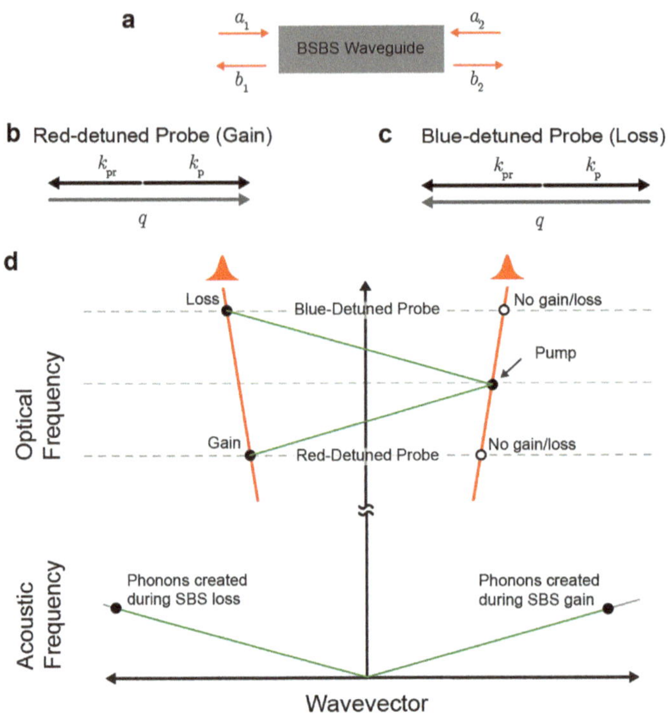

Fig. 5 Geometric picture of backward SBS (BSBS) in a guided-wave system. (A) Light is coupled from incoming (a_i) to outgoing (b_j) channels in the same spatial mode through a scattering matrix representing the SBS process (Eq. 13). (B,C) Simplified wavevector matching diagrams for Brillouin gain and loss through stimulated Stokes scattering. (D) Phase matching for BSBS plotted in frequency–wavevector space. A phonon mode couples light between forward- and backward-propagating wavevectors at two unique points on the optical dispersion relation. Forward scattering is not permitted through an interaction with this same phonon mode.

The nonreciprocal behavior of the signal wave can be summarized in a simplified scattering matrix representation. Through a stimulated Stokes process, energy transfer occurs from the higher frequency to lower frequency optical wave, with a relative gain G (the gain in power units is $|G|^2$, in comparison to scattering efficiency η in the scattering regime). When only backward SBS is allowed by phase matching, this results in the scattering relationship

$$\begin{bmatrix} b_1 \\ b_2 \end{bmatrix} = \begin{bmatrix} 0 & G \\ 1 & 0 \end{bmatrix} \begin{bmatrix} a_1 \\ a_2 \end{bmatrix} \tag{13}$$

where a_1, a_2, b_1, and b_2 are the field amplitudes of the incoming and outgoing waves as illustrated in Fig. 5A. Similarly, injecting a blue-detuned probe would result in a scattering matrix of the same form, except $|G| < 1$ due to Brillouin loss.

This analysis has assumed ideal energy conservation for the SBS process. In practice, the finite lifetime of the phonons mediating SBS leads to a finite frequency bandwidth over which SBS can occur, typically in the few kHz to ~100 MHz range depending on system and material parameters. This means that the process is nonreciprocal for a relatively narrow band of probe detunings around the phonon frequency that satisfies phase matching.

So far, we have seen cases where SBS occurs only between counter-propagating waves. Note that while SBS can occur between copropagating optical waves ("forward SBS"), it requires modal engineering of the acoustic dispersion to manifest elastic modes with suitably small wavevector. For instance, this can be realized in guided-wave mechanical systems where laterally confined cutoff elastic modes exist with high frequency but very small wavevector.

Based on these analyses, we find that nonreciprocal SBS manifests as unidirectional gain or loss for a suitable optical probe, which is in principle frequency resolvable. In practical systems, however, the separation of pump and probe may require narrowband filtering which can be technologically challenging to implement. As a result, interpolarization or intermodal versions of SBS are attractive to more easily produce useful nonreciprocal scattering matrices in practical systems, as polarization- (or mode-) multiplexing can be more straightforward to implement with standard linear optical components (Kang et al., 2011).

We present an example of such a system in Fig. 6. Here, we consider an optical waveguide that supports two optical modes with distinct transverse spatial profiles, but this picture can be readily translated to systems with other forms of distinct optical channels or polarizations. Pump light is injected into mode 1 of the waveguide, while the frequency detuned probe is coupled into mode 2 of the same waveguide.

This example considers forward stimulated intermodal scattering (F-SIMS, Fig. 6) where Brillouin coupling occurs when pump and probe waves are copropagating, provided that the phonon dispersion permits the interaction. In frequency–wavevector space, the pump and probe occupy points on two distinct optical dispersion branches. When we inject a red-detuned probe at frequency $\omega_s = \omega_p - \Omega$ where Ω is the frequency of a Brillouin-active phonon, the signal wave experiences strong gain. In this case, the phonon satisfies the standard phase matching relationship $\overline{k}_p = \overline{k}_s + q$. However, if the signal wave is injected in the opposite

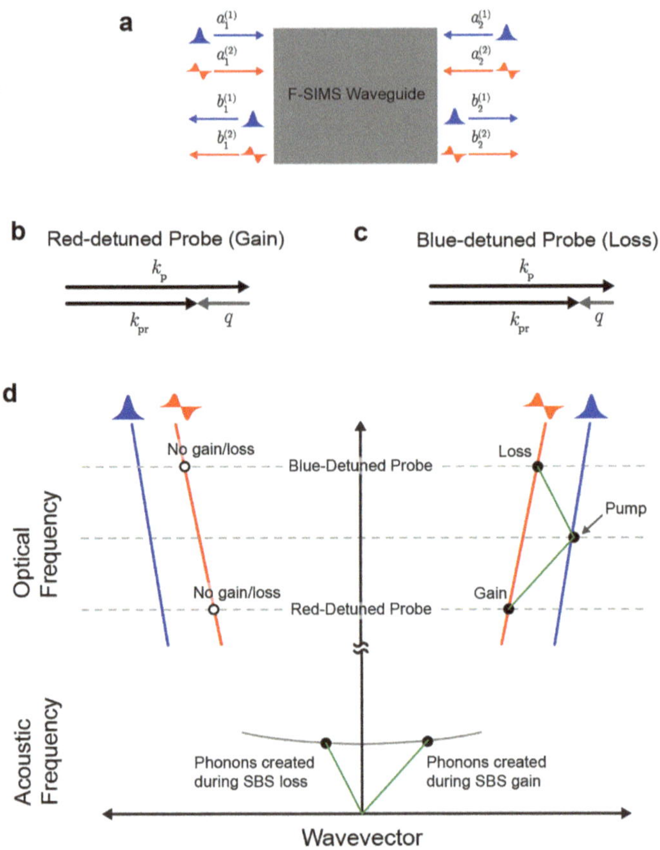

Fig. 6 Intermodal forward SBS in a guided-wave system. (A) Input–output diagram of intermodal forward SBS (F-SIMS). Light is coupled from incoming ($a_i^{(j)}$) to outgoing ($b_k^{(l)}$) channels through a scattering matrix representing the intermodal SBS process (Eq. 14). (B,C) Simplified wavevector-matching diagrams for Brillouin gain and loss through this process. (D) Phase matching for intermodal forward SBS plotted in frequency–wavevector space. A phonon couples light between two forward-propagating optical waves guided in distinct optical modes (or polarizations), corresponding to two unique points on the optical dispersion relations. Coupling is not permitted in the backward direction via this phonon because its wavevector is too small in magnitude.

direction, it passes through the waveguide unaffected, since there is no phonon satisfying the phase-matching condition for backward scattering. After the coupling, the optical waves are demultiplexed using, for example, a mode-selective directional coupler or a polarizing beam-splitter. For this two-mode (or two-polarization) system, the scattering matrix becomes:

$$\begin{bmatrix} b_1^{(1)} \\ b_1^{(2)} \\ b_2^{(1)} \\ b_2^{(2)} \end{bmatrix} = \begin{bmatrix} 0 & 0 & 1 & 0 \\ 0 & 0 & 0 & 1 \\ G & 0 & 0 & 0 \\ 0 & 1 & 0 & 0 \end{bmatrix} \begin{bmatrix} a_1^{(1)} \\ a_1^{(2)} \\ a_2^{(1)} \\ a_2^{(2)} \end{bmatrix} \tag{14}$$

where superscripts (1) and (2) represent the first and second optical modes, and subscripts 1 and 2 refer to waves on the lefthand (righthand) side of the interaction region and a (b) refers to incoming (outgoing) optical waves. Through this operation, the functional conclusion is essentially the same—the pump wave sets the allowable gain or loss process for the three-wave interaction based on the available phonon mode for coupling.

Through these many forms of SBS interactions, the achievable nonreciprocal contrast is set by the net strength of the Brillouin interaction–in other words by the amount of unidirectional gain or loss. Practically speaking, this is thus limited by the $G_B PL$ product, where G_B is the Brillouin gain coefficient in units of $1/(\text{W}\cdot\text{m})$, P is the pump power, and L is the interaction length. $G_B PL > 6.9$ is required for 99.9% attenuation (i.e., 30 dB isolation). Such large couplings are generally possible in fiber-based waveguide systems. Alternatively, we can consider optical or optomechanical resonators which create strong resonant enhancement of the nonlinear interaction. Using this approach, compact, low-power, and high-contrast nonreciprocal elements can be engineered.

In summary, stimulated Brillouin gain and loss are nonreciprocal due to directional (or chiral) biasing from the pump-wave and the required phase-matching condition. This directionality can be used to create optically controlled nonreciprocal components. Typically, these are narrowband effects due to the frequency bandwidth of the SBS interaction, in contrast to the linear scattering process discussed in the previous section. Note that, as with other strongly dispersive processes (Budker et al., 1999; Hau et al., 1999; Kasapi et al., 1995; Kash et al., 1999), SBS can be used to produce slow and fast light (Okawachi et al., 2005), which is also inherently nonreciprocal.

2.3 Induced transparency processes

The combination of nonreciprocal attenuation or gain with a resonant optical cavity can produce nonreciprocal transfer functions that are significantly enhanced compared against single-pass configurations. Two prominent examples are Brillouin scattering-induced transparency (BSIT) and Brillouin scattering-induced absorption (BSIA). Similar to the phenomena of electromagnetically induced transparency (EIT), the coherent interference between stimulated Brillouin attenuation (or gain) and an optical cavity response can give rise to a characteristic transparency (or absorption) window. While narrowband, these effects are very strong and can be leveraged for isolator or circulator applications.

The optically absorptive electronic transitions in atomic vapors are widely used as they are associated with states that have long coherence times, and can support quantum information storage (Liu et al., 2001; Phillips et al., 2001), optical delays (Budker et al., 1999; Hau et al., 1999; Kasapi et al., 1995; Kash et al., 1999), and interferometric experiments (Dimopoulos et al., 2007; Peters et al., 2001; Shin et al., 2004). In 1991, Boller et al. (1991) showed that this type of optical absorption could be suppressed by generating an interfering pathway via pumping of an auxiliary electronic transition. This family of mechanisms associated with optically pumped inhibited absorption is now known as EIT. The effect is typically generated in a window that is narrower than the primary absorptive transition and therefore leads to an extreme dispersion. For details and energy-level configurations for EIT, we refer the reader to the review by Fleischhauer et al. (2005). Of interest to us are the key requirements to generate EIT: (1) the pump must be on resonance with the detuning of the auxiliary state relative to the primary absorptive state, (2) the auxiliary state needs to ideally have a very long lifetime, that is, a much narrower linewidth than the original absorptive transition, and (3) transitions between the auxiliary state and the reference ground state should be disallowed.

Later advances in ultrahigh-Q optical resonators (Armani et al., 2003; Collot et al., 1993; Grudinin et al., 2006; Lee et al., 2012; Soltani et al., 2007) and the generation of large opto-mechanical coupling (Eichenfield et al., 2009; Kippenberg and Vahala, 2007) opened the door to EIT-like physics in solid-state systems through photon–phonon interactions. Optical resonators exhibited discrete narrow linewidth modes that could be engineered and used in the same way as electronic transitions in atoms. Mechanical vibrational modes of these resonators, in the sub-GHz regime, were (comparatively) extremely long-lived coherent states that could be coupled to the optical modes using radiation pressure. The first optomechanical demonstrations of EIT, that is, optomechanically induced transparency (OMIT) leveraged these new capabilities and were presented by Weis et al. (2010) and Safavi-Naeini et al. (2011).

Stimulated Brillouin scattering, as we have learned, is essentially an optomechanical coupling process, though the possibility of generating EIT-like effects initially seemed elusive since the propagating phonons associated with backward SBS are in the >10 GHz range with very short lifetimes (nanoseconds) compared to the photon lifetimes (microseconds). This situation is quite different in forward SBS where much lower frequency mechanical modes can be accessed, with frequencies in the MHz range and phonon lifetimes approaching milliseconds. The first experiments demonstrating such a forward BSIT were presented in 2015 (Dong et al., 2015; Kim et al., 2015) using silica whispering gallery resonators (WGRs) with a modal configuration shown in Fig. 7A and B. The distinctive feature of BSIT, in contrast to OMIT, is the phase-matching requirement for the interfering pathway to be generated, and as we have seen, this phase matching is key to producing nonreciprocal effects. As the phase matching diagram shows (Fig. 7B) the generation of BSIT requires a forward Brillouin scattering configuration, with two optical modes of the WGR (an upper mode at (ω_1, k_1) and a lower mode at (ω_2, k_2)) having separation in frequency–wavevector space that matches a co-located high-Q low-frequency mechanical mode (Ω, q). We emphasize that unlike the waveguide cases discussed previously, the modes participating in the Brillouin process within a resonator must occupy discrete points in the frequency–wavevector space. The lower optical mode (ω_2) is pumped via the waveguide in only one circulation direction, while the higher frequency mode (ω_1) can be probed in both directions to characterize the forward and backward transmission coefficients. Since a clockwise (counterclockwise) pump only induces the Brillouin coupling of light and phonons in the clockwise (counterclockwise) direction within the WGR, the BSIT effect is expected to be nonreciprocal and this nonreciprocity should be measurable via the waveguide. Ultimately, this waveguide nonreciprocity is of the greatest interest since in the strong coupling regime the transmission through the waveguide can approach unity.

Using coupled mode theory we can prepare a system of equations that describe the interacting fields within the system. Note that the pump equation of motion is not needed and its effect is more simply incorporated into the pump-enhanced Brillouin coupling rate (i.e., scattering rate) g.

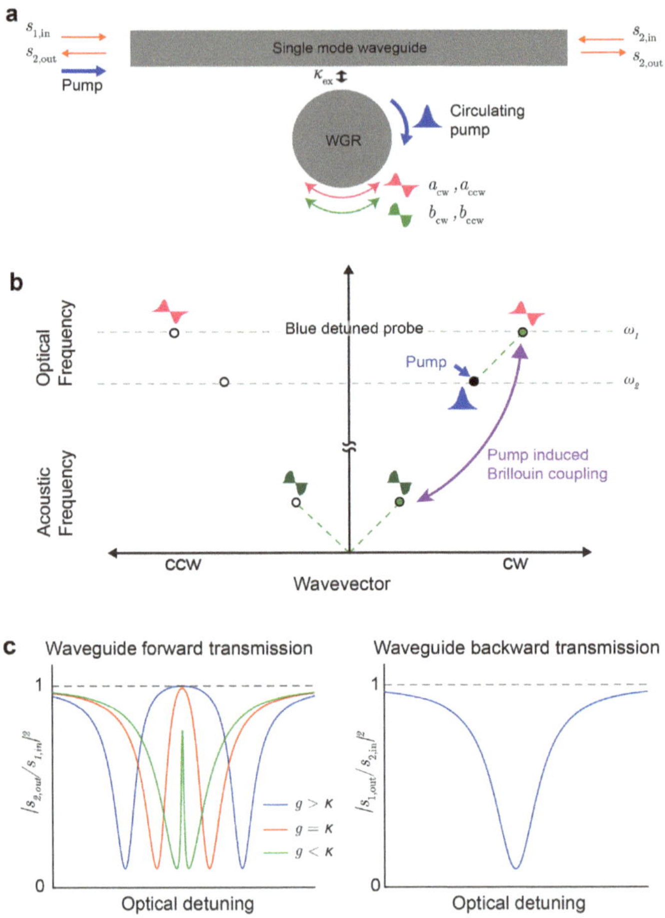

Fig. 7 Configuration for observation of Brillouin scattering-induced transparency (BSIT). (A) We consider a system composed of a single-mode waveguide and a two-mode whispering gallery resonator (WGR). The variables are described in detail in the accompanying text. (B) The optical modes of the WGR belonging to distinct mode families are configured for forward Brillouin phase matching with a phonon mode of the same WGR as shown. The circulating pump at ω_2 (here shown having a clockwise or cw wavevector) induces a coupling between light in the cw propagating optical mode at ω_1 and cw propagating phonons at Ω. As a result, light tuned to ω_1 propagating in left-to-right direction in the waveguide will interact with the cw modes of the WGR and experience the BSIT. Light propagating in the right-to-left direction in the waveguide will only experience absorption by the WGR. (C) Example curves showing how the nonreciprocal transparency arises near ω_1 when with coupling rate g.

$$\begin{bmatrix} \dot{a}_{ccw} \\ \dot{b}_{ccw} \\ \dot{a}_{cw} \\ \dot{b}_{cw} \end{bmatrix} = \begin{bmatrix} -i\omega_1 - \kappa/2 & ig/2 & 0 & 0 \\ ig^*/2 & -i\Omega - \Gamma/2 & 0 & 0 \\ 0 & 0 & -i\omega_1 - \kappa/2 & 0 \\ 0 & 0 & 0 & -i\Omega - \Gamma/2 \end{bmatrix} \begin{bmatrix} a_{ccw} \\ b_{ccw} \\ a_{cw} \\ b_{cw} \end{bmatrix} + \begin{bmatrix} \sqrt{\kappa_{ex}} s_{1,\ in} e^{-i\omega_l t} \\ 0 \\ \sqrt{\kappa_{ex}} s_{2,\ in} e^{-i\omega_l t} \\ 0 \end{bmatrix} \tag{15}$$

where ω_1 and Ω are the resonant frequencies of the optical probe mode and mechanical mode as described above; κ and Γ are their respective loss rates; ω_l is the optical probe frequency; and κ_{ex} is the evanescent coupling rate between the optical mode and the waveguide. Each mode participating in the process occurs in frequency degenerate clockwise and counterclockwise pairs; we use $a_{cw,\ ccw}$ to denote the intracavity optical fields and $b_{cw,\ ccw}$ to denote the intracavity mechanical fields. Since the probing is performed through the waveguide, we can additionally write the input–output relation from the perspective of the waveguide as

$$\begin{bmatrix} s_{2,\ out} \\ s_{1,\ out} \end{bmatrix} = \begin{bmatrix} s_{1,\ in} \\ s_{2,\ in} \end{bmatrix} - \begin{bmatrix} \sqrt{\kappa_{ex}} & 0 \\ 0 & \sqrt{\kappa_{ex}} \end{bmatrix} \begin{bmatrix} a_{cw} \\ a_{ccw} \end{bmatrix} \tag{16}$$

where $s_{1,\ in}$ ($s_{2,\ in}$) is the input field from Port 1 (Port 2) and $s_{1,\ out}$ ($s_{2,\ out}$) is the output field from Port 1 (Port 2). In Fig. 7C we present example optical transfer curves through the waveguide as a function of increasing g. For $g = 0$, we find that the response is completely symmetric and light is absorbed from the waveguide to the resonator. However, as the cw pump power increases, the nonreciprocal inhibited absorption becomes readily apparent. In the weak coupling regime, where the pump-enhanced Brillouin coupling rate is lower than the optical loss rate in the WGR ($g \ll \kappa$), the bandwidth of the transparency is limited to the mechanical mode loss rate (typically ~ kHz–MHz range). On the other hand, when the Brillouin coupling exceeds the optical loss rate ($g \gg \kappa$, i.e., the strong coupling regime) the cw optical mode is strongly coupled to the cw phonon mode, resulting in a large mode split and near-unity transmission at the center. Ultimately, even with strong coupling, the nonreciprocal isolation bandwidth is limited by the optical mode as its linewidth also sets the absorption band (~ GHz) in the opposite direction. Moreover, by engineering critical coupling between the WGR and waveguide, we can simultaneously create a large absorption for light propagating in the opposite direction (Kim et al., 2017a).

The BSIT process can also be intuited in the following way. The pump aligned with the lower WGR optical mode induces coupling (or scattering) between the higher optical mode and the mechanical mode; essentially a pump photon at ω_2 and a phonon at Ω can be annihilated to create a new photon at ω_1 in the blue-shifted mode, or via stimulated scattering a photon can be annihilated from the mode at ω_1 to create a new photon at ω_2 and a phonon at Ω. In this view, an additional pump-induced scattering pathway is generated for light from the interacting with the optical mode at ω_1, by scattering to the mechanical mode and back to the ω_1 mode with a π phase shift. This interfering effect naturally occurs only over the linewidth of the mechanical mode, and therefore produces a window of transparency where the absorption of light into the optical mode is suppressed (Fig. 7C).

Interestingly, if the pump and the probe are exchanged, that is, the higher mode is pumped and the lower mode is probed, then the interference generates an enhanced absorption rather than a transparency. By analogy with electromagnetically induced absorption (EIA) (Lezama et al., 1999), this effect is termed as BSIA (Kim et al., 2015). As with other strongly dispersive processes (Budker et al., 1999; Fiore et al., 2011; Hau et al., 1999; Kasapi et al., 1995; Kash et al., 1999; Safavi-Naeini et al., 2011; Weis et al., 2010), BSIT and BSIA can be used to produce slow and fast light with much greater group delays than with SBS in nonresonant systems, though over a much narrower bandwidth (the product of gain and bandwidth remains comparable).

3. Experimental platforms

Recent advancements in chip-scale integrated photonics have renewed the interest in developing compact nonreciprocal devices like isolators and circulators. Indeed, it is often not possible or convenient to adopt the conventional magneto-optic approach in foundries due to technical limitations in materials and processing. The stray magnetic fields emanating from magneto-optic devices are also not suitable for some modern applications, for example, in quantum devices and atom optics, which makes the search for nonmagnetic isolators even more important. Brillouin interactions represent a compelling candidate approach to realize these types of device technologies, as they can be extremely efficient in chip-scale optomechanical waveguides. Importantly, the pumping mechanism can be turned on and off as needed, which allows for more reconfigurability than magneto-optic systems. However, prior to the advent of chip-based Brillouin devices, it was already long understood that acousto-optic interactions could be used to produce nonreciprocal effects. Here, we briefly review the history of Brillouin based nonreciprocal systems from off-chip to chip-scale systems.

3.1 Bulk and fiber media

Brillouin scattering in bulk media, that is, in acousto-optic modulators (AOMs), has been used over many decades for optical isolation functions (Corcoran, 1978; Saleh and Teich, 2019) especially by the atom optics community. Similarly, Brillouin scattering processes have been effectively leveraged for nonreciprocal effects in fiber (Bashan et al., 2021; Hwang et al., 1997; Kang et al., 2011). These systems all employ variations of the spontaneous scattering principle as described in Section 2.1 as well as the stimulated scattering principle as described in Section 2.2, and are especially useful over wavelength ranges where magneto-optic materials are unavailable or in situations in which they are unsuitable. While these implementations require active input, such as a laser or mechanical actuator, and are therefore slightly less practical than magneto-optic isolators and circulators, they are noteworthy demonstrations of nonreciprocal dynamics that can also be adapted to chip-scale systems.

A simple frequency-shifting nonreciprocal device can be implemented using a conventional acousto-optic frequency shifter, as shown in Fig. 8A, by leveraging the spatial separation of momentum resolved scattering sidebands. Here, a piezoelectric transducer launches a propagating acoustic wave at frequency Ω through an optically transparent material, which results in a dynamic refractive index grating via the photoelastic effect. This grating diffracts and frequency-shifts the incident light (ω_l) with a phase-matching relationship as shown in Fig. 8A. The first-order scattering sideband (at $\omega_l + \Omega$) is angularly separated from other sidebands and can be captured as the frequency shifted output signal. We now consider how a returning signal at $\omega_l + \Omega$ propagates through the system in the backward direction, and find that any returning light experiences an additional frequency up-shift and acquires additional momentum. By then blocking the signal at $\omega_l + 2\Omega$ using a spectral filter or narrowband absorber, this AOM system can be effectively used as an optical isolator. Even without a filter, the photonic system may be insensitive to the frequency up-shifted light depending on the operating bandwidth of interest.

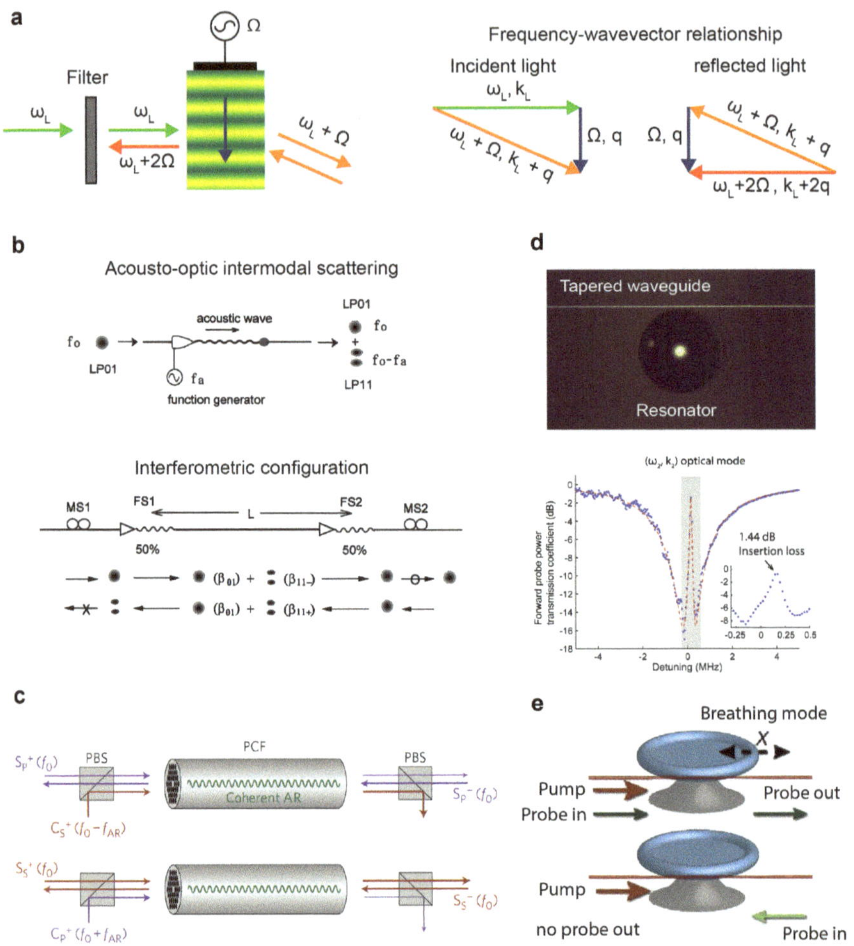

Fig. 8 Examples of bulk and fiber systems that exhibit Brillouin scattering based nonreciprocity. (A) Conventional acousto-optic modulators have been used for optical isolation for many decades (Corcoran, 1978; Saleh and Teich, 2019). (B) An all fiber optical isolator using intermodal scattering was demonstrated in Hwang et al. (1997). (C) Photonic crystal fiber demonstration of SBS based optical isolation (Kang et al., 2011). (D) Silica whispering gallery resonator for Brillouin scattering induced transparency (BSIT) based optical nonreciprocity (Kim et al., 2015, 2017a). (E) Optomechanically induced transparency (OMIT) in microtoroid resonators (Ruesink et al., 2016).

Similar concepts have been explored within optical fiber. The first explicit experiment on nonreciprocal light transmission in an optical fiber was performed by Hwang et al. in 1997 using a very simple intermodal scattering configuration similar to Fig. 3. Here, the researchers injected acoustic waves into the fiber by simply shaking the fiber with a piezoelectric actuator (Fig. 8B). As the fiber supported two optical modes, the light was scattered between the modes only for one direction of propagation. The authors also showed that an interferometric scheme composed of two such AOMs along with a phase delay line could be used to cause the frequency-shifted terms to destructively interfere and enhance the isolation effect (Fig. 8B). Notably, the bandwidth of this configuration is limited by the frequency band over which the phase matching exists, which can be several nanometers.

Experimental studies on using optically pumped stimulated Brillouin scattering for nonreciprocity have been relatively recent demonstrated. The first explicit experiments in this area were performed by Kang et al. (2011) with the help of a specially engineered photonic crystal fiber (PCF). The PCF was designed to support not only the required optical modes but also an acoustic mode confined in the transverse direction. The fiber's micrometer-scale core helped with strong co-confinement of light and vibration so that the stimulated Brillouin scattering process could be significantly enhanced. The researchers were able to demonstrate optical isolation using forward intramodal scattering and forward interpolarization scattering processes over 15 m of PCF fiber. More recently, these nonreciprocity concepts were also demonstrated in commercial polarization maintaining fiber (Bashan et al., 2021). For such designs the isolation bandwidth is limited by the gain/loss bandwidth for stimulated Brillouin scattering, which corresponds to the mechanical linewidth (typically kHz–MHz range).

3.2 Whispering gallery resonators

The most notable feature of whispering gallery micro-resonators (WGRs) is their ability to colocalize and recirculate phonons with high finesse in a very small volume, which significantly reduces the power threshold required for nonlinear optical processes. In the context of Brillouin scattering, the tight confinement and resonance feature of WGRs help enhance the interaction between the optical and mechanical modes by orders of magnitude compared to single-pass systems. Since optical waveguides (e.g., tapered fibers) are commonly used to access the circulating modes of a WGR, and since Brillouin coupling within the WGR can be configured to be nonreciprocal, the resulting input–output relations for the waveguide can be made nonreciprocal.

While stimulated Brillouin lasing readily appears in microresonators (Grudinin et al., 2009; Lee et al., 2012; Li et al., 2012, 2014; Tomes and Carmon, 2009), the first demonstrations of nonreciprocal effects with

WGRs in the form of BSIT and BSIA by Kim et al. (2015) and Dong et al. (2015). Typically, light in the waveguide that on resonance with a WGR mode is absorbed reciprocally and dissipated in the WGR. Both these experiments used ultrahigh-Q silica WGRs and showed that, if the resonator had suitable Brillouin-coupled modes (as described in Section 2.3), a unidirectional transparency could be induced for one direction of propagation within the waveguide. Kim et al. (2017a) later elaborated the conditions under which BSIT can achieve complete optical isolation, that is, ideal transparency simultaneously with complete absorption in the blocking direction.

Notably, this type of nonreciprocal-induced transparency can be generated not only in Brillouin systems in which there are whispering gallery acoustic modes but also in resonators having standing-wave vibrational modes with zero wavevector. This is because the momentum bias is set by the selection of the optical pump field, and the interaction between the pump and a forward probe is different from that with a backward probe. Such nonreciprocal OMIT was demonstrated using fused silica microsphere WGRs (Shen et al., 2016) and also in microtoroid resonators (Ruesink et al., 2016).

These WGR based nonreciprocal systems can also be used for circulator functionality by introducing additional add/drop waveguides. Notable experimental demonstrations have been performed using OMIT (Ruesink et al., 2018; Shen et al., 2018). An innovative BSIT-based circulator has also been theoretically proposed with the help of an additional Mach–Zehnder interferometer (Fu et al., 2015) but is yet to be realized in practice. Finally, as discussed in Section 2.1 of this chapter and in the next section, we should not overlook the possibility of biasing the interaction through phonon pumping as it is extremely effective in an on-chip WGR environment (Sohn et al., 2018, 2021; Tian et al., 2021).

While port-to-port nonreciprocal transmission effects are of interest for isolators and circulators, we should also look beyond this basic functionality. In particular, any Brillouin process breaks the underlying symmetry, of the clockwise and counterclockwise modes of the WGR, that is, imparts chirality to the modal properties, which can impart protection against undesirable scattering processes. For instance, it was shown in Kim et al. (2017b) that Brillouin anti-Stokes cooling (Bahl et al., 2012) imparts a chiral damping to the phonon population in a WGR, which can be used to suppress undesirable Rayleigh backscattering of the phonon population over the nonreciprocal band. Similarly, BSIT in a WGR also imparts a strong chirality to the optical modes, which can be used to shut down undesirable

Rayleigh backscattering of photons between cw and ccw directions (Kim et al., 2019). These ideas show that Brillouin scattering based nonreciprocity could be leveraged to produce photonic waveguiding systems that are intrinsically resilient against microscale defects.

3.3 Integrated photonic systems

Brillouin-based nonreciprocal effects are also of great interest in emerging chip-based platforms, where conventional approaches to nonreciprocity are challenging to implement. In particular, existing techniques to create integrated magneto-optic isolators and circulators require the inclusion of nonstandard magnetic materials which are difficult to introduce in foundries and even in boutique-integrated photonics platforms (Bi et al., 2011; Huang et al., 2016; Shoji et al., 2008; Sobu et al., 2013; Tien et al., 2011). Moreover, these materials are often optically lossy and only operate well in certain wavelength ranges. Techniques based on electro-optic modulation (Doerr et al., 2011, 2014; Ibrahim et al., 2004) and light-sound interactions are therefore compelling as alternative approaches.

As we have discussed, SBS is intrinsically nonreciprocal, and therefore represents a powerful approach to achieve nonmagnetic nonreciprocity on-chip (Fig. 9). There are many notable examples of backward and forward SBS in integrated photonic waveguides, many of which exhibit exceptionally large Brillouin couplings (Gundavarapu et al., 2019; Kittlaus et al., 2016, 2017; Liu et al., 2021; Merklein et al., 2021; Morrison et al., 2017; Pant et al., 2011; Shin et al., 2013; Van Laer et al., 2015). From a technological standpoint, to adapt SBS to a practical nonreciprocal device like an isolator, it is desirable to have a means to spatially separate the various participating optical fields, for example, pump and probe, through spectral or modal multiplexing. In photonic waveguides, this can be achieved using integrated mode multiplexers and intermodal SBS to produce nonreciprocal gain and loss in a multiport system. Demonstrations of this idea have been presented with both forward SBS in a silicon waveguide (Kittlaus et al., 2017) and backward SBS in a hybrid chalcogenide/silicon photonic platform (Liu et al., 2021).

To achieve large Brillouin gain, necessary for practical nonreciprocal components, in integrated waveguides typically requires some combination of long device lengths and relatively high pump powers. To reduce footprint and power requirements, resonant structures may be employed, as mentioned previously in Section 3.2. Demonstrations of SBS-based nonreciprocal effects

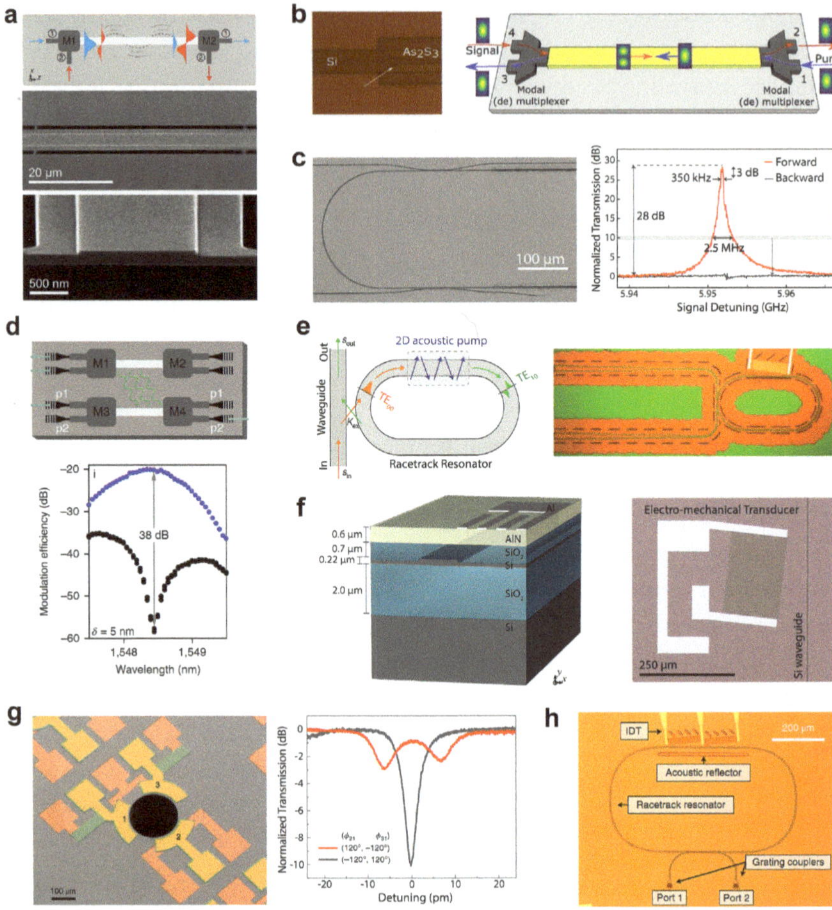

Fig. 9 Examples of nonreciprocal systems based on Brillouin scattering in integrated photonic waveguides. (A) Intermodal forward SBS in silicon (Kittlaus et al., 2017). (B) Intermodal backward SBS in As_2S_3 (Liu et al., 2021). (C) A resonant nonreciprocal SBS amplifier in silicon (Otterstrom et al., 2019). (D) Nonreciprocal scattering from an elastic wave driven through SBS (Kittlaus et al., 2018). (E) Nonreciprocal scattering in an AlN ring resonator using an electromechanically driven elastic wave (Sohn et al., 2018). (F) Traveling-wave nonreciprocal modulation in a silicon waveguide using an AlN transducer (Kittlaus et al., 2021). (G) Nonreciprocity in a Si_3N_4 resonator using modulation from bulk acoustic waves (Tian et al., 2021). (H) Nonreciprocal mode splitting in lithium-niobate-on-insulator (Sohn et al., 2021).

have been performed in a variety of resonator systems, including off-chip microsphere WGRs (Dong et al., 2015; Kim et al., 2015, 2017a, b, 2019; Shen et al., 2016, 2018) and on-chip microtoroid WGRs (Ma et al., 2020; Ruesink et al., 2016, 2018), nanocavities (Fang et al., 2017), and waveguide racetrack resonators (Otterstrom et al., 2019, 2020). The intrinsic nonreciprocal properties of SBS gain can also produce a form of backscatter immunity in Brillouin laser oscillators and injection locking physics that exhibits distinct nonreciprocal effects (Otterstrom et al., 2020).

Generally, nonreciprocal systems based on SBS are relatively narrowband, since they are limited by the bandwidth of stimulated Brillouin gain (typically <100 MHz), and/or by the linewidths of optical modes in the case of resonator devices. By contrast, techniques based on linear Brillouin light scattering or acousto-optic modulation (see Section 2.1.1) can produce very wideband operation limited only by phase-matching considerations.

The elastic wave used to mediate nonreciprocal Brillouin scattering in this manner may be optically or electro-mechanically generated. In the optical driving case, the elastic wave is initially produced using SBS at a wavelength away from the band of interest, and then induces broadband light scattering at a different wavelength where suitable phase matching is available. This approach was initially proposed in the context of fiber optics (Huang and Fan, 2011; Yu and Fan, 2009); Poulton et al. proposed a design that translates this approach to an integrated multimode waveguide system with an achievable operation bandwidth >25 nm (Poulton et al., 2012). This was experimentally realized in a suspended silicon waveguide system in Kittlaus et al. (2018). Notably, the operating wavelength of this approach was further tunable over a >35 nm range by changing the wavelength of the optical pump, leveraging the flexibility of wavevector-tunable optical driving of elastic waves.

Electro-mechanical driving, such as in the familiar case of an acousto-optic modulator, eliminates the need for optical pumping, offering potential benefits in terms of efficiency and practicality. Indeed, various forms of acousto-optic modulator and mode converter devices have been demonstrated in chip-based platforms as early as the 1970s (de Lima et al., 2006; Gorecki et al., 1997; Kuhn et al., 1970, 1971; Ohmachi and Noda, 1977; Sasaki et al., 1974; Tadesse and Li, 2014). As in bulk acousto-optic devices, these interactions can be highly efficient: For example, Sasaki et al. (1974) realized acousto-optic mode conversion with efficiency up to 88% in a ZnO waveguide using a surface acoustic wave (SAW). Over many decades, there have been significant advances in acousto-optic modulation using piezoelectric materials including quartz, gallium arsenide (de Lima et al., 2006), lithium niobate ($LiNbO_3$; LN) (Cheng and Tsai, 1992; Schmidt et al., 1973; White et al., 1974), and lithium tantalate ($LiTaO_3$) (Duffy et al., 1987). More recently, aluminum nitride (AlN) and LN have risen to prominence in the context of integrated photonics since both permit the fabrication of low-loss waveguides, offer relatively strong piezoelectric coefficients, have low acoustic attenuation at GHz frequencies, and have high refractive index which assists with elasto-optic modulation. AlN, in particular, has the added advantage of being CMOS compatible (Xiong et al., 2012). Leveraging these favorable characteristics, acousto-optic modulation of integrated AlN resonators at frequencies above 10 GHz has been demonstrated using sub-optical-wavelength surface acoustic waves driven by interdigital transducers (Tadesse and Li, 2014). Recent work has demonstrated a number of intriguing acousto-optic devices in LN (Jiang et al., 2019; Sarabalis et al., 2020; Shao et al., 2019, 2020) and AlN (Li et al., 2019; Liu et al., 2019).

All these acousto-optic systems can be adapted for nonreciprocal photonics based on propagating acoustic waves. However, as discussed in Section 2.1.1, nonreciprocity through Brillouin scattering requires that the interaction length is sufficient to resolve the nonreciprocal wavevector mismatch. This can be effectively realized using optical resonators, as has been applied, along with an electro-mechanically driven SAW, to produce nonreciprocal intermodal scattering in suspended (Sohn and Bahl, 2019; Sohn et al., 2018) and nonsuspended (Sohn et al., 2020) AlN ring resonators. Recently, these effects have been extended in lithium-niobate on insulator (Sohn et al., 2021) and Si_3N_4 (Tian et al., 2021) photonic integrated circuits, where acoustic waves are used to break the chiral symmetry of the optical resonator modes. Alternately, sufficient propagation lengths can be achieved directly by increasing the interaction region. For instance, in Kittlaus et al. (2021), a buried silicon waveguide modulated by propagating SAWs was configured in a serpentine pattern to increase the interaction region to many centimeters, as a means to demonstrate strong nonreciprocal modulation. This approach preserves the broad bandwidth of the Brillouin interaction, while using the relatively efficient elasto-optic scattering of silicon.

Looking forward, Brillouin-based nonreciprocity can leverage a variety of material platforms such as Si_3N_4 waveguides (Gundavarapu et al., 2019; Gyger et al., 2020; Tian et al., 2020), silicon, III–V semiconductors (Balram et al., 2017), AlN (Li et al., 2019; Liu et al., 2019; Tadesse and Li, 2014), lithium-niobate on insulator (Jiang et al., 2019; Shao et al., 2019), chalcogenide glasses (Liu et al., 2021; Pant et al., 2011), and others (Huang et al., 2022). Advances in material processing techniques, as well as new designs to control optical and acoustic waves, may further improve the performance and flexibility of Brillouin-based approaches to engineering optical nonreciprocity. These advances promise to enable a variety of nonreciprocal technologies that can be readily integrated within future photonic systems.

Acknowledgments

Part of this chapter was written at the Jet Propulsion Laboratory, California Institute of Technology, under a contract with the National Aeronautics and Space Administration (80NM0018D0004). Part of this chapter was written at Sandia National Laboratories (SNL). SNL is managed and operated by NTESS under DOE NNSA contract DE-NA-0003525. GB additionally acknowledges the support of the US Office of Naval Research grant N00014-17-1-2209 and the Presidential Early Career Award for Scientists and Engineers.

References

Adam, J.D., Davis, L.E., Dionne, G.F., Schloemann, E.F., Stitzer, S.N., 2002. Ferrite devices and materials. IEEE Trans. Microw. Theory. Tech. 50 (3), 721–737.

Armani, D.K., Kippenberg, T.J., Spillane, S.M., Vahala, K.J., 2003. Ultra-high-Q toroid microcavity on a chip. Nature 421 (6926), 925–928.

Bahl, G., Tomes, M., Marquardt, F., Carmon, T., 2012. Observation of spontaneous Brillouin cooling. Nat. Phys. 8 (3), 203–207.

Balram, K.C., Davanço, M.I., Ilic, B.R., Kyhm, J.-H., Song, J.D., Srinivasan, K., 2017. Acousto-optic modulation and optoacoustic gating in piezo-optomechanical circuits. Phys. Rev. Appl. 7 (2), 024008.

Bashan, G., Diamandi, H.H., London, Y., Sharma, K., Shemer, K., Zehavi, E., Zadok, A., 2021. Forward stimulated Brillouin scattering and opto-mechanical non-reciprocity in standard polarization maintaining fibres. Light Sci. Appl. 10 (1), 119.

Bi, L., Hu, J., Jiang, P., Kim, D.H., Dionne, G.F., Kimerling, L.C., Ross, C.A., 2011. On-chip optical isolation in monolithically integrated non-reciprocal optical resonators. Nat. Photonics 5, 758. https://doi.org/10.1038/nphoton.2011.270.

Boller, K.-J., Imamoğlu, A., Harris, S.E., 1991. Observation of electromagnetically induced transparency. Phys. Rev. Lett. 66, 2593–2596. https://doi.org/10.1103/PhysRevLett.66.2593.

Boyd, R.W., 2020. Nonlinear Optics. Academic Press.

Budker, D., Kimball, D.F., Rochester, S.M., Yashchuk, V.V., 1999. Nonlinear magneto-optics and reduced group velocity of light in atomic vapor with slow ground state relaxation. Phys. Rev. Lett. 83, 1767–1770. https://doi.org/10.1103/PhysRevLett.83.1767.

Cheng, Z.Y., Tsai, C.S., 1992. Baseband integrated acousto-optic frequency shifter. Appl. Phys. Lett. 60 (1), 12–14.

Collot, L., Lefevre-Seguin, V., Brune, M., Raimond, J.M., Haroche, S., 1993. Very high-Q whispering-gallery mode resonances observed on fused silica microspheres. EPL 23 (5), 327.

Corcoran, V.J., 1978. A new acousto-optic isolator for the stabilization of optically pumped submillimetre lasers. Infrared Phys. 18 (5-6), 635–645.

de Lima, M.M., Beck, M., Hey, R., Santos, P.V., 2006. Compact Mach-Zehnder acousto-optic modulator. Appl. Phys. Lett. 89 (12), 121104.

Dimopoulos, S., Graham, P.W., Hogan, J.M., Kasevich, M.A., 2007. Testing general relativity with atom interferometry. Phys. Rev. Lett. 98, 111102.

Doerr, C.R., Dupuis, N., Zhang, L., 2011. Optical isolator using two tandem phase modulators. Opt. Lett. 36 (21), 4293–4295. https://doi.org/10.1364/OL.36.004293. http://ol.osa.org/abstract.cfm?URI=ol-36-21-4293.

Doerr, C.R., Chen, L., Vermeulen, D., 2014. Silicon photonics broadband modulation-based isolator. Opt. Express 22 (4), 4493–4498. https://doi.org/10.1364/OE.22.004493.

Dong, C., Shen, Z., Zou, C., Zhang, Y., Fu, W., Guo, G., 2015. Brillouin-scattering-induced transparency and non-reciprocal light storage. Nat. Commun. 6, 6193.

Dötsch, H., Bahlmann, N., Zhuromskyy, O., Hammer, M., Wilkens, L., Gerhardt, R., Hertel, P., Popkov, A.F., 2005. Applications of magneto-optical waveguides in integrated optics. JOSA B 22 (1), 240–253.

Duffy, J.F., Al-Shukri, S.M., De La Rue, R.M., 1987. Guided-wave acousto-optic interaction on proton-exchanged lithium tantalate. Electron. Lett. 23 (16), 849–850.

Eichenfield, M., Chan, J., Camacho, R.M., Vahala, K.J., Painter, O., 2009. Optomechanical crystals. Nature 462 (7269), 78–82.

Fang, K., Luo, J., Metelmann, A., Matheny, M.H., Marquardt, F., Clerk, A.A., Painter, O., 2017. Generalized non-reciprocity in an optomechanical circuit via synthetic magnetism and reservoir engineering. Nat. Phys. 13 (5), 465–471.

Fiore, V., Yang, Y., Kuzyk, M.C., Barbour, R., Tian, L., Wang, H., 2011. Storing optical information as a mechanical excitation in a silica optomechanical resonator. Phys. Rev. Lett. 107 (13), 133601.

Fleischhauer, M., Imamoglu, A., Marangos, J.P., 2005. Electromagnetically induced transparency: optics in coherent media. Rev. Mod. Phys. 77 (2), 633.

Fu, W., Shu, F.-J., Zhang, Y.-L., Dong, C.-H., Zou, C.-L., Guo, G.-C., 2015. Integrated optical circulator by stimulated Brillouin scattering induced non-reciprocal phase shift. Opt. Express 23 (19), 25118–25127.

Gorecki, C., Chollet, F., Bonnotte, E., Kawakatsu, H., 1997. Silicon-based integrated interferometer with phase modulation driven by surface acoustic waves. Opt. Lett. 22 (23), 1784–1786.

Grudinin, I.S., Matsko, A.B., Savchenkov, A.A., Strekalov, D., Ilchenko, V.S., Maleki, L., 2006. Ultra high Q crystalline microcavities. Opt. Commun. 265 (1), 33–38.

Grudinin, I.S., Matsko, A.B., Maleki, L., 2009. Brillouin lasing with a CaF_2 whispering gallery mode resonator. Phys. Rev. Lett. 102, 043902. https://doi.org/10.1103/PhysRevLett.102.043902.

Gundavarapu, S., Brodnik, G.M., Puckett, M., Huffman, T., Bose, D., Behunin, R., Wu, J., Qiu, T., Pinho, C., Chauhan, N., et al., 2019. Sub-Hertz fundamental linewidth photonic integrated Brillouin laser. Nat. Photonics 13 (1), 60–67.

Gyger, F., Liu, J., Yang, F., He, J., Raja, A.S., Wang, R.N., Bhave, S.A., Kippenberg, T.J., Thévenaz, L., 2020. Observation of stimulated Brillouin scattering in silicon nitride integrated waveguides. Phys. Rev. Lett. 124 (1), 013902.

Hau, L.V., Harris, S.E., Dutton, Z., Behroozi, C.H., 1999. Light speed reduction to 17 metres per second in an ultracold atomic gas. Nature 397 (6720), 594–598.

Huang, X., Fan, S., 2011. Complete all-optical silica fiber isolator via stimulated Brillouin scattering. J. Lightwave Technol. 29 (15), 2267–2275. https://doi.org/10.1109/JLT.2011.2158886.

Huang, C., Shi, H., Yu, L., Wang, K., Cheng, M., Huang, Q., Jiao, W., Sun, J., 2022. Acousto-optic modulation in silicon waveguides based on piezoelectric aluminum scandium nitride film. Adv. Opt. Mater. 10 (6), 2102334.

Huang, D., Pintus, P., Zhang, C., Shoji, Y., Mizumoto, T., Bowers, J.E., 2016. Electrically driven and thermally tunable integrated optical isolators for silicon photonics. IEEE J. Sel. Top. Quantum Electron. 22 (6), 271–278. https://doi.org/10.1109/JSTQE.2016.2588778.

Hwang, I.K., Yun, S.H., Kim, B.Y., 1997. All-fiber-optic nonreciprocal modulator. Opt. Lett. 22 (8), 507–509.
Ibrahim, S.K., Bhandare, S., Sandel, D., Zhang, H., Noe, R., 2004. Non-magnetic 30 dB integrated optical isolator in III/V material. Electron. Lett. 40 (20), 1293–1294. https://doi.org/10.1049/el:20045901.
Jalas, D., Petrov, A., Eich, M., Freude, W., Fan, S., Yu, Z., Baets, R., Popović, M., Melloni, A., Joannopoulos, J.D., et al., 2013. What is—and what is not—an optical isolator. Nat. Photonics 7 (8), 579–582.
Jiang, W., Patel, R.N., Mayor, F.M., McKenna, T.P., Arrangoiz-Arriola, P., Sarabalis, C.J., Witmer, J.D., Van Laer, R., Safavi-Naeini, A.H., 2019. Lithium niobate piezo-optomechanical crystals. Optica 6 (7), 845–853.
Kang, M.S., Butsch, A., Russell, P.S.J., 2011. Reconfigurable light-driven opto-acoustic isolators in photonic crystal fibre. Nat. Photonics 5 (9), 549–553.
Kasapi, A., Jain, M., Yin, G.Y., Harris, S.E., 1995. Electromagnetically induced transparency: propagation dynamics. Phys. Rev. Lett. 74, 2447–2450. https://doi.org/10.1103/PhysRevLett.74.2447.
Kash, M.M., Sautenkov, V.A., Zibrov, A.S., Hollberg, L., Welch, G.R., Lukin, M.D., Rostovtsev, Y., Fry, E.S., Scully, M.O., 1999. Ultraslow group velocity and enhanced nonlinear optical effects in a coherently driven hot atomic gas. Phys. Rev. Lett. 82, 5229–5232. https://doi.org/10.1103/PhysRevLett.82.5229.
Kim, J., Kuzyk, M.C., Han, K., Wang, H., Bahl, G., 2015. Non-reciprocal Brillouin scattering induced transparency. Nat. Phys. 11 (3), 275–280.
Kim, J., Kim, S., Bahl, G., 2017a. Complete linear optical isolation at the microscale with ultralow loss. Sci. Rep. 7 (1), 1647. https://doi.org/10.1038/s41598-017-01494-w.
Kim, S., Xu, X., Taylor, J.M., Bahl, G., 2017b. Dynamically induced robust phonon transport and chiral cooling in an optomechanical system. Nat. Commun. 8 (1), 205.
Kim, S., Taylor, J.M., Bahl, G., 2019. Dynamic suppression of Rayleigh backscattering in dielectric resonators. Optica 6 (8), 1016–1022.
Kippenberg, T.J., Vahala, K.J., 2007. Cavity opto-mechanics. Opt. Express 15 (25), 17172–17205.
Kittlaus, E.A., Shin, H., Rakich, P.T., 2016. Large Brillouin amplification in silicon. Nat. Photonics 10 (7), 463–467.
Kittlaus, E.A., Otterstrom, N.T., Rakich, P.T., 2017. On-chip inter-modal Brillouin scattering. Nat. Commun. 8, 15819. https://doi.org/10.1038/ncomms15819.
Kittlaus, E.A., Otterstrom, N.T., Kharel, P., Gertler, S., Rakich, P.T., 2018. Non-reciprocal interband Brillouin modulation. Nat. Photonics 12 (10), 613–619.
Kittlaus, E.A., Jones, W.M., Rakich, P.T., Otterstrom, N.T., Muller, R.E., Rais-Zadeh, M., 2021. Electrically driven acousto-optics and broadband non-reciprocity in silicon photonics. Nat. Photonics 15 (1), 43–52.
Kuhn, L., Dakss, M.L., Heidrich, P.F., Scott, B.A., 1970. Deflection of an optical guided wave by a surface acoustic wave. Appl. Phys. Lett. 17 (6), 265–267.
Kuhn, L., Heidrich, P.F., Lean, E.G., 1971. Optical guided wave mode conversion by an acoustic surface wave. Appl. Phys. Lett. 19 (10), 428–430.
Lee, H., Chen, T., Li, J., Yang, K.Y., Jeon, S., Painter, O., Vahala, K.J., 2012. Chemically etched ultrahigh-Q wedge-resonator on a silicon chip. Nat. Photonics 6 (6), 369–373.
Lezama, A., Barreiro, S., Akulshin, A.M., 1999. Electromagnetically induced absorption. Phys. Rev. A 59, 4732–4735.
Li, J., Lee, H., Chen, T., Vahala, K.J., 2012. Characterization of a high coherence, Brillouin microcavity laser on silicon. Opt. Express 20 (18), 20170–20180.
Li, J., Lee, H., Vahala, K.J., 2014. Low-noise Brillouin laser on a chip at 1064 nm. Opt. Lett. 39 (2), 287–290.

Li, H., Liu, Q., Li, M., 2019. Electromechanical Brillouin scattering in integrated planar photonics. APL Photonics 4 (8), 080802.
Liu, C., Dutton, Z., Behroozi, C.H., Hau, L.V., 2001. Observation of coherent optical information storage in an atomic medium using halted light pulses. Nature 409 (6819), 490–493. https://doi.org/10.1038/35054017.
Liu, Q., Li, H., Li, M., 2019. Electromechanical Brillouin scattering in integrated optomechanical waveguides. Optica 6 (6), 778–785.
Liu, Y., Choudhary, A., Ren, G., Choi, D.-Y., Casas-Bedoya, A., Morrison, B., Ma, P., Nguyen, T.G., Mitchell, A., Madden, S.J., et al., 2021. Circulator-free Brillouin photonic planar circuit. Laser Photonics Rev. 15 (5), 2000481.
Ma, J., Wen, J., Ding, S., Li, S., Hu, Y., Jiang, X., Jiang, L., Xiao, M., 2020. Chip-based optical isolator and nonreciprocal parity-time symmetry induced by stimulated Brillouin scattering. Laser Photon. Rev. 14 (5), 1900278.
Merklein, M., Stiller, B., Vu, K., Ma, P., Madden, S.J., Eggleton, B.J., 2021. On-chip broadband nonreciprocal light storage. Nanophotonics 10 (1), 75–82.
Morrison, B., Casas-Bedoya, A., Ren, G., Vu, K., Liu, Y., Zarifi, A., Nguyen, T.G., Choi, D.-Y., Marpaung, D., Madden, S.J., Mitchell, A., Eggleton, B.J., 2017. Compact Brillouin devices through hybrid integration on silicon. Optica 4 (8), 847–854. https://doi.org/10.1364/OPTICA.4.000847. http://www.osapublishing.org/optica/abstract.cfm?URI=optica-4-8-847.
Ohmachi, Y., Noda, J., 1977. LiNbO3 TE-TM mode converter using collinear acoustooptic interaction. IEEE J. Quantum Electron. 13 (2), 43–46. https://doi.org/10.1109/JQE.1977.1069276.
Okawachi, Y., Bigelow, M.S., Sharping, J.E., Zhu, Z., Schweinsberg, A., Gauthier, D.J., Boyd, R.W., Gaeta, A.L., 2005. Tunable all-optical delays via Brillouin slow light in an optical fiber. Phys. Rev. Lett. 94 (15), 153902.
Otterstrom, N.T., Kittlaus, E.A., Gertler, S., Behunin, R.O., Lentine, A.L., Rakich, P.T., 2019. Resonantly enhanced nonreciprocal silicon Brillouin amplifier. Optica 6 (9), 1117–1123.
Otterstrom, N.T., Gertler, S., Zhou, Y., Kittlaus, E.A., Behunin, R.O., Gehl, M., Starbuck, A.L., Dallo, C.M., Pomerene, A.T., Trotter, D.C., et al., 2020. Backscatter-immune injection-locked Brillouin laser in silicon. Phys. Rev. Appl. 14 (4), 044042.
Pant, R., Poulton, C.G., Choi, D.-Y., Mcfarlane, H., Hile, S., Li, E., Thevenaz, L., Luther-Davies, B., Madden, S.J., Eggleton, B.J., 2011. On-chip stimulated Brillouin scattering. Opt. Express 19 (9), 8285–8290. https://doi.org/10.1364/OE.19.008285. http://www.opticsexpress.org/abstract.cfm?URI=oe-19-9-8285.
Peters, A., Chung, K.Y., Chu, S., 2001. High-precision gravity measurements using atom interferometry. 38 (1), 25–61.
Phillips, D.F., Fleischhauer, A., Mair, A., Walsworth, R.L., Lukin, M.D., 2001. Storage of light in atomic vapor. Phys. Rev. Lett. 86, 783–786. https://doi.org/10.1103/PhysRevLett.86.783.
Poulton, C.G., Pant, R., Byrnes, A., Fan, S., Steel, M.J., Eggleton, B.J., 2012. Design for broadband on-chip isolator using stimulated Brillouin scattering in dispersion-engineered chalcogenide waveguides. Opt. Express 20 (19), 21235–21246. https://doi.org/10.1364/OE.20.021235. http://www.opticsexpress.org/abstract.cfm?URI=oe-20-19-21235.
Ruesink, F., Miri, M.-A., Alù, A., Verhagen, E., 2016. Nonreciprocity and magnetic-free isolation based on optomechanical interactions. Nat. Commun. 7, 13662.
Ruesink, F., Mathew, J.P., Miri, M.-A., Alù, A., Verhagen, E., 2018. Optical circulation in a multimode optomechanical resonator. Nat. Commun. 9 (1), 1798.

Safavi-Naeini, A.H., Alegre, T.M., Chan, J., Eichenfield, M., Winger, M., Lin, Q., Hill, J.T., Chang, D.E., Painter, O., 2011. Electromagnetically induced transparency and slow light with optomechanics. Nature 472 (7341), 69–73.

Saleh, B.E.A., Teich, M.C., 2019. Fundamentals of Photonics. John Wiley & Sons.

Sarabalis, C.J., McKenna, T.P., Patel, R.N., Van Laer, R., Safavi-Naeini, A.H., 2020. Acousto-optic modulation in lithium niobate on sapphire. APL Photonics 5 (8), 086104.

Sasaki, H., Kushibiki, J., Chubachi, N., 1974. Efficient acousto-optic TE⇌TM mode conversion in ZnO films. Appl. Phys. Lett. 25 (9), 476–477. https://doi.org/10.1063/1.1655554.

Schmidt, R.V., Kaminow, I.P., Carruthers, J.R., 1973. Acousto-optic diffraction of guided optical waves in LiNbO3. Appl. Phys. Lett. 23 (8), 417–419.

Shao, L., Yu, M., Maity, S., Sinclair, N., Zheng, L., Chia, C., Shams-Ansari, A., Wang, C., Zhang, M., Lai, K., et al., 2019. Microwave-to-optical conversion using lithium niobate thin-film acoustic resonators. Optica 6 (12), 1498–1505.

Shao, L., Sinclair, N., Leatham, J., Hu, Y., Yu, M., Turpin, T., Crowe, D., Lončar, M., 2020. Integrated microwave acousto-optic frequency shifter on thin-film lithium niobate. Opt. Express 28 (16), 23728–23738.

Shen, Z., Zhang, Y.-L., Chen, Y., Zou, C.-L., Xiao, Y.-F., Zou, X.-B., Sun, F.-W., Guo, G.-C., Dong, C.-H., 2016. Experimental realization of optomechanically induced non-reciprocity. Nat. Photonics 10 (10), 657–661.

Shen, Z., Zhang, Y., Chen, Y., Sun, F., Zou, X., Guo, G., Zou, C., Dong, C., 2018. Reconfigurable optomechanical circulator and directional amplifier. Nat. Commun. 9 (1), 1797.

Shin, Y., Saba, M., Pasquini, T.A., Ketterle, W., Pritchard, D.E., Leanhardt, A.E., 2004. Atom interferometry with Bose-Einstein condensates in a double-well potential. Phys. Rev. Lett. 92, 050405.

Shin, H., Qiu, W., Jarecki, R., Cox, J.A., Olsson, R.H., Starbuck, A., Wang, Z., Rakich, P.T., 2013. Tailorable stimulated Brillouin scattering in nanoscale silicon waveguides. Nat. Commun. 4, 1944. https://doi.org/10.1038/ncomms2943.

Shoji, Y., Mizumoto, T., Yokoi, H., Hsieh, I.-W., Osgood, R.M., Jr., 2008. Magneto-optical isolator with silicon waveguides fabricated by direct bonding. Appl. Phys. Lett. 92 (7), 071117. https://doi.org/10.1063/1.2884855.

Sobu, Y., Shoji, Y., Sakurai, K., Mizumoto, T., 2013. GaInAsP/InP MZI waveguide optical isolator integrated with spot size converter. Opt. Express 21 (13), 15373. https://doi.org/10.1364/OE.21.015373. http://www.opticsexpress.org/abstract.cfm?URI=oe-21-13-15373.

Sohn, D.B., Bahl, G., 2019. Direction reconfigurable nonreciprocal acousto-optic modulator on chip. APL Photonics 4 (12), 126103.

Sohn, D.B., Kim, S., Bahl, G., 2018. Time-reversal symmetry breaking with acoustic pumping of nanophotonic circuits. Nat. Photonics 12 (2), 91–97.

Sohn, D., Kim, S., Bahl, G., 2020. Unreleased on-chip frequency-shifting optical isolator. In: CLEO: Science and Innovations, Optical Society of America. STh4J–5.

Sohn, D.B., Örsel, O.E., Bahl, G., 2021. Electrically driven optical isolation through phonon-mediated photonic Autler-Townes splitting. Nat. Photonics 15, 822–827.

Soltani, M., Yegnanarayanan, S., Adibi, A., 2007. Ultra-high Q planar silicon microdisk resonators for chip-scale silicon photonics. Opt. Express 15 (8), 4694–4704.

Sounas, D.L., Alù, A., 2017. Non-reciprocal photonics based on time modulation. Nat. Photonics 11 (12), 774–783.

Tadesse, S.A., Li, M., 2014. Sub-optical wavelength acoustic wave modulation of integrated photonic resonators at microwave frequencies. Nat. Commun. 5, 5402.

Tian, H., Liu, J., Dong, B., Skehan, J.C., Zervas, M., Kippenberg, T.J., Bhave, S.A., 2020. Hybrid integrated photonics using bulk acoustic resonators. Nat. Commun. 11 (1), 3073.

Tian, H., Liu, J., Siddharth, A., Wang, R.N., Blésin, T., He, J., Kippenberg, T.J., Bhave, S.A., 2021. Magnetic-free silicon nitride integrated optical isolator. Nat. Photonics 15, 828–836.

Tien, M.-C., Mizumoto, T., Pintus, P., Kromer, H., Bowers, J.E., 2011. Silicon ring isolators with bonded nonreciprocal magneto-optic garnets. Opt. Express 19 (12), 11740. https://doi.org/10.1364/OE.19.011740. http://www.opticsexpress.org/abstract.cfm?URI=oe-19-12-11740.

Tomes, M., Carmon, T., 2009. Photonic micro-electromechanical systems vibrating at *X*-band (11-GHz) rates. Phys. Rev. Lett. 102, 113601. https://doi.org/10.1103/PhysRevLett.102.113601.

Van Laer, R., Kuyken, B., Van Thourhout, D., Baets, R., 2015. Interaction between light and highly confined hypersound in a silicon photonic nanowire. Nat. Photonics 9 (3), 199–203. https://doi.org/10.1038/nphoton.2015.11.

Weis, S., Rivière, R., Deléglise, S., Gavartin, E., Arcizet, O., Schliesser, A., Kippenberg, T.J., 2010. Optomechanically induced transparency. Science 330 (6010), 1520–1523.

White, J.M., Heidrich, P.F., Lean, E.G., 1974. Thin-film acousto-optic interaction in LiNbO3. Electron. Lett. 10 (24), 510–511.

Xiong, C., Pernice, W.H., Sun, X., Schuck, C., Fong, K.Y., Tang, H.X., 2012. Aluminum nitride as a new material for chip-scale optomechanics and nonlinear optics. New J. Phys. 14 (9), 095014.

Yu, Z., Fan, S., 2009. Complete optical isolation created by indirect interband photonic transitions. Nat. Photonics 3 (2), 91–94. https://doi.org/10.1038/nphoton.2008.273.

CHAPTER FOURTEEN

Electromechanical Brillouin scattering

Huan Li[a,*], Omar Florez[b], Bingcheng Pan[a], Guilhem Madiot[b], Clivia M. Sotomayor Torres[b,c], and Mo Li[d,*]

[a]State Key Laboratory for Modern Optical Instrumentation, College of Optical Science and Engineering, International Research Center for Advanced Photonics, Zhejiang University, Hangzhou, China
[b]Catalan Institute of Nanoscience and Nanotechnology (ICN2), CSIC and BIST, Bellaterra, Spain
[c]ICREA, Pg. Lluís Companys 23, Barcelona, Spain
[d]University of Washington, Seattle, WA, United States
[*]Corresponding authors: e-mail address: lihuan20@zju.edu.cn; moli96@uw.edu

Contents

In his 1922 paper (Brillouin, 1922), L. Brillouin theoretically predicted light being scattered by an elastic wave, or a sound wave, at the Bragg condition, which is named after him as Brillouin scattering. Brillouin's description of the scattering process is specific for the elastic wave of thermal origin. However, Brillouin scattering has evolved into a broad concept over time, covering elastic waves of diverse origins. The elastic wave can be spontaneously produced by thermal agitation of the environment (i.e., spontaneous Brillouin scattering), or stimulated (narrowly defined to be optically excited) by a strong light source (i.e., stimulated Brillouin scattering (SBS)). The elastic wave can also be excited by external stimuli, such as optical pulses, thermal shocks, and electrical and magnetic fields. This chapter discusses

Semiconductors and Semimetals, Volume 110
ISSN 0080-8784
https://doi.org/10.1016/bs.semsem.2022.05.007

Brillouin scattering of light by electromechanically excited elastic waves, a process that can be dubbed as electromechanical Brillouin scattering (EBS). The main difference between EBS and SBS is how the elastic waves (i.e., sound or phonons) are generated. In EBS, the sound wave is generated by an electromechanical transducer using piezoelectric materials, which convert radio frequency (RF) and microwave (MW) electromagnetic waves to propagating elastic waves, or in some cases, localized mechanical modes. The electromechanical conversion efficiency of the transducers can be very high, approaching unity in optimized devices. In contrast, in SBS the elastic wave is stimulated by the optical field through optical forces in the materials, including radiation pressure and electrostriction.

The operation principles of the EBS devices essentially cover those of the conventional acousto-optic (AO) devices (Tsai, 1990), which were extensively investigated from 1960s to 1990s. Oftentimes, the terms of EBS and acousto-optics are used interchangeably in the literature. In the 2000s and 2010s, due to the availability of new thin-film materials and rapid advancement in nanofabrication technologies, the interest in integrated acousto-optics revived. Such a revival is also fueled by the concomitant advancement in the general field of integrated nano-optomechanical systems, especially cavity optomechanics (Aspelmeyer et al., 2014a), as well as integrated SBS devices (Eggleton et al., 2019). The latest development of integrated EBS devices feature unprecedented high acoustic frequencies and strong co-confinement of the acoustic and optical modes within wavelength-scale structures, leading to significantly enhanced performance compared to their conventional counterparts. The new-generation integrated EBS devices are promising to reinvent guided-wave acousto-optics for "conventional" applications in optical beam steerers, tunable filters, frequency shifters, and single-sideband modulators.

Furthermore, considering that the RF/MW electromagnetic wave has the same frequency as the excited elastic wave, the inherent energy conversion efficiency can be high. This compares favorably with SBS, which involves five orders of magnitude frequency down-conversion from the optical frequency of 100s of THz to acoustic frequencies of MHz to GHz. Therefore, by engineering efficient acousto-optic interaction in integrated devices, the EBS process is promising to achieve efficient conversion between RF/MW and optical frequencies, with a wide range of novel applications from microwave photonics to quantum transduction between optical and microwave qubits.

More interestingly, the newly-emerging phononic circuits is anticipated to complement the functionalities of the photonic and electronic circuits, leading to integrated nano-opto-electro-mechanical systems (NOEMS), which exploit the interaction of all the three "x-ons," namely, photons, electrons, and phonons. The exciting prospect of integrated NOEMS circuits that implement sophisticated sensing and information processing functionalities through EBS mechanisms in the classical and quantum regimes is attracting increasing research efforts.

In this chapter, the electromechanical excitation of acoustic waves based on piezoelectric IDTs is introduced first in Section 1, followed by a brief overview of EBS processes, configurations, and devices in Section 2. Finally, the chapter is concluded with the prospective applications of EBS devices in Section 3.

1. Electromechanical excitation of acoustic waves

To excite acoustic waves electromechanically in integrated photonic devices, materials with piezoelectricity (Nye, 1985) are needed. While silicon photonics is prevailing in integrated photonics, silicon lacks piezoelectricity. It is difficult to exploit silicon in EBS applications, despite its widespread adoption in nanofabrication and semiconductors. This is reminiscent of the absence of direct bandgap in silicon which limits its use in photonics particularly for the realization of active optoelectronic devices. Within this context, a possible strategy consists of integrating piezoelectric materials onto silicon chips, enabling the introduction of desired functionalities. Besides silicon, photonic devices can also be built completely with materials with strong piezoelectricity. Those piezoelectric materials that have been explored for integrated EBS devices include lithium niobate ($LiNbO_3$), zinc oxide (ZnO), and III-V materials, especially aluminum nitride (AlN).

1.1 Common piezoelectric materials

1.1.1 Lithium niobate

Lithium niobate ($LiNbO_3$) (Weis and Gaylord, 1985) has been the most prevalent material for guided-wave acousto-optics since the early 1970s. Below its Curie temperature at $\sim$1210 °C, $LiNbO_3$ is a ferroelectric material with a trigonal crystal structure and is characterized by large piezoelectric, electro-optic, photoelastic coefficients, and acousto-optic figure-of-merit.

$LiNbO_3$ is a uniaxial crystal with a wide transparency window from 0.4 to 5 μm. Its ordinary and extraordinary refractive indices are $n_o=2.21$ and $n_e=2.14$ in the telecom C-band. The richness of its favorable optical and acoustic properties and large susceptibilities has made lithium niobate one of the most widely used materials in applications such as integrated dielectric waveguides, acoustic wave transducers, acoustic delay lines, acoustic filters, acousto-optic Q-switches, beam deflectors, and others.

Conventional $LiNbO_3$ guided-wave acousto-optic devices (Tsai, 1990) are based on bulk crystal substrates, where surface acoustic waves (SAWs) and diffusion-based optical waveguides are typically employed to achieve acousto-optic interactions. In the early 2000s, lithium-niobate-on-insulator (LNOI), fabricated with Smart-Cut technology, emerged as a promising material platform to rejuvenate the then matured field of $LiNbO_3$ devices. The last decade has witnessed the rapid commercialization of LNOI wafers and major breakthroughs in the argon-based Inductively Coupled Plasma Reactive Ion Etching (ICP-RIE) technologies for LNOI, which enable the development of the next-generation integrated EBS devices (Sarabalis et al., 2021; Shao et al., 2019). Compared with their conventional counterparts prevalent in the last century, such EBS devices feature strong co-confinement of acoustic and optical modes in high-quality LNOI waveguides with wavelength-scale cross-sections, leading to unprecedented compact footprint and significantly enhanced EBS processes.

1.1.2 Zinc oxide

Zinc oxide (ZnO) (Tsai, 1990) has been extensively investigated as the material platform for guided-wave acousto-optics since the 1970s. ZnO is a semiconductor material with a direct band gap of 3.37 eV. As a member of the hexagonal wurtzite class, ZnO features a low dielectric constant and a high electromechanical coupling factor, as well as a low-loss transparency window from 0.4 to 2 μm, with ordinary and extraordinary refractive indices of $n_o= 1.93$ and $n_e=1.94$ in the telecom C-band, rendering it a well-suited material for EBS devices.

Various kinds of fabrication techniques for ZnO thin films have been proposed and demonstrated, including DC and RF sputtering, chemical vapor deposition (CVD), and planar magnetron sputtering. The c-axis oriented piezoelectric ZnO thin films deposited onto non-piezoelectric substrates, such as fused-silica, sapphire, and silicon-on-insulator (SOI), can be further patterned with standard lithographic and CH_4/H_2-based

ICP-RIE processes. Alternatively, ZnO thin films can also be patterned with lift-off or wet etching processes.

1.1.3 III–V materials

III–V materials feature diverse electronic band structures, rich optoelectronic properties, piezoelectric and electro-optic effects, as well as a wide range of functionalities in quantum photonics, including single- and entangled-photon generation by radiative recombination in quantum dots, low-loss routing, electro-optic modulation and single-photon detection (Dietrich et al., 2016; Wang et al., 2014). Inspired by the prospects in combining optical, electronic and mechanical properties on the same platform (see Section 3), an increasing research effort is made to use III–V materials to generate, guide and interact with acoustic waves. Commonly used III-V materials for electromechanical actuation of phononic circuits include aluminum nitride (AlN) (Liu et al., 2019; Xiong et al., 2012), gallium arsenide (GaAs) and/or aluminum gallium arsenide (AlGaAs or $Al_xGa_{1-x}N$) (Balram et al., 2016; Crespo-Poveda et al., 2013; Modica et al., 2020), gallium nitride (GaN) (Fu et al., 2019), and gallium phosphide (GaP) (Hönl et al., 2022), among others. Both epitaxy processes and bonding techniques have been used to implement the desired material stacks.

AlN (Rumyantsev et al., 2004) is the most representative and widely used III-V material for EBS devices. It is a semiconductor material with a wide bandgap of 6.2 eV. It has a wurtzite crystalline structure. With magnetron sputter deposition technology, c-axis oriented polycrystalline AlN thin films can be sputtered onto a variety of substrates, including Si and SiO_2, with retained piezoelectricity. The sputtered AlN thin films can be further patterned with standard lithographic and chlorine-based ICP-RIE processes.

Due to its excellent piezoelectric and acoustic properties, fabrication compatibility with CMOS processes, and availability of wafers from foundries, AlN has been widely employed in acoustic RF filters (Ruppel, 2017) based on film bulk acoustic resonator (FBAR) in the MHz and GHz frequency range. Featuring superior performance to their electronic counterparts, such acoustic RF filters have been widely deployed in cellular phones and networks.

Furthermore, in the past decade, AlN thin films have been extensively investigated as a material platform for EBS (Tadesse and Li, 2014), due to their excellent optical properties. AlN is a uniaxial crystal with a wide transparency window from 0.2 to 13.6 μm. Its ordinary and extraordinary refractive indices are $n_o = 2.12$ and $n_e = 2.16$ in the telecom C-band, providing sufficient refractive index contrast to SiO_2 to form well confined optical waveguide modes. Sputtered AlN thin films can be patterned into high-quality waveguides and resonators, rendering them an excellent platform for EBS devices. In addition, AlN possesses a moderate electro-optic effect, which contributes to the EBS process.

1.2 Generation of acoustic waves: Electromechanical transducers

On bulk or thin films of piezoelectric materials, various types (or modes) of acoustic waves can be generated using various types of electromechanical transducers. Those acoustic transducers have significant applications in RF/MW signal processing as filters, delay lines, and correlators. The readers are referred to the many monographs on acoustic devices for RF/MW signal processing (Campbell, 1998; Datta, 1986; Hashimoto, 2000; Morgan, 2007). In this section, we briefly introduce the types of transducers that are commonly used in EBS devices.

1.2.1 Interdigital transducers

The interdigital transducers (IDTs) (Fig. 1A) are the most straightforward and efficient way to excite surface acoustic waves (SAWs) on a piezoelectric substrate. The IDTs are conveniently patterned in thin metal films deposited on the substrate. Sending RF signals to the countering electrodes in the IDT

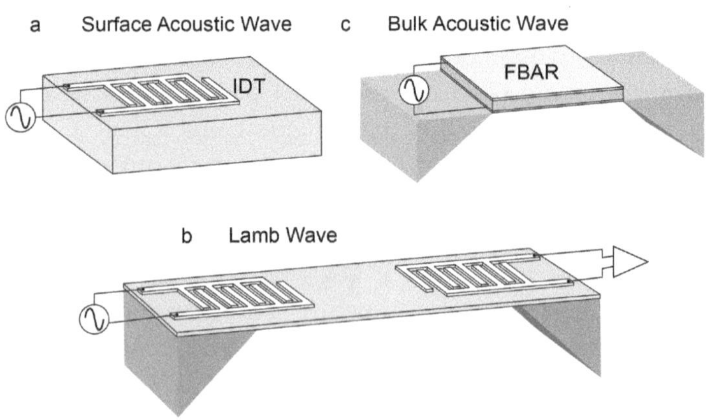

Fig. 1 Types of electromechanical transducers commonly used to generate acoustic waves in devices. (A) Surface acoustic wave generated on a piezoelectric substrate with IDTs. (B) Lamb wave generated on a free-standing piezoelectric thin film with IDTs. (C) Bulk acoustic wave generated in a piezoelectric film vertically sandwiched between two electrodes. FBAR: film bulk acoustic resonator.

will thus excite mechanical vibrations through the piezoelectricity of the material. Basically, IDTs enable acoustic wave generation via both the spatial periodicity resulting from the IDT geometry and the temporal periodicity of the electric potential applied to it. On a bulk substrate, the excited wave typically is in the Rayleigh mode, which is an elliptically polarized (in the propagation plane) mode as a mixture of longitudinal and transverse mode. In a similar fashion, IDTs are also capable of exciting Lamb waves on free-standing piezoelectric thin films (Fig. 1B). Typical Lamb modes are transverse modes with major displacement direction normal to the thin film. The wavelength λ, and thus the frequency f, of the acoustic wave is determined by the periodicity (Λ) of the IDT. The most common IDT has a uniform periodicity, so it is symmetric and bidirectional, emitting acoustic waves in both directions, which in most scenarios is undesirable. Unidirectional IDTs can be realized with more advanced designs (Malocha, 2004; Morgan, 2007).

It is important to understand that, as the acoustic wave is generated by the IDT and propagating, the invariant of the system is the frequency, determined by the RF/MW source. The acoustic wavelength, and therefore the wavevector, however, is different in different regions of the device where the phase (and group) velocity of the acoustic mode is different. For example, in the IDT region, the substrate is mass-loaded with the metal fingers, so the acoustic phase velocity is lower than the region without the IDT (the open region). As the acoustic wavelength in the IDT region is set by the IDT periodicity, the acoustic wavelength in the open region will be longer because of the higher phase velocity. Understanding the changing velocity, i.e., the dispersion relation, in different regions of the device is very important when considering the phase-matching conditions of the Brillouin scattering process (see Section 1).

IDTs are natural candidates to answer the low-power and compactness requirements of integrated EBS devices. In the last decade, fundamental investigation of phonons and their interactions with light, electrons, or other quantum degrees of freedom, has motivated a strong interest in developing new designs performing in the 100s MHz to GHz frequency range. Such a frequency range coincides with the optical linewidths of telecom domain nanocavities, resulting in pronounced cavity optomechanical dynamical backaction. Also, the acoustic waves within this frequency range possess wavelengths (hence wavevectors) that are comparable to their optical counterparts in the telecom domain, satisfying the phase-matching conditions for Brillouin scattering.

1.2.2 Bulk acoustic wave transducers

The bulk acoustic wave (BAW) mode is another type of acoustic mode that is straightforward to generate in piezoelectric devices. A BAW transducer (Fig. 1C) is constructed by sandwiching the piezoelectric film between two electrodes. The RF/MW signal applied to the electrodes excites the acoustic wave propagating across the thickness of the film. Only the modes that can form the standing wave in the film will resonate in the device, generating a significant response. The resonant condition is satisfied when the film thickness d and the acoustic wavelength are related as: $d = n\lambda/2$, where $n = 1, 2, 3 \ldots$. Therefore, the frequency of the BAW mode is determined by the thickness of the piezoelectric film, which contrasts with the IDT's lateral periodicity. Using thinner films will increase the frequency but at the cost of lowered electromechanical coupling efficiency.

In the context of acousto-optic interaction, the BAW transducer is less convenient than the SAW transducers because the acoustic mode is in the vertical direction. To achieve acousto-optic coupling, the photonic devices need to be integrated below or above the transducers, requiring multilayer configurations. Nevertheless, recently there have been several remarkable works on integrating AlN-based BAW transducers with silicon and silicon nitride photonic devices (Tian et al., 2020, 2021).

2. Brillouin scattering by electromechanically excited acoustic wave

2.1 Acoustic modes

Oftentimes, the acoustic waves propagate and induce EBS processes in the same piezoelectric material where they are electromechanically excited. Due to piezoelectricity, the mechanical and electric field quantities are linearly coupled and co-propagate to form the acoustic modes. The mechanical field quantities should obey the equation of motion, that is the Newton's second law. The electric field quantities should be quasi-electrostatic, described by the Gauss's law, because the acoustic wavelength is generally five orders of magnitude smaller than that of the electromagnetic wavelength at the same frequency. The mechanical and electric field quantities are linearly coupled through the elastic, piezoelectric, and dielectric properties of the material. For non-piezoelectric materials, the mechanical and electric field quantities are decoupled, leading to a purely mechanical acoustic mode. Furthermore, in acoustic modes, the potential energy density includes both the mechanical potential energy and electric energy, while the kinetic energy includes the mechanical contribution only. The total energy flux is the sum of the mechanical and electric contributions. In most cases, however, the mechanical contribution dominates.

The acoustic modes in piezoelectric waveguides and cavities can be readily solved using the Finite Element Method (FEM), with commercial software such as COMSOL Multiphysics. Besides the mechanical field, note that the electric field also contributes to the EBS process through the linear electro-optic effects in materials such as AlN and $LiNbO_3$. Therefore, the electric field must be solved for accurately in such cases. Due to anisotropy, the acoustic mode profiles and dispersion curves depend on the propagation direction with respect to the crystal orientation.

The acoustic modes can be categorized by the confinement dimensionality. The bulk acoustic modes include longitudinal and transverse (shear) modes, which are not confined in any dimension. The surface and film acoustic modes are confined in one dimension, forming 2D waveguides. Acoustic modes in strip waveguides and phononic crystal line defect waveguides are confined in two dimensions, forming 1D waveguides. Finally, acoustic cavity modes, including phononic crystal cavity and film bulk acoustic resonator (FBAR) modes, are confined in all three dimensions, forming 0D cavities.

Low loss is a key metric for acoustic modes to implement sophisticated EBS circuits and functionalities. Within the frequency range of 100s MHz to GHz, the pertinent loss mechanisms (Slobodnik et al., 1970) include the interaction with thermal phonons (temperature dependent), air damping (pressure dependent), and the scattering by crystalline defects, impurities, and surface roughness (fabrication dependent). As a rule of thumb, at room temperature and under atmospheric pressure, the acoustic propagation loss is on the order of a few dB/mm, which increases with frequency squared. At cryogenic temperature and under vacuum, however, the acoustic propagation loss can be a few dB/m or even lower (MacCabe et al., 2020; Sakuma and Nakayama, 1974), which has profound implications for quantum applications.

2.1.1 Acoustic waveguide and cavity

On the surface of a homogeneous bulk material, generally there is only one type of acoustic mode supported, termed Rayleigh SAW modes (Fig. 2A). Each differential volume in the SAW mode moves in an elliptical pattern, and the displacement decays exponentially from the surface into the bulk. The depth of the SAW modes is generally one wavelength. Intuitively, the free-to-move surface is "softer" than the bulk, hence an acoustic mode

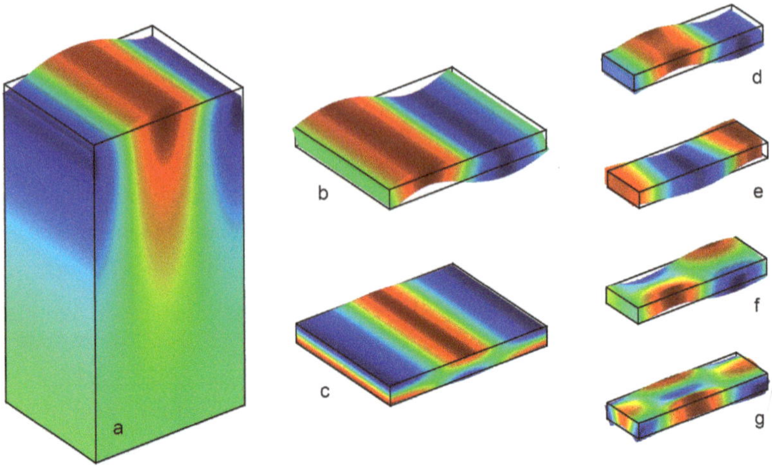

Fig. 2 Representative acoustic waveguide modes. (A) Surface acoustic wave (SAW) mode. Antisymmetric (B) and symmetric (C) Lamb wave modes. SV_0 (D), SH_0 (E), SV_1 (F), and SV_2 (G) modes in a suspended strip waveguide, where "S" stands for "shear-wave," "V" stands for "vertical," "H" stands for "horizontal," and the subscript stands for the mode order. For example, SV_1 is the 1st-order shear-wave mode with dominantly vertical displacement.

slower than any of the bulk acoustic modes can be guided at the surface without leaking into the bulk. In a sense, SAW modes are 2D acoustic waveguide modes that only require flat surfaces. Due to such fabrication convenience, SAW modes have been widely employed in conventional guided-wave acousto-optic devices (Tsai, 1990) on bulk $LiNbO_3$ wafers and ZnO thin films on non-piezoelectric substrates. They have also been widely employed in recent demonstrations of integrated EBS devices on diverse material platforms, including LNOI (Shao et al., 2020; Yu and Sun, 2020), GaAs (Crespo-Poveda et al., 2015; de Lima et al., 2006), and AlN (Li et al., 2015).

The acoustic modes in thin films are collectively termed Lamb modes. Lamb modes have more varieties, including the antisymmetric (Fig. 2B) and symmetric (Fig. 2C) modes, categorized by the symmetry of the out-of-plane displacement with respect to the dissecting plane of the thin film. The thin films are generally free-standing, forming the 2D acoustic waveguides where the acoustic modes are confined simply due to the complete reflection at the boundary between solid and air/vacuum. To fabricate such suspended thin films, the sacrificial layer beneath the thin film needs to be under-cut by isotropic etching through the releasing windows patterned in the thin film. Alternatively, the substrate can be removed from the backside, leaving only a frame to support the suspended thin films. Such acoustic waveguides have been demonstrated on suspended GaAs (Kapfinger et al., 2015), AlN (Kim et al., 2021; Li et al., 2019), and $LiNbO_3$ (Shao et al., 2019) thin films.

Strip waveguides possess continuous translational invariance along the propagation direction, mostly with rectangular or trapezoidal cross-sections in integrated EBS devices. Strip waveguides can be suspended, or on top of a substrate in which all the bulk acoustic modes have higher acoustic phase velocities than that in the strip waveguide. In the latter case, total internal reflection prevents the acoustic mode from leaking into the substrate. The acoustic modes within such 1D acoustic waveguides can be generally categorized by the major displacement direction and numbered by the presence of zeros in the displacement profile (Fig. 2D–G), in a fashion similar to that for optical modes. Typically, the major displacement directions include longitudinal, horizontal transverse (shear) and vertical transverse (shear). However, due to trapezoidal cross-section and/or material anisotropy, the mode profiles and dispersion curves may become nontrivial with complicated mode hybridization. The strip waveguides are usually lithographically patterned and etched with ICP-RIE processes, followed by an optional under-cut etching process to suspend them. Strip acoustic waveguides have been demonstrated in recent integrated EBS devices, including suspended AlN (Liu et al., 2019) and $LiNbO_3$ (Sarabalis et al., 2021) strip waveguides. Notably, unsuspended GaN strip acoustic circuits, which consist of ring resonators and coupling waveguides, have been demonstrated on a sapphire substrate (Fu et al., 2019).

In addition to the acoustic waveguide above, one of the simplest forms of an acoustic cavity is a thin film, plate, or membrane, which are also employed for integrated EBS devices (Tian et al., 2020, 2021). Basically, these FBAR modes are the standing wave formed by the bulk acoustic modes in the thin films due to the multi-reflection and interference between the two surfaces. The acoustic phonons in free-standing membranes of silicon-on-insulator (SOI) were first observed by Raman scattering (Sotomayor Torres et al., 2004), and later further confirmed and investigated by Brillouin scattering (Cuffe et al., 2012, 2013; Neogi et al., 2015).

2.1.2 Phononic crystal slab waveguide and cavity

The spatial periodicity of materials in a phononic crystal offers the possibility to engineer the mechanical band dispersion and its phononic properties.

In the context of integrated EBS devices, only 2D phononic crystal slabs and 1D phononic crystal beams are relevant due to convenient fabrication. Through careful design of the patterned structure (Kushwaha et al., 1993; Sigalas and Economou, 1993), acoustic band gaps appear over a certain range of frequencies (Martínez-Sala et al., 1995). This phenomenon is due to the destructive interference of scattered waves at Bragg wavelengths which prevents the propagation of sound waves through the crystal for all incident angles/directions. The frequency at which these forbidden zones occur depends on the size of the crystal and is inversely proportional to the structure periodicity, i.e., the smaller the period, the higher the frequency. The essential characteristic that the unit cell should possess to open a broad mechanical gap is the distribution of mass leading to large masses connected by small necks. Such mass contrast opens the phononic bandgap in a fashion similar to how the refractive index contrast opens the photonic bandgap. Therefore, the higher the mass contrast, the wider the mechanical gap becomes. The fabrication processes of suspended phononic crystal slabs and waveguides are similar to those of suspended strip waveguides described in Section 2.1.1.

Complete phononic bandgaps have been experimentally demonstrated on suspended silicon slabs with a hexagonal lattice of cylindrical air holes (Fig. 3A) (Mohammadi et al., 2008), a triangular lattice of snow-flake unit

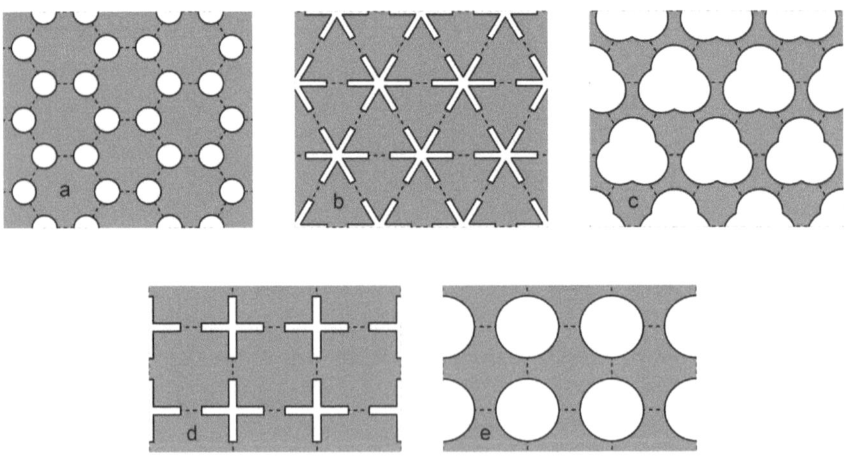

Fig. 3 Representative phononic crystal slab lattices and unit cells. (A) Hexagonal lattice of cylindrical air holes. (B) Triangular lattice of snow-flake unit cells. (C) Triangular lattice of shamrock unit cells. (D) Rectangular lattice of cross unit cells. (E) Rectangular lattice of cylindrical air holes or gold pillars deposited on the top.

cells (Fig. 3B) (Safavi-Naeini et al., 2014), and a triangular lattice of shamrock unit cells (Fig. 3C) (Florez et al., 2022; Söllner et al., 2016), as well as on suspended GaAs slabs with a rectangular lattice of cross unit cells (Fig. 3D) (Balram et al., 2016). Other forms of phononic crystal slabs have also been studied, including suspended silicon slabs with rectangular lattices of cylindrical air holes and gold pillars deposited on the top (Fig. 3E) (Graczykowski et al., 2015).

Within an otherwise perfect phononic crystal slab, 1D acoustic waveguides and 0D cavities can be implemented by creating line or point defects, respectively. This is a versatile approach to constructing acoustic waveguides and cavities. The profiles and band dispersion of the supported Bloch modes can be fully engineered for specific purposes, such as a slow group velocity, or enhanced EBS processes. The phononic modes are laterally confined by the phononic bandgaps and vertically confined by the reflection at the boundaries between the slab and air or the substrate. The phononic modes in the GHz frequency range can be guided and manipulated within such structures.

Because the acoustic wavelength at GHz frequency becomes similar to the optical wavelength, it is possible to design periodic structures with both phononic and photonic gaps. Such structures are dubbed as *optomechanical* crystals (Eichenfield et al., 2009) or *phoxonic* crystals (Djafari-Rouhani et al., 2016). The spatial co-localization of photonic and phononic modes in nanobeam optomechanical crystals enables engineering phonon-photon interaction in the strongly coupled regime, which leads to novel phenomena and quantum applications such as optomechanically induced transparency and quantum transduction (Mirhosseini et al., 2020).

2.2 Electromechanical Brillouin scattering

2.2.1 Physical effects

The Brillouin scattering in materials with strong piezoelectric, electro-optic effects, and/or optical anisotropy, such as $LiNbO_3$, involves many physical effects (Nye, 1985; Tsai, 1990) that have not been fully and explicitly accounted for in the recent literature. Here a complete account of all the physical effects involved in the EBS processes in a general non-magnetic material is provided, as shown in Fig. 4.

The acoustic waves change the material refractive indices through the primary photoelastic effect, the secondary photoelastic effect, the roto-optic effect, and the dielectric tensor redistribution. The primary photoelastic effect is the change of the dielectric tensor directly by the strain in the

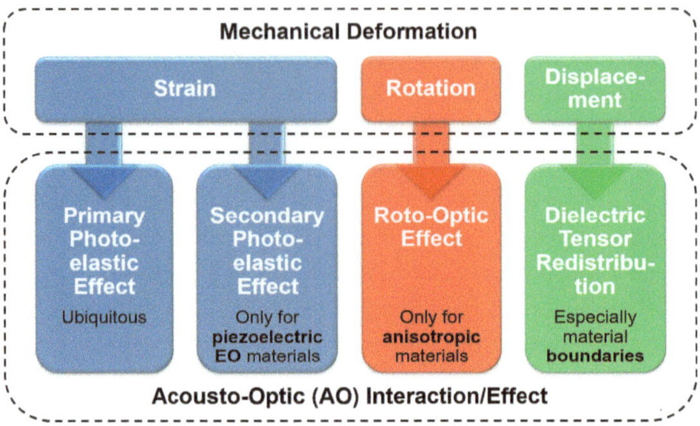

Fig. 4 The physical effects involved in the EBS processes in a general non-magnetic material.

acoustic mode. It is ubiquitous in any material. The secondary photoelastic effect is the change of the dielectric tensor indirectly by the strain in the acoustic mode through the piezoelectric and electro-optic effects. It is only present in materials with strong piezoelectric and electro-optic effects. The roto-optic effect (Nelson and Lax, 1971) is the change of the dielectric tensor, observed in the original reference frame, by the local pure rotation in the acoustic mode of an optically anisotropic material. Its contribution is non-negligible in materials with strong optical anisotropy. The dielectric tensor redistribution refers to the change of the dielectric tensor distribution by the local displacement in the acoustic mode, which is ubiquitous in any material. However, such an effect is only pronounced at the discontinuities of the dielectric tensor distribution, that is the material boundaries (Aspelmeyer et al., 2014a, 2014b). Otherwise, its contribution is of higher-order perturbation. Both perturbative (Sarabalis et al., 2021) and non-perturbative (Liu et al., 2019) approaches can be employed to quantify all the physical effects described above.

2.2.2 Phase-matching conditions

The phase-matching conditions of EBS processes include both the frequency and wavevector matching, which has been elaborated in Section 1 for SBS processes. However, there are two major differences between the phase-matching conditions of EBS and SBS processes.

First, for SBS, the process is sustained only if phonons are emitted, therefore anti-Stokes processes can only occur spontaneously, that is through thermal phonons. The SBS process stops when the pump is fully depleted, and the Stokes photons do not convert back to the pump photons. In contrast, for EBS, due to the abundance of electromechanically excited phonons, both Stokes and anti-Stokes processes can occur, and the conversion between the pump and Stokes/anti-Stokes photons can occur in both ways.

Another major difference is that SBS has only been observed in the collinear configuration (Fig. 5A and B), while EBS has also been demonstrated in the transverse (Fig. 5C) and 2D-free-space (Fig. 5D) configurations. The three configurations are categorized by the dimensionalities of their respective acoustic and optical waveguides. In the collinear configuration, the acoustic and optical waves co-propagate in the same 1D waveguide, such as a strip waveguide. The phase-matching conditions, in this case, is the same as those for SBS. On the other hand, in the 2D-free-space configuration, the acoustic and optical waves co-propagate in the same 2D waveguide, such as a slab waveguide, but in arbitrary directions. Hence, their wavevector-matching conditions must be generalized into 2D to account for the different directions of the wavevectors. In the transverse configuration, the acoustic waves in a 2D waveguide, such as a slab waveguide, traverse the optical waves in a 1D waveguide, such as a ridge waveguide, at an oblique or right angle, hence only the projection of the acoustic wavevector along the optical waveguide should be considered in the wavevector-matching conditions.

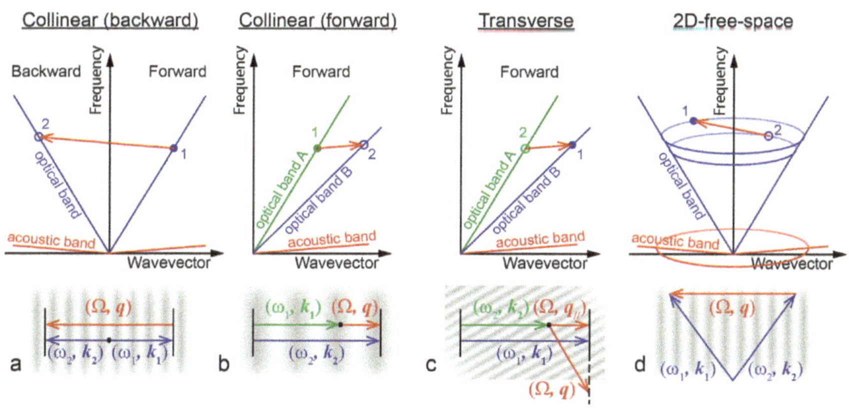

Fig. 5 Phase-matching conditions in representative EBS configurations. (A) Collinear backward intraband EBS. (B) Collinear forward interband EBS. (C) Transverse forward interband EBS. (D) 2D-free-space any-direction intraband.

2.2.3 Electromechanical Brillouin scattering in different configurations

In Section 2.2.2, three feasible EBS configurations have been introduced, including the collinear, transverse and 2D-free-space configurations. For each configuration, the EBS process can be further categorized according to the direction (forward, backward, or any-direction) and band (interband or intraband) of the scattered optical mode, with respect to the pump optical mode. In the following, representative works on different EBS configurations are briefly reviewed. Note that there is so far no demonstration of backward interband EBS in any configuration, even though it is theoretically feasible.

Collinear backward intraband EBS has been demonstrated in suspended thin-film AlN strip waveguides (Liu et al., 2019), as shown in Fig. 6A, which is so far the first and only demonstration of this EBS configuration. The challenge lies in the ultrahigh acoustic frequency required for the wavevector matching. In this work, an acoustic wave of ~16 GHz is employed, whose wavelength is only half of that of the optical TE_0 mode in the AlN waveguide, thereby forming an effective traveling Bragg grating that reflects the TE_0 mode. The scattered TE_0 mode also experiences a Doppler blue shift at ~16 GHz, which naturally implements single-sideband modulation. Using this device as the modulator, a prototype single-sideband RF photonic link has been demonstrated with homodyne and heterodyne demodulation schemes. The acoustic frequency of ~16 GHz in this work has been so far the highest in EBS devices.

Collinear forward interband EBS has been demonstrated in suspended thin-film $LiNbO_3$ strip waveguides (Sarabalis et al., 2021), as shown in Fig. 6B, which features so far the highest EBS efficiency of 18%. Collinear forward interband EBS has been demonstrated using conventional guided-wave AO devices (Brot et al., 1995; Wagner et al., 1994), but (Sarabalis et al., 2021) is so far the first and only demonstration of such an EBS configuration in suspended, wavelength-scale waveguides. The suspended waveguide is 250 μm long, with a measured acoustic loss of 11.7 dB/mm and it uses acoustic power of 30 μW to achieve 18% conversion efficiency. The EBS device has also been used to demonstrate nonreciprocal direction-dependent Doppler shift. Overall, this EBS configuration is optimal in terms of AO interaction enhancement, hence promising to achieve near-unity optical mode conversion efficiency with only sub-mW acoustic power.

Transverse forward interband EBS has been demonstrated in a series of works (Kim et al., 2021; Sohn et al., 2018; Sohn and Bahl, 2019) on

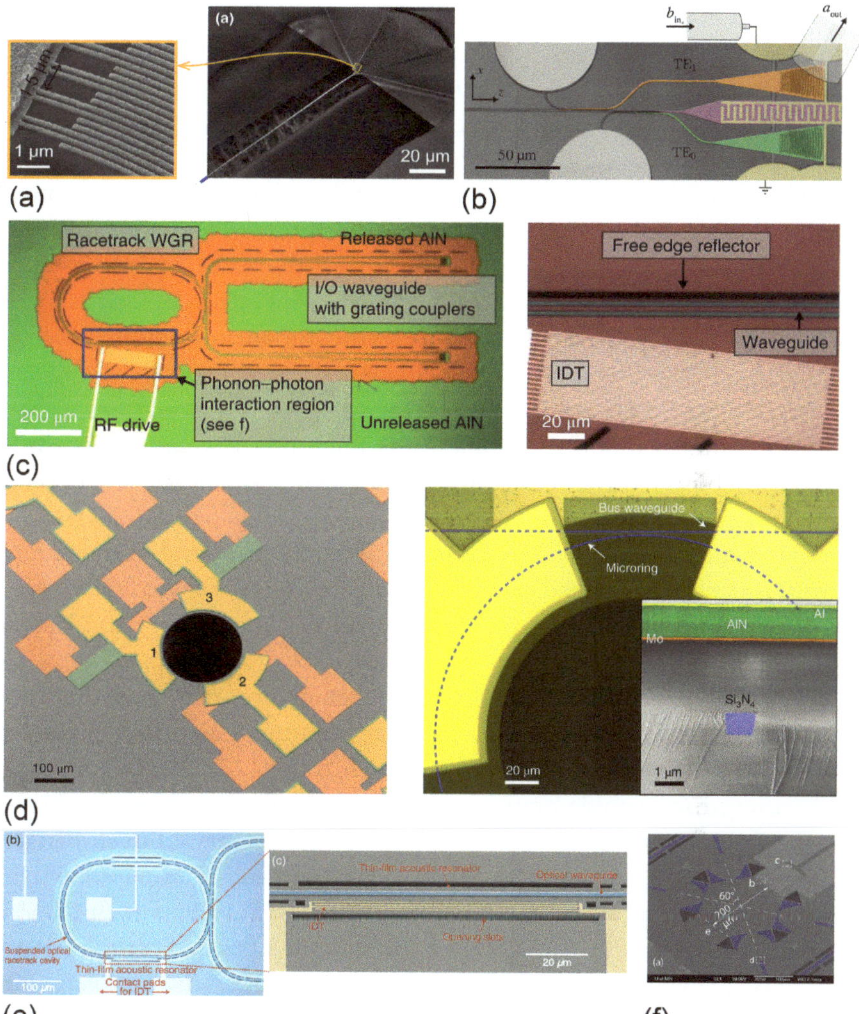

Fig. 6 Electromechanical Brillouin scattering in different configurations. (A) Collinear backward intraband EBS. (B) Collinear forward interband EBS. (C) Transverse forward interband EBS with Lamb waves. (D) Transverse forward interband EBS with bulk acoustic waves. (E) Transverse forward intraband EBS. (F) 2D-free-space any-direction intraband EBS. *Panel (A): Adapted with permission from Liu, Q., Li, H., Li, M., 2019. Electromechanical Brillouin scattering in integrated optomechanical waveguides. Optica 6, 778. doi:10.1364/OPTICA.6.000778. © The Optical Society. Panel (B): Adapted with permission from Sarabalis, C.J., Van Laer, R., Patel, R.N., Dahmani, Y.D., Jiang, W., Mayor, F.M., Safavi-Naeini, A.H., 2021. Acousto-optic modulation of a wavelength-scale waveguide. Optica 8, 477. doi:10.1364/OPTICA.413401. © The Optical Society. Panel (C): Adapted with permission from Sohn, D.B., Kim, S., Bahl, G., 2018. Time-reversal symmetry*

(Continued)

suspended thin-film AlN ridge waveguides, as shown in Fig. 6C, which features the demonstration of broken time-reversal symmetry, hence nonreciprocity, in EBS devices. On one side of the optical waveguide, the IDT fingers, hence the acoustic wavefront, make a small oblique angle with respect to the optical waveguide to tailor the phase-matching conditions. On the other side of the optical waveguide, a free edge reflector is used to achieve the optimal acoustic standing wave profile. The optical waveguide is embedded within a high-Q racetrack resonator to enhance the EBS efficiency. Furthermore, a magnetic-free integrated optical isolator has been demonstrated on AlN piezoelectric modulators monolithically integrated on low-loss Si_3N_4 photonic integrated circuits (Tian et al., 2020, 2021), as shown in Fig. 6D. The transmission reciprocity is broken by spatio-temporal modulation of a Si_3N_4 microring resonator with three AlN bulk acoustic wave resonators driven with a rotational phase. This design creates an effective rotating acoustic wave that allows transverse forward interband EBS in only one direction along the microring. A maximum of 10 dB isolation is achieved using 300 mW total RF power applied to three actuators, with minimum insertion loss of 0.1 dB and an isolation bandwidth of 700 MHz.

Transverse forward intraband EBS has been demonstrated in numerous works, as shown in Fig. 6E, including acousto-optic Mach-Zehnder modulators (Crespo-Poveda et al., 2013; Shao et al., 2019; Yu and Sun, 2020), arrayed waveguide gratings (AWGs) (Crespo-Poveda et al., 2015), racetrack resonators (Shao et al., 2019), photonic crystal nanocavities (Kapfinger et al., 2015; Li et al., 2015), and optomechanical adiabatic frequency shifters (Fan et al., 2016), among others. However, it should be noted that in most of the cases above, the acousto-optic interaction length is not sufficiently long to enable precise phase-matching, which is analogous to the Raman-Nath regime in conventional AO devices, instead of the

Fig. 6—Cont'd *breaking with acoustic pumping of nanophotonic circuits. Nat. Photonics 12, 91–97. doi:10.1038/s41566-017-0075-2. © Springer Nature. Panel (D): Adapted with permission from Tian, H., Liu, J., Siddharth, A., Wang, R.N., Blésin, T., He, J., Kippenberg, T.J., Bhave, S.A., 2021. Magnetic-free silicon nitride integrated optical isolator. Nat. Photonics 15, 828–836. doi:10.1038/s41566-021-00882-z. © Springer Nature. Panel (E): Adapted with permission from Shao, L., Yu, M., Maity, S., Sinclair, N., Zheng, L., Chia, C., Shams-Ansari, A., Wang, C., Zhang, M., Lai, K., Lončar, M., 2019. Microwave-to-optical conversion using lithium niobate thin-film acoustic resonators. Optica 6, 1498. doi:10.1364/OPTICA.6.001498. © The Optical Society. Panel (F): Adapted from Li, H., Liu, Q., Li, M., 2019. Electromechanical Brillouin scattering in integrated planar photonics. APL Photonics 4, 080802. doi:10.1063/1.5108672.*

Bragg regime. This is often manifested in that these EBS processes result in both the Stokes and anti-Stokes sidebands of comparable amplitudes, which can be simply considered as a phase modulation of the optical carrier (pump) mode. Within a high-Q cavity, either the Stokes or the anti-Stokes sideband may be significantly suppressed or enhanced due to the cavity resonance.

Finally, 2D-free-space any-direction intraband EBS has been demonstrated on suspended AlN slab waveguides (Li et al., 2019) and LNOI slab waveguides (Shao et al., 2020), as shown in Fig. 6F . The former work implemented AO beam deflectors featuring an unprecedented large deflection angle of ~60° and high frequency shift of ~11 GHz, while the latter work implemented an integrated 3-GHz acousto-optic frequency shifter (AOFS) with a carrier suppression over 30 dB, as well as a gigahertz spaced optical frequency comb featuring more than 200 lines over a 0.6-THz optical bandwidth. The performance of such devices is superior compared with their conventional counterparts.

3. Applications

3.1 Next-generation acousto-optics

The integrated EBS devices are well poised to comprehensively reinvent the conventional guided-wave acousto-optics (AO), with unprecedented compact footprints, enhanced EBS processes, and high acoustic frequencies. Many of the applications demonstrated in the conventional AO devices (Tsai, 1990) are still relevant for the next-generation AO, including optical beam steerers, tunable filters, frequency shifters, and single-sideband modulators.

Following is a simplified argument of scaling the AO devices: if the acoustic power density is kept the same, so is the refractive index change it induces, then the total acoustic power consumption to achieve the same function is proportional to the cross-sectional area of the acoustic wave for the AO interactions. From bulk-wave AO devices to guided-wave AO devices, the acoustic wave cross-sectional area is reduced from a few mm^2 to 100s μm^2, resulting in acoustic power reduction from a few W to a few mW or sub-mW. The next-generation AO devices are expected to feature an acoustic wave cross-section of a few μm^2 or sub-μm^2, leading to an acoustic power reduction of another two or three orders of magnitude.

Other than the significant reduction of the acoustic power consumption, the performance metrics of the most representative AO devices will also be greatly improved. For optical beam steerers, the steering angle is expected to increase from only a few degrees to potentially 180°, that is, arbitrary angles are feasible. For optical tunable filters, the operation wavelength can be extended to the telecom bands and even the mid-infrared, and the entire device can be monolithically integrated with other peripheral devices on an LNOI chip to implement a complete system for signal processing. For frequency shifters and single-sideband modulators, the operation frequency is expected to increase from 100 s MHz to as high as 10s GHz. However, a crucial premise for the performance improvement above is that the acoustic loss remains sufficiently low, especially at GHz frequency and above. Such a premise may be invalid at room temperature, due to the abundance of thermal phonons that interact with the acoustic wave of interest. Nevertheless, at cryogenic temperatures, the acoustic loss has been experimentally demonstrated to be low enough to support meter-scale propagation length and above (MacCabe et al., 2020), promising for quantum applications.

3.2 Nonreciprocal devices and circuits

Unless the acoustic wavevector is perpendicular to the optical wavevector, EBS processes induced by propagating acoustic modes are intrinsically nonreciprocal. In such processes, the time-reversal symmetry of light propagation is broken in the sense that when the time is reversed, the propagating acoustic mode is not accordingly reversed. In other words, the optical modes that propagate along and against the acoustic mode "sees" slightly different systems. This is an intuitive explanation for the nonreciprocity. In integrated EBS devices, such nonreciprocity can be leveraged to implement optical isolation or circulation devices, with potentially only sub-mW acoustic power consumption. Even more intriguingly, because the acoustic modes can be excited and manipulated with phononic circuits, it is potentially possible to achieve reconfigurable nonreciprocal devices and circuits. Such reconfigurability cannot be conveniently achieved with magneto-optic thin-film devices.

3.3 Quantum transduction

The ideal platform to achieve RF-to-optical conversion and vice versa would combine optomechanics with both photonic and electromechanical controls. Thus, a new approach consists in combining photonic and phononic circuits on the same platform. For example, the realization of on-chip quantum coherent interconnections between superconducting quantum processors remains an important challenge to scaling up the computing systems while limiting the power consumption in cryogenic systems. In this context, optomechanical systems can be a candidate to perform microwave-to-optical conversion of quantum states (Balram et al., 2016;

Bochmann et al., 2013; Mirhosseini et al., 2020). Such a hybrid integrated platform that incorporates an optomechanical cavity with electromechanical actuation has been proposed (Hönl et al., 2022) and fabricated using gallium phosphide (GaP) bonded onto a silicon chip. The electric field generated by two niobate electrodes placed below the optomechanical nanobeam couples to the mechanical motion due to the piezoelectric properties of GaP. Then, relying on standard optomechanical characterization techniques, the spectral analysis of the nanobeam thermal phonon modes can be performed using the fact that these GHz mechanical modes can be piezoelectrically driven. In this preliminary demonstration, the platform is only exploited at room temperature as a preliminary demonstration prior to future investigations at cryogenic temperatures, in the quantum regime.

3.4 Nano-opto-electro-mechanical (NOEM) platforms

The association of optical forces and electromechanical actuations in a unique Nano-Opto-Electro-Mechanical (NOEM) platform is now rendered possible by the convergence of the respective fields of cavity optomechanics and nano-electromechanics (Midolo et al., 2018; Xu et al., 2021). The term "NOEM" is also known as "Nano-Electro-Opto-Mechanical (NEOM)," which will be used interchangeably hereafter.

Such configuration has recently been achieved in SOI using nanocrystalline Si (Navarro-Urrios et al., 2022). This device (see Fig. 7) incorporates an AlN layer and Al interdigitated electrodes for piezoelectric generation of SAWs. It already incorporates mechanical lasing in the central part of the circuit, electrical actuation by IDTs, optomechanical interaction via an

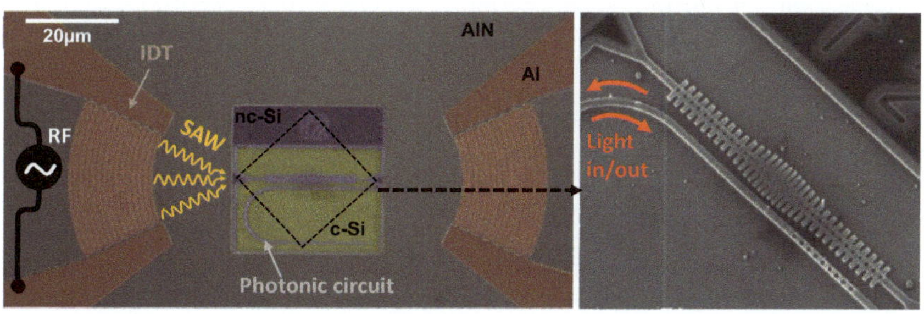

Fig. 7 NEOMS circuit. Colorized SEM micrograph highlighting the AlN IDTs (in red) enabling SAW launching into a 1D optomechanical nanobeam (in violet). Optical access is enabled by an integrated photonic circuitry (blue). *Adapted with permission from Navarro-Urrios, D., Colombano, M.F., Arregui, G., Madiot, G., Pitanti, A., Griol, A., Makkonen, T., Ahopelto, J., Sotomayor-Torres, C.M., Martínez, A., 2022. Room-temperature silicon platform for GHz-frequency nanoelectro-opto-mechanical systems. ACS Photonics 9, 413–419. doi:10.1021/acsphotonics.1c01614. © American Chemical Society.*

integrated photonic waveguide, coherent phonon propagation from the IDTs to the nanobeams and vice versa and, to be demonstrated, phonon detection by a second IDT. This constitutes the first demonstration of a CMOS-compatible NEOM device at room temperature. Such phononic circuit promises a range of applications in all-optical processing in wireless networks such as RF-to-optical conversion.

Acknowledgments

HL acknowledges the faculty start-up fund provided by Zhejiang University. ICN2 is supported by the Severo Ochoa program of the Spanish Research Agency (AEI, grant no. SEV-2017-0706) and by the CERCA Programme/Generalitat de Catalunya. The ICN2 authors acknowledge support from the Spanish MICINN project SIP (PGC2018-101743-B-I00), the European Union grants TOCHA (GA Nr. 824140), PhENOMEN (GA Nr. 713450) and LEIT (GA Nr. 885689). OF acknowledges the support of an EU MS Curie COFUND PREBIST PhD fellowship (GA Nr. 754558). ML acknowledges the funding support from National Science Foundation (NSF).

References

Aspelmeyer, M., Kippenberg, T.J., Marquardt, F., 2014a. Cavity optomechanics. Rev. Mod. Phys. 86, 1391–1452. https://doi.org/10.1103/RevModPhys.86.1391.

Aspelmeyer, M., Kippenberg, T.J., Marquardt, F. (Eds.), 2014b. Cavity Optomechanics. Springer Berlin Heidelberg, Berlin, Heidelberg, https://doi.org/10.1007/978-3-642-55312-7.

Balram, K.C., Davanço, M.I., Song, J.D., Srinivasan, K., 2016. Coherent coupling between radiofrequency, optical and acoustic waves in piezo-optomechanical circuits. Nat. Photonics 10, 346–352. https://doi.org/10.1038/nphoton.2016.46.

Bochmann, J., Vainsencher, A., Awschalom, D.D., Cleland, A.N., 2013. Nanomechanical coupling between microwave and optical photons. Nat. Phys. 9, 712–716. https://doi.org/10.1038/nphys2748.

Brillouin, L., 1922. Diffusion de la lumière et des rayons X par un corps transparent homogène. Ann. Phys. (Paris). 9, 88–122. https://doi.org/10.1051/anphys/192209170088.

Brot, C., Duchet, C., Di Maggio, M., 1995. Interdigital transducer for acousto-optic tunable filter on LiNbO3. Electron. Lett. 31, 1235–1237. https://doi.org/10.1049/el:19950851.

Campbell, C.K., 1998. Surface Acoustic Wave Devices for Mobile and Wireless Communications. Academic Press, San Diego, CA.

Crespo-Poveda, A., Hey, R., Biermann, K., Tahraoui, A., Santos, P.V., Gargallo, B., Muñoz, P., Cantarero, A., de Lima, M.M., 2013. Synchronized photonic modulators driven by surface acoustic waves. Opt. Express 21, 21669. https://doi.org/10.1364/OE.21.021669.

Crespo-Poveda, A., Hernández-Mínguez, A., Gargallo, B., Biermann, K., Tahraoui, A., Santos, P.V., Muñoz, P., Cantarero, A., de Lima, M.M., 2015. Acoustically driven arrayed waveguide grating. Opt. Express 23, 21213. https://doi.org/10.1364/OE.23.021213.

Cuffe, J., Chávez, E., Shchepetov, A., Chapuis, P.-O., El Boudouti, E.H., Alzina, F., Kehoe, T., Gomis-Bresco, J., Dudek, D., Pennec, Y., Djafari-Rouhani, B., Prunnila, M., Ahopelto, J., Sotomayor Torres, C.M., 2012. Phonons in slow motion: dispersion relations in ultrathin Si membranes. Nano Lett. 12, 3569–3573. https://doi.org/10.1021/nl301204u.

Cuffe, J., Ristow, O., Chávez, E., Shchepetov, A., Chapuis, P.-O., Alzina, F., Hettich, M., Prunnila, M., Ahopelto, J., Dekorsy, T., Sotomayor Torres, C.M., 2013. Lifetimes of confined acoustic phonons in ultrathin silicon membranes. Phys. Rev. Lett. 110, 095503. https://doi.org/10.1103/PhysRevLett.110.095503.

Datta, S., 1986. Surface Acoustic Wave Devices. Prentice-Hall, Englewood Cliffs, NJ.

de Lima, M.M., Beck, M., Hey, R., Santos, P.V., 2006. Compact Mach-Zehnder acousto-optic modulator. Appl. Phys. Lett. 89, 121104. https://doi.org/10.1063/1.2354411.

Dietrich, C.P., Fiore, A., Thompson, M.G., Kamp, M., Höfling, S., 2016. GaAs integrated quantum photonics: towards compact and multi-functional quantum photonic integrated circuits. Laser Photonics Rev. 10, 870–894. https://doi.org/10.1002/lpor.201500321.

Djafari-Rouhani, B., El-Jallal, S., Pennec, Y., 2016. Phoxonic crystals and cavity optomechanics. Comptes Rendus Phys. 17, 555–564. https://doi.org/10.1016/j.crhy.2016.02.001.

Eggleton, B.J., Poulton, C.G., Rakich, P.T., Steel, M.J., Bahl, G., 2019. Brillouin integrated photonics. Nat. Photonics. https://doi.org/10.1038/s41566-019-0498-z.

Eichenfield, M., Chan, J., Camacho, R.M., Vahala, K.J., Painter, O., 2009. Optomechanical crystals. Nature 462, 78–82. https://doi.org/10.1038/nature08524.

Fan, L., Zou, C.-L., Poot, M., Cheng, R., Guo, X., Han, X., Tang, H.X., 2016. Integrated optomechanical single-photon frequency shifter. Nat. Photonics 10, 766–770. https://doi.org/10.1038/nphoton.2016.206.

Florez, O., Arregui, G., Albrechtsen, M., Ng, R.C., Gomis-Bresco, J., Stobbe, S., Sotomayor-Torres, C.M., García, P.D., 2022. Engineering nanoscale hypersonic phonon transport. Nat. Nanotechnol. 17. https://doi.org/10.1038/s41565-01178-1.

Fu, W., Shen, Z., Xu, Y., Zou, C.-L., Cheng, R., Han, X., Tang, H.X., 2019. Phononic integrated circuitry and spin–orbit interaction of phonons. Nat. Commun. 10, 2743. https://doi.org/10.1038/s41467-019-10852-3.

Graczykowski, B., Sledzinska, M., Alzina, F., Gomis-Bresco, J., Reparaz, J.S., Wagner, M.R., Sotomayor Torres, C.M., 2015. Phonon dispersion in hypersonic two-dimensional phononic crystal membranes. Phys. Rev. B 91, 075414. https://doi.org/10.1103/PhysRevB.91.075414.

Hashimoto, K., 2000. Surface Acoustic Wave Devices in Telecommunications: Modelling and Simulation. Springer, Berlin.

Hönl, S., Popoff, Y., Caimi, D., Beccari, A., Kippenberg, T.J., Seidler, P., 2022. Microwave-to-optical conversion with a gallium phosphide photonic crystal cavity. Nat. Commun. 13, 2065. https://doi.org/10.1038/s41467-022-28670-5.

Kapfinger, S., Reichert, T., Lichtmannecker, S., Müller, K., Finley, J.J., Wixforth, A., Kaniber, M., Krenner, H.J., 2015. Dynamic acousto-optic control of a strongly coupled photonic molecule. Nat. Commun. 6, 8540. https://doi.org/10.1038/ncomms9540.

Kim, S., Sohn, D.B., Peterson, C.W., Bahl, G., 2021. On-chip optical non-reciprocity through a synthetic hall effect for photons. APL Photonics 6, 011301. https://doi.org/10.1063/5.0034291.

Kushwaha, M.S., Halevi, P., Dobrzynski, L., Djafari-Rouhani, B., 1993. Acoustic band structure of periodic elastic composites. Phys. Rev. Lett. 71, 2022–2025. https://doi.org/10.1103/PhysRevLett.71.2022.

Li, H., Tadesse, S.A., Liu, Q., Li, M., 2015. Nanophotonic cavity optomechanics with propagating acoustic waves at frequencies up to 12 GHz. Optica 2, 826. https://doi.org/10.1364/OPTICA.2.000826.

Li, H., Liu, Q., Li, M., 2019. Electromechanical Brillouin scattering in integrated planar photonics. APL Photonics 4, 080802. https://doi.org/10.1063/1.5108672.

Liu, Q., Li, H., Li, M., 2019. Electromechanical Brillouin scattering in integrated optomechanical waveguides. Optica 6, 778. https://doi.org/10.1364/OPTICA.6.000778.

MacCabe, G.S., Ren, H., Luo, J., Cohen, J.D., Zhou, H., Sipahigil, A., Mirhosseini, M., Painter, O., 2020. Nano-acoustic resonator with ultralong phonon lifetime. Science (80) 370, 840–843. https://doi.org/10.1126/science.abc7312.

Malocha, D.C., 2004. Evolution of the SAW transducer for communication systems. In: IEEE Ultrasonics Symposium, 2004. IEEE, pp. 302–310, https://doi.org/10.1109/ULTSYM.2004.1417726.

Martínez-Sala, R., Sancho, J., Sánchez, J.V., Gómez, V., Llinares, J., Meseguer, F., 1995. Sound attenuation by sculpture. Nature 378, 241. https://doi.org/10.1038/378241a0.

Midolo, L., Schliesser, A., Fiore, A., 2018. Nano-opto-electro-mechanical systems. Nat. Nanotechnol. 13, 11–18. https://doi.org/10.1038/s41565-017-0039-1.

Mirhosseini, M., Sipahigil, A., Kalaee, M., Painter, O., 2020. Superconducting qubit to optical photon transduction. Nature 588, 599–603. https://doi.org/10.1038/s41586-020-3038-6.

Modica, G., Zhu, R., Horvath, R., Beaudoin, G., Sagnes, I., Braive, R., 2020. Slow propagation of 2 GHz acoustical waves in a suspended GaAs phononic waveguide on insulator. Appl. Phys. Lett. 117, 193501. https://doi.org/10.1063/5.0019949.

Mohammadi, S., Eftekhar, A.A., Khelif, A., Hunt, W.D., Adibi, A., 2008. Evidence of large high frequency complete phononic band gaps in silicon phononic crystal plates. Appl. Phys. Lett. 92, 221905. https://doi.org/10.1063/1.2939097.

Morgan, D.P., 2007. Surface Acoustic Wave Filters with Applications to Electronic Communications and Signal Processing, second ed. Academic Press.

Navarro-Urrios, D., Colombano, M.F., Arregui, G., Madiot, G., Pitanti, A., Griol, A., Makkonen, T., Ahopelto, J., Sotomayor-Torres, C.M., Martínez, A., 2022. Room-temperature silicon platform for GHz-frequency nanoelectro-opto-mechanical systems. ACS Photonics 9, 413–419. https://doi.org/10.1021/acsphotonics.1c01614.

Nelson, D.F., Lax, M., 1971. Theory of the photoelastic interaction. Phys. Rev. B 3, 2778–2794. https://doi.org/10.1103/PhysRevB.3.2778.

Neogi, S., Reparaz, J.S., Pereira, L.F.C., Graczykowski, B., Wagner, M.R., Sledzinska, M., Shchepetov, A., Prunnila, M., Ahopelto, J., Sotomayor-Torres, C.M., Donadio, D., 2015. Tuning thermal transport in ultrathin silicon membranes by surface nanoscale engineering. ACS Nano 9, 3820–3828. https://doi.org/10.1021/nn506792d.

Nye, J.F., 1985. Physical Properties of Crystals: Their Representation by Tensors and Matrices. Clarendon Press, Oxford [Oxfordshire], New York.

Rumyantsev, S.L., Shur, M.S., Levinshtein, M.E., 2004. Materials properties of nitrides: summary. Int. J. High Speed Electron. Syst. 14, 1–19. https://doi.org/10.1142/S012915640400220X.

Ruppel, C.C.W., 2017. Acoustic wave filter technology—a review. IEEE Trans. Ultrason. Ferroelectr. Freq. Control 64, 1390–1400. https://doi.org/10.1109/TUFFC.2017.2690905.

Safavi-Naeini, A.H., Hill, J.T., Meenehan, S., Chan, J., Gröblacher, S., Painter, O., 2014. Two-dimensional phononic-photonic band gap optomechanical crystal cavity. Phys. Rev. Lett. 112, 153603. https://doi.org/10.1103/PhysRevLett.112.153603.

Sakuma, T., Nakayama, T., 1974. Attenuation of elastic surface waves by anharmonic interactions at low temperatures. Appl. Phys. Lett. 25, 176–177. https://doi.org/10.1063/1.1655428.

Sarabalis, C.J., Van Laer, R., Patel, R.N., Dahmani, Y.D., Jiang, W., Mayor, F.M., Safavi-Naeini, A.H., 2021. Acousto-optic modulation of a wavelength-scale waveguide. Optica 8, 477. https://doi.org/10.1364/OPTICA.413401.

Shao, L., Yu, M., Maity, S., Sinclair, N., Zheng, L., Chia, C., Shams-Ansari, A., Wang, C., Zhang, M., Lai, K., Lončar, M., 2019. Microwave-to-optical conversion using lithium niobate thin-film acoustic resonators. Optica 6, 1498. https://doi.org/10.1364/OPTICA.6.001498.

Shao, L., Sinclair, N., Leatham, J., Hu, Y., Yu, M., Turpin, T., Crowe, D., Lončar, M., 2020. Integrated microwave acousto-optic frequency shifter on thin-film lithium niobate. Opt. Express 28, 23728. https://doi.org/10.1364/OE.397138.

Sigalas, M., Economou, E.N., 1993. Band structure of elastic waves in two dimensional systems. Solid State Commun. 86, 141–143. https://doi.org/10.1016/0038-1098(93)90888-T.

Slobodnik, A.J., Carr, P.H., Budreau, A.J., 1970. Microwave frequency acoustic surface-wave loss mechanisms on $LiNbO_3$. J. Appl. Phys. 41, 4380–4387. https://doi.org/10.1063/1.1658471.

Sohn, D.B., Bahl, G., 2019. Direction reconfigurable nonreciprocal acousto-optic modulator on chip. APL Photonics 4, 126103. https://doi.org/10.1063/1.5123497.

Sohn, D.B., Kim, S., Bahl, G., 2018. Time-reversal symmetry breaking with acoustic pumping of nanophotonic circuits. Nat. Photonics 12, 91–97. https://doi.org/10.1038/s41566-017-0075-2.

Söllner, I., Midolo, L., Lodahl, P., 2016. Deterministic single-phonon source triggered by a single photon. Phys. Rev. Lett. 116, 234301. https://doi.org/10.1103/PhysRevLett.116.234301.

Sotomayor Torres, C.M., Zwick, A., Poinsotte, F., Groenen, J., Prunnila, M., Ahopelto, J., Mlayah, A., Paillard, V., 2004. Observations of confined acoustic phonons in silicon membranes. Phys. Status Solidi 1, 2609–2612. https://doi.org/10.1002/pssc.200405313.

Tadesse, S.A., Li, M., 2014. Sub-optical wavelength acoustic wave modulation of integrated photonic resonators at microwave frequencies. Nat. Commun. 5, 5402. https://doi.org/10.1038/ncomms6402.

Tian, H., Liu, J., Dong, B., Skehan, J.C., Zervas, M., Kippenberg, T.J., Bhave, S.A., 2020. Hybrid integrated photonics using bulk acoustic resonators. Nat. Commun. 11, 3073. https://doi.org/10.1038/s41467-020-16812-6.

Tian, H., Liu, J., Siddharth, A., Wang, R.N., Blésin, T., He, J., Kippenberg, T.J., Bhave, S.A., 2021. Magnetic-free silicon nitride integrated optical isolator. Nat. Photonics 15, 828–836. https://doi.org/10.1038/s41566-021-00882-z.

Tsai, C.S. (Ed.), 1990. Guided-Wave Acousto-Optics, Springer Series in Electronics and Photonics. Springer Berlin Heidelberg, Berlin, Heidelberg, https://doi.org/10.1007/978-3-642-75225-4.

Wagner, E., Hinkov, I., Hinkov, V., 1994. Low power integrated acousto-optical tunable filters in first telecommunication window. Electron. Lett. 30, 1884–1885. https://doi.org/10.1049/el.19941240.

Wang, J., Santamato, A., Jiang, P., Bonneau, D., Engin, E., Silverstone, J.W., Lermer, M., Beetz, J., Kamp, M., Höfling, S., Tanner, M.G., Natarajan, C.M., Hadfield, R.H., Dorenbos, S.N., Zwiller, V., O'Brien, J.L., Thompson, M.G., 2014. Gallium arsenide (GaAs) quantum photonic waveguide circuits. Opt. Commun. 327, 49–55. https://doi.org/10.1016/j.optcom.2014.02.040.

Weis, R.S., Gaylord, T.K., 1985. Lithium niobate: summary of physical properties and crystal structure. Appl. Phys. A Solids Surfaces 37, 191–203. https://doi.org/10.1007/BF00614817.

Xiong, C., Pernice, W.H.P., Sun, X., Schuck, C., Fong, K.Y., Tang, H.X., 2012. Aluminum nitride as a new material for chip-scale optomechanics and nonlinear optics. New J. Phys. 14, 095014. https://doi.org/10.1088/1367-2630/14/9/095014.

Xu, N., Cheng, Z.-D., Tang, J.-D., Lv, X.-M., Li, T., Guo, M.-L., Wang, Y., Song, H.-Z., Zhou, Q., Deng, G.-W., 2021. Recent advances in nano-opto-electro-mechanical systems. Nanophotonics 10, 2265–2281. https://doi.org/10.1515/nanoph-2021-0082.

Yu, Z., Sun, X., 2020. Acousto-optic modulation of photonic bound state in the continuum. Light Sci. Appl. 9, 1. https://doi.org/10.1038/s41377-019-0231-1.

CHAPTER FIFTEEN

Brillouin light scattering in biological systems

Irina Kabakova[a,*], Giuliano Scarcelli[b], and Seok-Hyun Yun[c,d]
[a]School of Mathematical and Physical Sciences, University of Technology Sydney, Broadway, NSW, Australia
[b]Fischell Department of Bioengineering, University of Maryland, College Park, MD, United States
[c]Harvard Medical School and Wellman Center for Photomedicine, Massachusetts General Hospital, Boston, MA, United States
[d]Harvard-MIT Health Sciences and Technology, Cambridge, MA, United States
*Corresponding author: e-mail address: irina.kabakova@uts.edu.au

Contents

1. Introduction to Brillouin light scattering in biological systems

The importance of the mechanical properties of tissues and cells and their interplay with the biological systems have long been recognized. A century ago it was observed that the trabeculae (porous bone tissue found at the end of long bones) in bones coordinated along with the principal stress lines of daily physical loading (Wolff, 1892). This led to the scientific understanding of the amazing ability of tissues to adapt to their mechanical

Semiconductors and Semimetals, Volume 110
ISSN 0080-8784
https://doi.org/10.1016/bs.semsem.2022.05.008

environment. Today, the critical role of extracellular mechanical environment has been well established in various branches of biology, including developmental biology (Jansen et al., 2015) and stem cell differentiation (Engler et al., 2006a; Wolfenson et al., 2019), as well as general cellular physiology (Weaver, 2017). In medical fields, abnormal mechanical properties of tissues are implicated in numerous diseases including cardiovascular diseases and hypertension to cancer. Lung pathologies, such as acute respiratory distress syndrome, fibrosis and emphysema, accompany the viscoelastic alterations of the airway tissues, which severely inhibit lung functions (Faffe and Zin, 2009).

Over the past decade or so, there have been growing interests in applying Brillouin light scattering (BLS) to measure the mechanical properties of tissues and cells to help understand the roles of biomechanics and to apply the knowledge for advancing life sciences and medical diagnosis and treatments. Unlike traditional mechanical testing techniques such as shear rheology, extensiometry, compression, and indentation, BLS does not require application of extrinsic forces nor the use of artificially introduced labels (see Fig. 1), making this technology particularly suitable for in situ and in vivo settings.

As in the solid-state applications given elsewhere in this volume, the two fundamental parameters that describe BLS in biological matter are the Brillouin frequency shift, BFS, (Ω) and the Brillouin linewidth (Γ). The BFS is proportional to the propagation velocity of hypersonic (GHz) longitudinal elastic waves c_s as $\Omega = \pm\, c_s q/(2\pi)$, where $q = \frac{4\pi n}{\lambda} \sin\Theta/2$ is the wavenumber of the elastic wave, n is the material refractive index, λ is the wavelength of light in vacuum, and Θ is the scattering angle between the optical and hypersonic waves. The sound speed is related to the complex longitudinal modulus $M = M' + jM''$ (also known as p-wave modulus) and mass density ρ as $c_s = \sqrt{M'/\rho}$. Thus, changes in BFS indicate changes in material's longitudinal modulus, density, and/or refractive index.

The propagation speed of the longitudinal elastic waves in solids and liquids falls in relatively narrow ranges, approximately, from 1000 to 1600 m/s for most liquids (~1500 m/s for water), and from 1500 m/s for soft rubber to 12,000 m/s for the hardest diamond. As a result, the difference in the BFS of liquids and solids is no larger than a factor of 10. Soft tissues are composed of solid constituents, such as macro-molecules, and liquid water with about 30% solids to 70% liquids ratio. With such a high content and the lower longitudinal modulus of water, the BFS of soft tissues is only slightly (<10%)

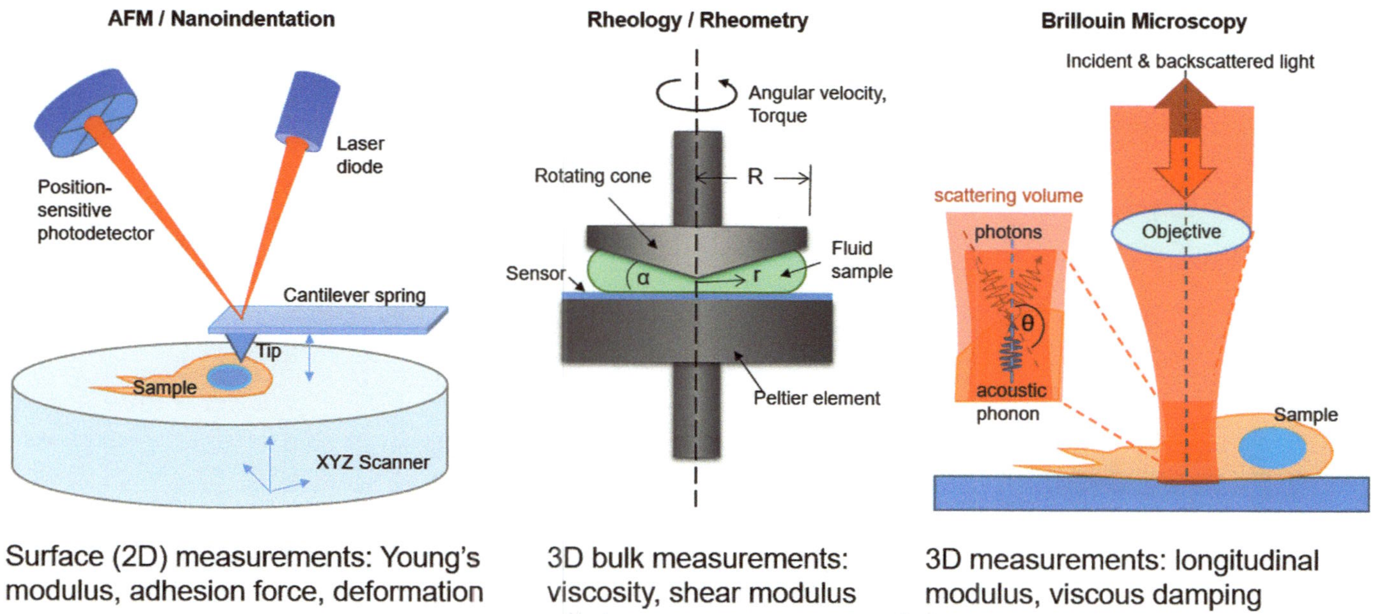

Fig. 1 Schematic illustration of comparative methods for measurements of mechanical properties in biological materials. *Left to right*: atomic force microscopy nanoindentation, rheology/rheometry, Brillouin microscopy. BLS relies on intrinsic thermal fluctuations and does not require any extrinsic mechanical load.

higher than the BFS of water. On the other hand, the BFS of soft tissues and cells (Scarcelli and Yun, 2008) are distinctly lower, by a factor 2 to 4, than the BFS of solid constituents, such as dehydrated fibrous components of muscles and tendons (Cusack and Miller, 1979; Palombo et al., 2014) and compact bone (Fukui et al., 2012).

The Brillouin linewidth $\Gamma = 1/\tau$ is inversely proportional to the finite interaction time between optical and elastic waves. In relatively homogeneous samples, τ is given by the lifetime of elastic waves (phonons) primarily by viscous damping in the water and solid constituents, and thus Γ can serve as a measure of local dynamic viscosity η at hypersonic frequencies, i.e., $\Gamma \propto \eta$. In highly heterogeneous samples or at the interface between different materials, further heterogeneous spectral broadening of BLS can occur. Moreover, in high-resolution microscopy or in materials with small penetration depth, the finite length of the optical beam can generate additional spectral broadening. The typical widths of Brillouin peaks in tissues, biological liquids, and gels are a few hundred MHz to a GHz, depending on the material composition and structure, which is at least an order of magnitude larger than Brillouin linewidths found in solid-state crystalline materials (Cusack and Miller, 1979; Palombo et al., 2014). This difference may not be that surprising, considering that biological materials typically do not possess any long-range structural order (unlike crystalline atomic lattices) and are composite materials containing nano- to micro-scale components of different material properties (e.g., micro-filaments versus lipids). Hydration, visco-elasticity, and heterogeneity are among common characteristics of biological materials, which set specific requirements for Brillouin instrumentation and applications.

Optical absorption and scattering in samples play an important role in the application of BLS in biological systems. The optical absorption is highly wavelength dependent and characteristic of specific molecules, but in general drops significantly at longer wavelength in the visible light spectrum. The Rayleigh and Mie scattering of light in tissue tends also to decrease with increasing optical wavelength λ. The intensity of BLS follows a $1/\lambda^4$ dependence. As a result, the optimal wavelength that makes a trade-off between absorption and scattering is considered to be in the region of red to near-infrared (NIR) light (600–850 nm).

More details of BLS in biological materials are discussed in the following sections of this chapter. In brief, a historical perspective of the BioBrillouin field is presented in Section 2. The technological development is linked to the motivating scientific questions and the driving needs from the applications. In Section 3, the relationship of key BLS parameters to elastic modulus and water content is discussed. In Section 4, the biomedical applications of BLS to mechanobiology and diagnostics are described. Finally, a discussion on current technical limitations and future directions of biomedical Brillouin research is presented.

2. History of Brillouin technology from spectroscopy to imaging

The first studies using Brillouin spectroscopy on biologically relevant samples date back to the 1970s. Bedborough and Jackson (1976) applied Brillouin spectroscopy to collagen gels at various concentrations and investigated the effect of water content in collagen gels. Harley et al. (1977) measured elastic modulus in both type-I collagen fibers and muscle proteins with microscopic resolution, which they related to the forces of hydrogen bonding along the protein backbone. Later studies by Randall et al. (1979) reported the elastic wave velocities and attenuation coefficients in rat tail collagen and rabbit muscle fibers. Cusack and Miller (1979) measured the angle dependence of both lateral and transverse modes in wet and dry collagen fibers. Randall and Vaughan (1982) measured the longitudinal elastic modulus of ocular tissues (Fig. 2).

All the early experiments were conducted using high-finesse Fabry–Perot type interferometry, adapted for BLS measurements. The instruments addressed two main challenges of BLS measurements: i.e., the weak BLS intensity, which is about 10^{-10} to 10^{-9} of the input laser intensity, and the small frequency shift of 1–20 GHz, which is ~10^{-5} to 10^{-6} of the frequency of light itself (Sandercock, 1976, 1978). The Fabry–Perot interferometers rely on the interference between multiple partial reflections at the surfaces of an etalon, which can be either a solid type or an air-spaced pair of mirrors. For a given thickness of the etalon, only the resonant, narrowband frequency component is transmitted while all other spectral components are reflected. Nowadays, a scanning tandem Fabry–Perot interferometer is still used for spontaneous BLS experiments due to its excellent performance in terms of its high spectral contrast of over 10^{11} and tunable free spectral range suitable for a wide range of materials (Mattana et al., 2018). The downside of the Fabry–Perot interferometers is, however, their high insertion loss and need for scanning the resonant frequency, resulting in long acquisition time. Typically, the time required to collect spectral data at a single location in a sample ranges 1–30 s depending on the material transparency and the laser

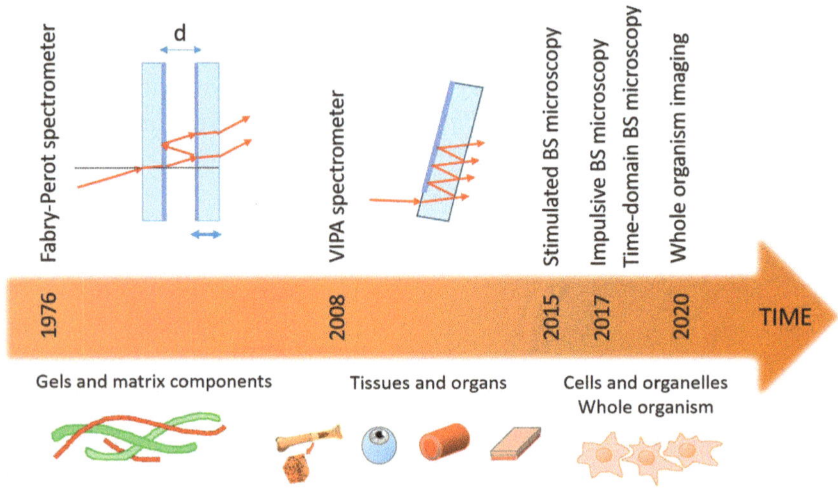

Fig. 2 Timeline of Brillouin light scattering technology in biomedical field. 1970s to 1990s have seen the first measurements of biologically relevant materials (gels and matrix fibers) using scanning Fabry–Perot spectrometers (Sandercock, 1976). Since 2008, VIPA spectrometers have been widely used (Scarcelli and Yun, 2008) to speed up the signal acquisition time and enable 2D and 3D scanning of tissues and organs. The technological development has exploded since 2015 with a number of specialized techniques based on time-domain and stimulated Brillouin scattering to improve SNR and achieve full organism imaging (Ballmann et al., 2017; Remer et al., 2020).

power. To avoid scanning, a Fabry–Perot etalon with a diverging incoming beam may be used, for which a different incident angle encounters a different effective thickness of the etalon and, thus, a different resonant frequency. As a result, every frequency within the spectrum is transmitted at a different angle, so that the whole spectrum can be acquired using a 2D camera in a single shot without mechanical scanning (Itoh, 1998; Koski and Yarger, 2005). However, this configuration does not improve the measurement throughput because at each angle only one frequency component is resonant and transmitted, while all other frequency components of the spectrum are reflected and lost. In both scanning and angle-resolved interferometers, there is a tradeoff between spectral resolution and throughput. The narrower the spectral transmission window is, the larger is the portion of the spectrum that is lost in reflection, and the longer it takes to scan over a desired spectral range. A modified configuration to recirculate the frequency components reflected by the etalon provides some improvement in throughput but with added complexity (Stephen et al., 2015).

A breakthrough in measurement speed, which opened a door for BLS technology to enter the domain of biomedical imaging and microscopy, occurred with the development of virtually imaged phased array (VIPA) spectrometers (Scarcelli and Yun, 2008). VIPA was first introduced by Shirasaki (1996) for optical telecommunication applications (Willits et al., 2012). The VIPA is a modified etalon in which the back surface is totally reflective except for a narrow window for the input optical beam. As in a tilted etalon the interference between multiple reflected components leads to different frequency components being transmitted at slightly different angles, thus enabling single-shot measurements. The difference is that in VIPAs no light leakage is allowed by the reflective back surface, and thus no frequency component is lost. This results in a dramatic improvement in throughput, allowing single spectral measurement to be acquired in 10–100 ms (Nikolic and Scarcelli, 2019). Typical VIPA offers a low finesse of <100 and therefore a spectral resolution of >0.2–0.5 GHz. A single VIPA provides a relatively low spectral contrast of about 10^3 to 10^4, which makes it difficult to separate Brillouin components from elastic scattering or other stray light. The contrast can been significantly enhanced by using a cascade of 2 to 3 VIPAs in series (Fiore and Scarcelli, 2021; Scarcelli and Yun, 2011) or introducing additional low-loss spatial filters (Antonacci et al., 2016; Edrei et al., 2017). State-of-the-art VIPA spectrometers achieve 10^7 to 10^8 spectral extinction (Zhang and Scarcelli, 2021), which is adequate for Brillouin measurements of biological cells, tissues and biomaterials (Fiore et al., 2016).

At a fundamental level, stimulated Brillouin scattering (SBS) is an attractive way to obtain greater scattering signals than spontaneous BLS. In SBS, elastic waves are generated via nonlinear electrostrictive modulation of the material density (and thus the refractive index) by high-power light beams. SBS can in principle be achieved using a single high-intensity pump laser, but the threshold intensity for meaningful SBS is too high to achieve in practical settings in biomedical applications. A simpler approach is using a second counterpropagating beam with its frequency detuned from the first beam by exactly the phase-matching phonon frequency. The beating of the two counter-propagating electric fields generates phonons or the elastic wave. A major advantage of SBS over spontaneous BLS is that the scattering efficiency can be much higher with sufficient intensities. In addition, the SBS process could allow more efficient filtering strategy of stray light or elastically scattered light than spontaneous BLS setups (Ballmann et al., 2015; Remer and Bilenca, 2016b). Last but not least, the spectral resolution of SBS measurement is not limited by spectrometers but depends on the linewidth of the lasers, which can be much narrower.

However, there are practical difficulties to the use of SBS in biomedical applications. High power lasers with tunable narrowband radiation are not routinely available. Access to a sample from both sides is needed to allow for counterpropagating pump-probe beams. In addition, laser stability is a concern given the precision in spectral and spatial overlap/tuning needed between pump and probe beams.

High laser stability and tunability has been achieved by using GHz electro-optic modulators (EOMs) to produce both pump and probe light from a single laser (Ballmann et al., 2015). Recently, biomechanical imaging of the entire organism such as *Caenorhabditis elegans* was demonstrated by using two-laser SBS in a high-resolution microscopy setting (Remer et al., 2020). A low spectral integration time of 20 ms in tissue and 2 ms in water were achieved using a relatively high continuous-wave (CW) laser power of 256 mW at the sample (Fig. 3). The relatively modest improvement in Brillouin signal strength over spontaneous BLS of equivalent power is due to the CW operation rather than pulses. A high spectral resolution (<50 MHz) enabled Brillouin linewidth measurements at high precision for label-free readout of material viscosity (Remer et al., 2020). Although no visible photo-damage was registered on the sample, the high power requirement of SBS microscopy still imposes a limit to straightforward translation of this technology in many biomedical applications.

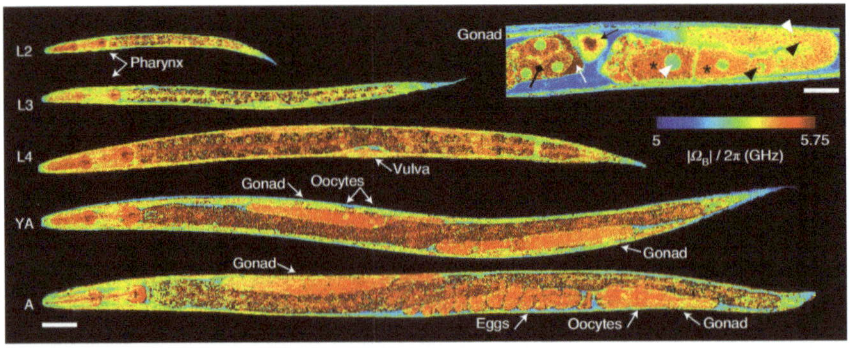

Fig. 3 SBS imaging of live *C. elegans* nematodes. *Adopted from Remer, I., Shaashoua, R., Shemesh, N., Ben-Zvi, A., Bilenca, A., 2020. High-sensitivity and high-specificity biomechanical imaging by stimulated Brillouin scattering microscopy. Nat. Methods 17 (9), 913–916. https://doi.org/10.1038/s41592-020-0882-0.*

When the required laser frequency stability and high coherence are not fulfilled, one might opt for a subclass technology within a broader class of SBS techniques, such as impulsive Brillouin scattering. Unlike conventional SBS imaging that uses a single pump and a single probe beam, in impulsive Brillouin scattering the elastic waves are created by two crossed pulsed pump beams (Ballmann et al., 2017; Krug et al., 2019). The interference pattern formed by such beams modulates the refractive index through the combination of electrostriction, photo-elastic and thermo-elastic effects. This traveling index grating, in turn, scatters a continuous wave probe beam, producing a scattered beam frequency-shifted in proportion to the elastic wave velocity. The advantage of this technique is relaxed requirements for the pump lasers in terms of laser linewidth and stability. Relatively inexpensive pulsed laser sources of sufficient peak power can be used without affecting the frequency resolution of the measurement (the latter is determined by the detector bandwidth and can be within a few MHz range). A downside of this technology is relatively low signal-to-noise ratio (Ballmann et al., 2017; Krug et al., 2019). Another complication may arise from sample heating at the measurement point due to high optical intensities.

Another alternative approach to creation of coherent elastic waves or acoustic phonons can be found via thermo-elastic excitation by ultra-short optical pulses in a thin metallic layer deposited onto the back surface of a transparent material. The phonons can be then probed by an optical beam with high sensitivity and precision using a heterodyne detection. In the case of semi-transparent materials where the detection of the elastic waves within the material is not possible, the elastic wave propagating on the surface of the sample may be observed. This subclass of SBS techniques is known as time-domain Brillouin microscopy. Although this technique has not yet provided a significant improvement in acquisition speed, it can offer high spatial resolution (Gusev and Ruello, 2018; La Cavera et al., 2021).

To summarize, a number of competitive approaches have been developed to achieve high spectral contrast, acquisition speed, and spatial resolution for biomedical Brillouin measurement. Each technique has its advantages and disadvantages and is currently applied in a number of research labs around the world. In the next section, we will overview some fundamental characteristics of BLS signals in biological samples.

3. Key aspects of Brillouin light scattering in biological matter

3.1 The relation of Brillouin frequency shift to mechanical properties

The acoustic phonons that give rise to Brillouin scattering interactions can be described as microscopic pressure/density waves traveling in a material. From a mechanical standpoint, Brillouin scattering probes the local response of a sample to a one-dimensional dynamic perturbation at the phonon frequency in the GHz range. All of the measurements in hydrated materials reported to date probe the longitudinal acoustic velocity, and therefore the response of the material governed by complex longitudinal modulus. The real part of the latter expresses the elastic response, and the imaginary part expresses the viscous response, i.e., the loss of acoustic energy in the sample.

From the relationships expressed in the previous section, it is evident that the conversion from measured Brillouin data to the local longitudinal elastic modulus requires the knowledge of the refractive index and density of the material. However, both index of refraction and density of the material may not be known, and their values may be different from one sample to another, and even vary within a sample. This uncertainty must be taken into account when interpreting the Brillouin measurement data in terms of longitudinal modulus. This is one reason why most studies report Brillouin frequency shift values rather than longitudinal modulus values when the refractive index and/or density are not precisely known. Interestingly, in the case of most tissue and cells, the factor containing density and index can be estimated to a good approximation. This is because each of the quantities does not vary much among soft tissues. Moreover, both density and index are dependent on the relative amount of water vs solid components of tissues in such a correlated way that almost cancels out their contributions (Scarcelli et al., 2011, 2012, 2015b). One exception is for lipid-rich samples, since lipids and water have different ratios of density/index. Care is needed when interpreting Brillouin frequency shift as longitudinal modulus for adipocytes, lipid or liquid–liquid phase separations (Kim and Guck, 2020; Schlüßler et al., 2020).

The longitudinal elastic modulus has a mechanical definition different from shear modulus and Young's modulus Prevedel and Antonacci (2019). The shear and Young's moduli are related to deformation or elastic waves without volume constraint, such as shear waves and torsional waves. By contrast, the longitudinal modulus also involves volume compression or compressional elastic waves. For soft biological tissues, the magnitude of longitudinal modulus is substantially different (higher) than shear and Young's modulus values. It is useful to revisit the relationship between these different elastic moduli for isotropic materials. The Young's modulus (E) is defined as the ratio between an applied axial stress (force over area) and resulting axial strain (relative elongation/compression over initial length). The Young's modulus is generally measured by compressive or tensile stress–strain tests with stress-free surface boundaries. In these tests, the Poisson's ratio, ν, describes the ratio between transverse and axial strain under axial loading. The shear modulus (G) is defined as the ratio between shear stress and shear strain and can be measured using rheometers. G is equal to $E/[2(1+\nu)]$; for incompressible materials, $\nu = 0.5$ and $G = E/3$. The bulk modulus (K) is the inverse of the volume compressibility, and is especially useful for characterizing liquids as it describes the ratio of applied pressure to volume change. The longitudinal modulus is defined as the ratio of axial stress to axial strain while the boundaries are constrained so that the material deformation is strictly uniaxial. This perturbation involves a volume change, which require large stress (>GPa) for nearly incompressible materials. Water and soft tissues are nearly incompressible as their molecular structure are packed with atoms without voids. (Their mass density ranges narrowly between 1000 and 1100 kg m^{-3}.)

The longitudinal modulus (M) is related to bulk and shear moduli as $M = K + 4/3G$. At GHz frequencies, the shear modulus term is not necessarily small and, in fact, approaches to an order of GPa. This is because at the subnanosecond time-scale all the slow relaxation processes that give rise to small shear modulus for soft tissues are essentially frozen, making the material nearly as hard as solid-state materials. This can be understood from the frequency-dependent modulus of viscoelastic materials (Mofrad and Kamm, 2006). Note that standard mechanical tests measure shear or Young's modulus at near DC to 100 Hz. Most physiological movements of tissues and cells have time scales in this low frequency range (Bailey et al., 2020).

Finally, the spatial length scale of Brillouin measurement is typically different from traditional mechanical tests. Traditional mechanical tests obtain mechanical properties of bulk samples that are mm to cm in size, giving average properties averaged over the entire length. Brillouin interactions instead can occur at the scale of acoustic phonon wavelengths, which is on the micron level (Caponi et al., 2020; Prevedel et al., 2019). This length scale sets the fundamental limit of the spatial resolution of Brillouin microscopy. An optical beam can be focused to a shorter focal length by using a lens with a high numerical aperture. However, this would cause considerable geometry-induced broadening of the BLS spectrum, making it difficult to measure the Brillouin frequency shift reliably.

For all these reasons, the longitudinal modulus measured by Brillouin measurements and traditional shear (or Young's) modulus are fundamentally different properties. There is no definitive relationship. However, when it comes to soft biological materials, many experiments in the past decade have shown that the two types of moduli are not entirely random and have relatively strong quantitative correlations (Scarcelli et al., 2011, 2015b; Webb et al., 2017). This empirical relationship is probably because of the general organic compositions of soft materials with high water content. From a practical standpoint, the quantitative correlation can be useful when comparing the softness and hardness (as defined by shear modulus) of samples and also measuring time-dependent changes in stiffness. Nonetheless, care is needed in any attempt to convert longitudinal to shear or Young's moduli, as is discussed in the following section.

3.2 The influence of hydration on Brillouin scattering

Liquid water in biological materials is primarily responsible for the softness of the materials. Human brain tissue has a relatively high water content of up to 80%–85%, giving a low shear modulus of 1–5 kPa. The skin and muscles have about 70%–75% water. The magnitude of the shear modulus is typically nonlinear with the water content, often accompanying a phase change from the gel-like behavior to solid-like behavior. Hard compact bone has low water content and exhibits high shear modulus in the order of several GPa.

When it comes to longitudinal modulus, all volume-incompressible materials behave like solids, with typical values in the order of GPa even for soft tissues. The presence of water affects the longitudinal modulus and elastic wave speeds. The effect of hydration on the Brillouin frequency shift (Ω) and linewidth (Γ) has been investigated with various fibrous materials (Cusack and Miller, 1979), collagen gels (Zhao and Vanderwal, 1997), hydrogels (Johnson, 1982; Scarcelli and Yun, 2018; Wu et al., 2018), and human corneas in vivo (Shao et al., 2018). Already in 1980s, a simple biphasic model was suggested to explain the effects of hydration on the material's complex longitudinal modulus ($M = M' + jM''$) and its compressibility ($\beta \approx 1/M'$) (Johnson, 1982). According to this model, the material is treated as a two-phase aggregate: a liquid phase with fraction ϵ and a solid one with fraction $(1 - \epsilon)$. This allows to write the compressibility of the aggregate material as a superposition of its components: $\beta_{agg} = \epsilon\beta_f + (1 - \epsilon)\beta_s$, where β_{agg}, β_f, β_s are compressibilities of the aggregate, fluid and solid components, respectively.

Since the Brillouin frequency shift measures material's storage longitudinal modulus $\Omega \propto \sqrt{M'} \approx 1/\sqrt{\beta}$ (valid for highly hydrated materials with negligibly small shear modulus), its value is therefore sensitive to changes in the material content including hydration. Using the first-order approximation valid for small fluctuations in hydration $\Delta\epsilon \ll \epsilon$, we write:

$$\frac{\Delta\beta_{agg}}{\beta_{agg}} = \left(1 - \frac{\beta_s}{\beta_{agg}}\right)\frac{\Delta\epsilon}{\epsilon} + (1 - \epsilon)\frac{\Delta\beta_s}{\beta_{agg}}. \tag{1}$$

For tissues, assuming $\epsilon \approx 0.75$ and $\dfrac{\beta_s}{\beta_{agg}} \approx 0.25$ (Scarcelli and Yun, 2018), we find the relative change in the Brillouin frequency shift of the aggregate to be:

$$\frac{\Delta\Omega_{agg}}{\Omega_{agg}} = -0.4\frac{\Delta\epsilon}{\epsilon} - 0.04\frac{\Delta\beta_s}{\beta_{agg}}. \tag{2}$$

Eq. (2) signifies that the Brillouin frequency shift is 10 times more sensitive to the changes in the material hydration than those in compressibility of the solid network. The higher sensitivity of Brillouin frequency shift to hydration is an important aspect that needs to be considered in the preparation of samples and interpretation of the measured data.

As mentioned above, it has been well established that increased hydration of hydrogels and tissues makes the materials softer (Hatami-Marbini and Pachenari, 2020). Crosslinking of tissues results in stiffening of the materials not only via reinforcing the solid network but also via reduction in hydration (Scarcelli et al., 2013). Indeed in a side by side comparison in corneal tissue, Seiler et al. (2018) recently showed that the Young's modulus was even more sensitive to hydration changes than longitudinal modulus. In terms of Brillouin measurements, Eq. (2) establishes a limit of hydration fluctuation, above which the uncertainty in the measurement of the fluid portion of a sample becomes comparable with the measurement of the solid network contributions. When the hydration is uncontrolled or uncertain by more than the limit, Brillouin measurements do not provide reliable information about the solid network. Otherwise, Brillouin measurements can provide useful information related to the solid constituents. In general, care needs to be taken when designing Brillouin imaging experiments to ensure that the level of hydration is known or can be taken into account for example by allowing tissues to swell over time and reach equilibrium level of hydration.

In reality, this can be a challenging task as changes in osmotic pressure and pathological processes affect cell and tissue hydration. Additionally, hydration of tissues cannot be easily controlled in in vivo and in situ settings. In the in vivo experiment by Shao et al. (2018), for example, corneas of healthy volunteers were scanned continuously over the course of 9 hours and the variations in the Brillouin frequency shift and corneal thickness were recorded by Brillouin imaging and optical coherence tomography. It was found that the Brillouin frequency shift was on average 25 MHz (0.9% of the mean value) lower in corneas immediately after awakening compared to later times during that day. In addition, the corneal hydration was found to correlate negatively with the Brillouin frequency shift, as predicted by the biphasic model (Shao et al., 2018). Depending on the instrumentation and the choice of the spectrometer, Brillouin imaging on average can resolve differences in the Brillouin frequency shift down to 0.1%–0.5% making it possible to resolve small changes in Ω due to environmental factors (temperature, pressure) and hydration. Overall, these considerations reveal the important role of hydration in cells, tissues, and biomaterials, and its effect on the material mechanical properties and Brillouin measurements. For a given series of measurements, it may be useful to monitor the sample hydration independently for extracting the mechanical properties of the solid component with greater accuracy.

4. Applications of Brillouin scattering to biology and medicine

4.1 Mechanobiology

The significance of the micromechanical interaction between biological cells and their extracellular microenvironment has been established with many phenomenological observations (Discher et al., 2005): cells migrate toward stiffer substrate regions (Lo et al., 2000; Pelham and Wang, 1997); substrate stiffness regulates cell proliferation (Huang and Ingber, 1999); stem cells differentiate into different lineages according to substrate elasticity

(Engler et al., 2006b); long-term gene expression is affected by matrix elasticity (Chiquet, 1999); also at the system level micromechanical interactions have been shown to be critical for tissue morphogenesis and angiogenesis (Huang and Ingber, 2005; Ingber, 2002).

At the center of the micromechanical interaction between cell and microenvironment is the biological process of mechanotransduction. Mechanotransduction accounts for how cells sense mechanical stimuli from external forces or stiffness of the extracellular matrix (ECM), and how cells convert the mechanical stimuli into intracellular signaling responses (Janmey and McCulloch, 2007; Tzima et al., 2005). Brillouin microscopy is useful in the studies of mechanotransduction by enabling the probing of the mechanical properties in the outside and inside of cells with high spatial resolution, in a noncontact and nondisturbing fashion, within both 2D and 3D microenvironments. In the past several years, several groups have shown the capability of Brillouin microscopy to measure intracellular mechanical properties and their perturbations. In the cytoplasm, the complex interaction of liquid components and cytoskeletal components (actin, microtubules, and intermediate filaments) regulate cell mechanical properties (Moeendarbary et al., 2013). Scarcelli et al. observed a significant decrease in Brillouin modulus when inhibiting actin polymerization and demonstrated sensitivity to actin polymerization and branching, two major mechanisms used by cells to modulate their internal stiffness (Scarcelli et al., 2015b). Antonacci et al. observed reduced modulus inhibiting cytoskeletal assembly (Antonacci and Braakman, 2016) and correlated regions of cells at higher moduli with cell's ability to adhere and produce forces (Coppola et al., 2019). Brillouin microscopy also showed sensitivity to perturbations due to surface stiffness changes and environment osmolality (the measure of solute concentration), in agreement with previous observations with gold-standard mechanical tests (Conrad et al., 2019). Brillouin microscopy was also recently applied to the measurement of nuclear mechanical properties, which is uniquely challenging for other techniques as the nucleus is inaccessible from outside the cell. Zhang et al. demonstrated that nuclear mechanics is not only regulated by internal components such as chromatin condensation and nuclear envelope, but can also be modulated by cytoskeletal perturbation, in agreement with chemomechanical modeling predictions (Zhang et al., 2017, 2020b). Interestingly, Brillouin microscopy at high resolution can measure condensates and aggregates within cells and nuclei, which have recently received a lot of attention for their links to human diseases (Antonacci et al., 2018; Fasciani et al., 2020; Schlüßler et al., 2020). In plant cells, spatially nonuniform variation in longitudinal modulus has been observed and linked to cell growth patterns (Altartouri et al., 2019; Elsayad et al., 2016). All these observations show that Brillouin microscopy has useful sensitivity to detect biologically relevant changes in mechanical properties of cells and have provided data and insights consistent with gold-standard techniques, but beyond those, it can enable biomechanical studies at high resolution, without perturbation or contact, and in a vast variety of experimental settings.

4.2 Medical diagnostics

4.2.1 Ophthalmology

The first biomedical field where Brillouin microscopy has made an impact is in ophthalmology, for mechanical examination of the eye tissues of patients in the clinic. The cornea has been most extensively studied so far (Scarcelli et al., 2012). The cornea is the outermost tissue of the eye and is kept under tension by the intraocular pressure. The collagen fiber network in the cornea provides the critical mechanical strength needed to maintain normal corneal shape. Abnormal loss of corneal mechanical strength can result in degenerative ectatic disorders, such as keratoconus (Roy and Dupps Jr, 2011) or post-laser-assisted in situ keratomileusis ectasia, a complication of refractive surgery (Binder, 2007). For these conditions, early diagnosis is crucial. Keratoconus progression could be halted by promising interventions, such as corneal collagen crosslinking (CXL) (Wollensak, 2006). Early-stage keratoconus can be a cause for post-LASIK ectasia. Early diagnosis may dramatically reduce the occurrence of this complication (Rabinowitz, 2006). Current diagnosis, however, primarily relies on corneal geometrical features, such as curvature and thickness. The morphological information alone has so far proven to be insufficient for definitive diagnosis of early stages of keratoconus (Binder et al., 2005; Rabinowitz, 2006). In recent years, Brillouin microscopy showed the potential to be useful in this space. Brillouin microscopy has provided the first and only direct mechanical evidence of decreased modulus in keratoconus corneas in vivo and found remarkable region-dependent modulus changes (Scarcelli et al., 2014, 2015a). Importantly, it has recently been shown that the analysis of spatial distribution of modulus within the cornea compares favorably against traditional metrics to detect keratoconus (Shao et al., 2019). Brillouin microscopy has also established a strong potential in the characterization of CXL procedures, displaying high sensitivity to quantify corneal modulus changes induced by CXL protocols (Randleman et al., 2017; Scarcelli et al., 2013; Webb et al., 2017; Zhang et al., 2019, 2020a) with 3D resolution (Kwok et al., 2016; Webb et al., 2019). Beyond ectatic disorders, Brillouin microscopy has shown potential for diagnostics and therapy monitoring also for other cornea diseases (Eltony et al., 2020a) and for the characterization of other ocular tissue such as crystalline lens (Ambekar et al., 2020a; Besner et al., 2016) and retina (Ambekar et al., 2020b; Weber et al., 2017).

4.2.2 Cancer research and diagnostics

In the field of cancer research, Brillouin microscopy has made important contributions. The critical role of biomechanics in tumor progression and metastasis is largely established (Kumar and Weaver, 2009): stiff ECM promotes tumor progression by enhancing integrin signaling, which is a known pathway related to biomechanics; mammary cells exposed to high stiffness in the ECM undergo malignant transformation (Provenzano et al., 2009) while ECM softening lowers malignancy transformation (Levental et al., 2009; Payne, 2005). Importantly, from a diagnostic standpoint, tumor masses have higher elastic moduli than healthy tissue (Nikolic et al., 2018), and thus could be a promising diagnostic target. At the cellular level, atomic force microscopy (AFM) was extensively used and showed that metastatic cells are significantly softer than normal cells (Cross et al., 2007). This finding has been confirmed with other techniques (Agus et al., 2013; Remmerbach et al., 2009; Tse et al., 2013), and recently also by Brillouin microscopy (Mattana et al., 2018; Nikolic et al., 2018).

Brillouin microscopy makes it possible to probe the biomechanics of cancer with subcellular resolution, for example, to investigate cell behavior during tumor progression and metastatic cascade in vivo. In experimental model of tumor cell extravasation, Brillouin longitudinal modulus of cells was measured to markedly decrease during transmigration (Roberts et al., 2021), as predicted by mechanics-based models of optimal transmigration (Cao et al., 2016). In ovarian cancer, where ascitic fluid currents have been associated with poor prognosis, the ascitic flow markedly reduces the longitudinal modulus of tumor nodules (Conrad et al., 2021), a known signature of tumor aggressiveness. In confined migration, the measurement of nuclear longitudinal modulus suggested that nuclear stiffness is a main mechanical barrier for efficient cell motility (Wisniewski et al., 2020). Across several tumorigenic and metastatic cascade steps, the longitudinal modulus measured by Brillouin microscopy has been a useful indicator.

4.2.3 Bone and cartilage health

Bone and cartilage are specialized forms of connective tissue that provide structural support for muscles and organs within the human body. Bone is a hard, calcified connective tissue that comprises the skeleton, whereas cartilage is a flexible, avascular tissue that serves to absorb shock in the joints between bones, or forms rigid, low-load bearing structures such as within the nose, ear, and larynx. The structure and mechanical properties of bone and cartilage are integral to their function, and disruption of these properties are implicated in injury and degenerative conditions such as arthritis and osteoporosis. Earlier detection of changes in the structure and strength of bone and cartilage will facilitate earlier diagnosis, which is key to effective treatment and improved patient outcomes.

The function-dependent microscopic structure of bone trabeculae and its anisotropy present challenges to mechanical measurements by standard macroscopic mechanical tests. Matsukawa et al. (2014) performed a comparative study between BLS and scanning acoustic microscopy (SAM) to determine the sensitivity of BLS in detecting anisotropy and decalcification, hallmarks of bone diseases. Brillouin spectra were site-matched with SAM; following decalcification, a significant reduction in Brillouin frequency shift (expressed as wave velocity), as well as increase of BLS intensity due to optical clearing effects of demineralization were measured. This suggests that BLS may be applied for characterizing bone quality in disease and healing processes.

Bone remodeling phenomena, particularly in the presence of implants, remain poorly understood due to the lack of means of evaluating implant stability and osseointegration in a noninvasive manner during healing postimplantation. To address this, Mathieu et al. (2011) investigated the capability of BLS in differentiating between mature and newly formed bone during postimplant healing based on viscoelastic properties and histological analysis. Results indicated lower mineralization in newly formed bone compared to mature cortical bone, along with higher heterogeneity which reflects the initial lack of organization of ECM proteins during healing. This work confirmed the sensitivity of BLS in detecting mechanically relevant compositional changes in bone during remodeling.

Akilbekova et al. (2018) similarly employed Brillouin microscopy to investigate the load dependency and viscoelastic properties of bone during healing. The efficacy of bone morphogenetic proteins (BMPs) and stem cell-loaded heparin-conjugated fibrin hydrogels in the repair of critical bone defects in a rabbit model was assessed by measuring changes in Brillouin frequency shift and linewidth, which are indicative of the rate of bone formation and healing processes at the site of the defects. Results were correlated to X-ray images, where Brillouin frequency shift and linewidth increased with formation of an organized endosteal callous and showed that BMPs coupled with stem cells facilitated bone regeneration, which is in line with previous findings.

Additionally, Brillouin microscopy was applied to study micromechanical anisotropy in articular cartilage (Ryu et al., 2021) and to detect the loss of ECM proteins in cartilage as occurs in the pathogenesis of osteoarthritis (Wu et al., 2019). In the study by Ryu et al. (2021), ECM was found to have a significantly higher longitudinal modulus than chondrocytes, with Brillouin frequency shift being the highest in the transitional zone and the lowest in the superficial zone of cartilage tissue, supported by previous studies of the elastic modulus variation with depth (Fig. 4). Wu et al. (2019) used ex vivo porcine articular cartilage subjected to trypsin digestion for 4 hours to mimic ECM degradation typical of that during onset of osteoarthritis. Significant changes in Brillouin frequency shift were detected and attributed to the degradation of cartilage ECM and 4% variation in tissue water content. Overall, these studies support the potential of Brillouin microscopy as a noninvasive tool for characterizing mechanical properties of bone and cartilage during formation and healing, as well as a method for early detection of changes in these properties in disease.

4.2.4 Diagnostics of plaques

Plaques are characteristic of respective pathological conditions; examples include dental plaque, amyloid proteins in Alzheimer's disease, psoriatic skin lesions, and atherosclerotic plaque. As plaques accumulate during disease progression, they tend to become semi-hardened over time and disrupt the elasticity and functionality of the tissue in situ. Mattana et al. (2017) applied Brillouin Microscopy in combination with Raman spectroscopy to map the viscoelastic profiles of individual amyloid beta protein plaques in ex vivo histological sections of mouse hippocampus tissue and showed that amyloid plaques are comprised of a rigid core of β-folded (β-amyloid) proteins in a sheet conformation surrounded by a softer ring-shaped region rich in lipids and other protein conformations Mattana et al. (2018). Due to the shallow imaging depth of Brillouin microscopy, in vivo monitoring of plaque formation is challenging and would require a catheter-based approach. However, Brillouin microscopy is likely to prove useful in understanding the biomechanics of plaque formation in biopsy samples as well as in evaluating the efficacy of treatments.

Atherosclerotic plaque is the hallmark presentation of atherosclerosis in coronary arteries. A vulnerable plaque is developed when the endothelial

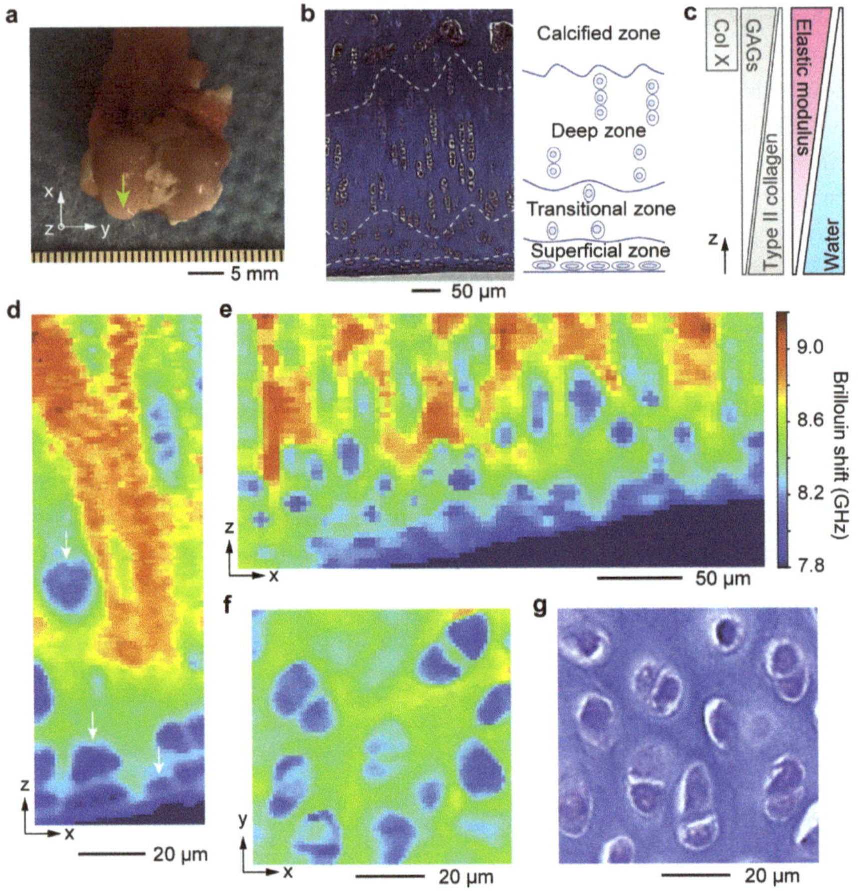

Fig. 4 Brillouin histology of articular cartilage tissues from 9-month-old female New Zealand white rabbit. (A) Articular cartilage at the femoral head (*arrow*) was scanned with Brillouin microscopy followed by histological staining for a morphological comparison. (B) Masson trichrome staining image of a coronal section of the articular cartilage. *Dashed lines* and an anatomical schematic indicate the four sublayer zones with chondrocytes. (C) Spatial gradients of collagen and GAGs proteins, elastic modulus, and water content. (D and E) Brillouin images of an intact articular cartilage tissue in the *x–z* plane. Arrows indicate chondrocytes ECM in the transitional zone has higher Brillouin shifts than that in the superficial zone. (F) Brillouin image in the *x–y* plane at the transitional zone. Chondrocytes and the stiffer ECM are clearly distinguished. (G) Toluidine blue stained image of a *x–y* cross-sectional tissue section of the transitional zone. *Adopted from Ryu, S., Martino, N., Kwok, S. J. J., Bernstein, L., Yun, S.-H., 2021. Label-free histological imaging of tissues using Brillouin light scattering contrast. Biomed. Opt. Express 12 (3), 1437–1448. https://doi.org/10.1364/BOE.414474.*

wall of blood vessels is disrupted, allowing infiltration of lipoproteins that progressively accumulate and form a necrotic core. A rupture of the atherosclerotic lesion can lead to acute cardiac events and heart attack. Early detection and treatment of the rupture-prone plaques is key to patient survival. Antonacci et al. (2015) investigated Brillouin microscopy for giving quantitative measures of stiffness across rupture-prone atherosclerotic plaques characterized by a thin, stiff fibrous cap covering a soft lipid-rich necrotic core within an arterial wall (Fig. 5). They measured distinct differences in the mean Brillouin frequency shifts of ex vivo plaque (15.79 ± 0.09 GHz) compared with normal vessel sections (16.24 ± 0.15 GHz) and showed a direct correlation between the small Brillouin frequency shifts in lipid-rich regions and the larger Brillouin frequency shifts in collagenous areas of the fibrous cap (Antonacci et al., 2015). This study supports the combination of Brillouin spectroscopy with intravascular imaging to improve detection of vulnerable plaques and predict rupture potential in atherosclerotic patients.

5. Challenges and future directions of BioBrillouin

5.1 Signal-to-noise limit of spontaneous BLS

Spontaneous Brillouin scattering involves two distinctive photon-phonon interactions: the Stokes and Anti-Stokes processes. The Stokes process generates a forward propagating phonon with a frequency Ω. For energy and momentum conservation, one forward propagating photon is annihilated while one backward scattered photon is created with a frequency down-shifted by Ω. The Anti-Stokes process involves the annihilation of a backward propagating phonon that is already present in the sample, and as a result, one backward scattered photon with a frequency up-shift of Ω is produced. The intrinsic phonon involved in the Anti-Stokes process is present by thermal fluctuations. At zero absolute temperature, there are no thermally excited phonons. Therefore, only Stokes BLS would occur at zero temperature. The average number of thermal phonons N_{phon} at frequency Ω and temperature T can be estimated, by applying the Planck distribution in high temperature limit, to be:

$$N_{\text{phon}} \approx \frac{k_B T}{\hbar\Omega}. \tag{3}$$

Here $k_B = 1.38 \times 10^{-23}$ J/K and $h = 6.63 \times 10^{-34}$ J/Hz are Boltzmann and Planck constants, respectively. For phonons at a frequency of $\Omega = 6$ GHz

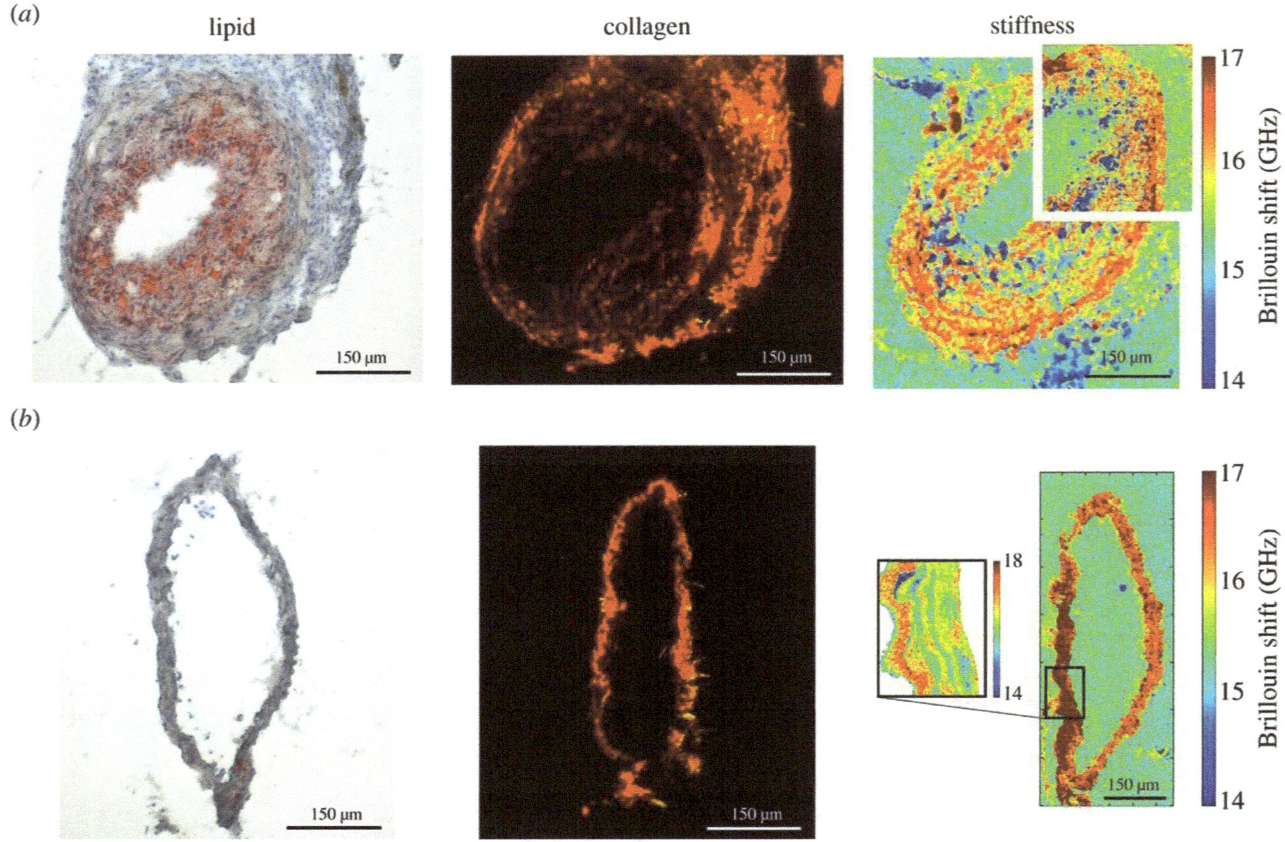

Fig. 5 Images of (*left*) lipid (imaged with light microscopy), (*center*) collagen (imaged with polarized light microscopy) and (*right*) Brillouin shift (imaged with Brillouin microscopy), which is related to stiffness, obtained from representative neighboring sections of the (A) instrumented and (B) control vessels. Scale bars, 150 µm. *Adopted from Antonacci, G., Pedrigi, R. M., Kondiboyina, A., Mehta, V. V., de Silva, R., Paterson, C., Krams, R., Torok, P., 2015. Quantification of plaque stiffness by Brillouin microscopy in experimental thin cap fibroatheroma. J. R. Soc. Interface 12, 20150843. https://doi.org/10.1098/rsif.2015.0843.*

(phonon frequency in water when probed with red monochromatic light) and at room temperature $T = 293$ K, we find a total $N_{phon} \approx 1000$ thermal phonons (per mode) available for the spontaneous BLS process to occur.

The scattering cross section can be calculated in the case of an isotropic homogeneous infinite medium and is approximately given by

$$\sigma = \frac{V k_B T}{6\pi\rho c_s^2 \lambda^4}, \tag{4}$$

where c is the speed of light in vacuum, V is the scattering volume and c_s is the acoustic phonon speed (Ching et al., 1990). The cross-section is the same for the Anti-Stokes and Stokes processes of spontaneous Brillouin scattering at room temperature. The cross-section has the $1/\lambda^4$ dependence, as discussed in Section 1.

Let us now consider a typical Brillouin scattering experiment that uses a 10 mW beam of monochromatic visible light at $\lambda = 632$ nm and $t = 1$ ms signal integration time per point. The total number of input photons within this integration time is $N_{input} = 3 \times 10^{14}$. For simplicity we consider water as the sample, with the refractive index of $n = 1.33$ and density of $\rho = 10^3$ kg m^{-3}. Based on Eq. (4) the number of BLS-generated photons is $N_{scat} = \frac{\sigma}{A} N_{input} \approx 3 \times 10^{-10} N_{input}$, where $A \approx \pi\lambda^2$ is an approximate beam area at the diffraction limit. This means that $N_{scat} \approx 10^5$ photons are generated from 10 mW of input light during 1 ms. This calculation sets the fundamental upper limit of SNR in Brillouin measurement. In actual measurements, only a fraction of this number would be collected and detected.

At the quantum limit of detection, known as the shot noise limit, the variance in the number of detected photons, or the shot noise photons, scales as $\sqrt{N_{det}} = \sqrt{N_{scat}}$, and the shot-noise limited SNR is given by

$$SNR = \frac{N_{det}}{\sqrt{N_{det}}} = \sqrt{N_{det}}. \tag{5}$$

For the chosen parameters of the experiment, we find a theoretical SNR limit to be $\sqrt{N_{det}} \approx 316$ per 10 mW beam power and 1 ms exposure. In reality, this performance is never reached due to additional noise sources such as the laser and electronic noises, as well as the losses in the optical system as mentioned above. In practice, SNRs of 10–20 have been routinely achieved for acquisition times of 10–100 ms (Scarcelli et al., 2015b; Wu et al., 2019).

5.2 Improvements in SNR and measurement speed

Recent years have witnessed tremendous improvements from Fabry–Perot interferometers (Antonacci et al., 2016; Mattana et al., 2018) to VIPA spectrometers (Zhang and Scarcelli, 2021) that greatly increased the spectral detection efficiency and have enabled in vivo measurements at safe power levels (Scarcelli et al., 2015a) and 2D/3D imaging of biological cells and tissues (Schlüßler et al., 2018; Wisniewski et al., 2020; Zhang et al., 2020b). However, because of the small cross-section of spontaneous Brillouin scattering, Brillouin microscopy remains a slow technique compared to other imaging modalities. For many applications, such as imaging of large areas or monitoring of dynamic events, reducing the data acquisition time is critical. One avenue that many labs are beginning to explore is to take advantage of nonlinear/stimulated Brillouin interactions rather than the spontaneous processes described in the previous section, as they potentially provide much higher energy conversion efficiency.

Several techniques based on stimulated scattering phenomena have been demonstrated (Ballmann et al., 2015; Krug et al., 2019; Remer and Bilenca, 2016a), as we described in Section 2. However, only modest improvements in measurement speed have been shown so far (Remer et al., 2020). While numerous nonlinear imaging techniques, such as two-photon excited fluorescence microscopy and stimulated Raman microscopy, have been used in biomedical sciences, the requirement of sub-GHz bandwidth lasers poses a unique challenge for SBS setups. Because of the small bandwidth, it is difficult to achieve a very high peak intensity required for efficient nonlinear processes, which in other modalities is relatively easily achieved using pico- to femto-seconds lasers. While technical advances in SBS are certainly expected in the future, spontaneous Brillouin scattering may remain as a useful, reliable, and low-cost solution for many biomedical applications.

A considerable improvement in speed can come from multiplexed spectral measurements. Traditional Brillouin spectrometers, whether based on etalons or VIPA, can only measure one spatial point of the sample at a time, and a 2D/3D image is obtained by coupling the spectrometer with a confocal microscope and scanning the volume of the object one point at a time. To multiplex spectral analysis, one approach is using a VIPA to simultaneously process scattered light from an extended line in a sample, which consists of a few hundreds of spatial points. The first demonstration of such a line-scan setup has demonstrated two orders of magnitude improvement in line-acquisition time over conventional point-scan Brillouin microscopes

(Zhang et al., 2016). A 2D Brillouin image is obtained by scanning the sample only in one dimension orthogonal to the linear beam path.

Another promising approach for multiplexing is using absorptive filters instead of dispersive filters. Atomic vapor absorption filters are available at several laser wavelengths with narrow absorption linewidths, high spectral extinction, and low insertion loss. Atomic vapor filters using Faraday anomalous dispersion have been used in Brillouin spectroscopy for environmental sensing (Popescu and Walther, 2005), but their linewidth and instrumentation have been suboptimal for biomedical applications. Practical implementation of narrowband tunable filters could dramatically simplify Brillouin microscope architecture and enable wide-field imaging (in a similar way that bulk dichroic filters are used in wide-field fluorescence imaging.) Such absorptive, tunable filters could also enable light-sheet illumination configurations for rapid acquisition of 2D images with optical sectioning capability. Much progress is expected in multiplexed Brillouin techniques (Scarcelli et al., 2022).

The collection efficiency of Brillouin scattered light depends on several geometrical parameters including the interaction volume and collection solid angle. Epi-detection configurations, in which the same objective lens is used for the sample illumination and the scattered light collection, are optimal for high collection efficiency, while line-scan and light-sheet settings need to be carefully designed to maximize throughput. In terms of background rejection, epi-detection is at a disadvantage compared to angled illumination because of additional back reflections and stray light from optical components along the beam path and from the sample surface. On the other hand, the multiplexed spectral analysis has intrinsically lower spectral contrast, and therefore future research is expected to focus on improving spectral contrast possibly with narrow band-pass filtering or vapor-based absorption filtering (Meng et al., 2014).

5.3 Reconstruction of the full elastic modulus

We have so far considered isotropic materials. In general, materials have direction-dependent mechanical properties. Then, all the elastic moduli should be described as direction-dependent parameters. In continuum mechanics, the elastic properties of a material is described with a stiffness tensor or its inverse, compliance tensor. In general, such a tensor contains 21 independent elements relating stress and strain along different directions. The tensor can be much simplified based on the symmetry properties of a specific material. For isotropic material it reduces to only two independent parameters that are directly related to longitudinal and shear moduli. Brillouin measurements provide longitudinal modulus along the propagation direction of phase matched acoustic phonons. Therefore, by varying the scattering direction, various components of the stiffness tensor can be extracted.

In addition, when incident and scattered light have orthogonal polarization directions, Brillouin measurements can also provide information regarding the shear components of the elastic tensor. Shear Brillouin measurements have been demonstrated for solid-state samples (Kim et al., 2016; Koski et al., 2013). Along with conventional longitudinal BLS, shear BLS clearly represents an intriguing research direction. Together, they provide the unique capability to directly measure the full elastic tensor. This possibility has been recently demonstrated in two interesting studies on silk and collagen samples (Koski et al., 2013; Palombo et al., 2014). Shear Brillouin measurement techniques in soft tissues and hydrogels, however, have yet to be developed.

Two recent studies have demonstrated the ability to measure material anisotropy in the cornea using direction-dependent longitudinal Brillouin microscopy (Eltony et al., 2020b; Webb et al., 2020). The cornea can be approximated as a transversely isotropic material, Elsheikh and Alhasso (2009), i.e., its mechanical properties are relatively uniform across the lateral plane but change along the depth in the cornea (Singh et al., 2016). In particular, corneal collagen fibers have strong directionality parallel to the cornea surface, which gives rise to higher stiffness along the plane than along the depth orthogonal to the corneal surface (Hatami-Marbini, 2018; McKee et al., 2011). To a first-order approximation, when the direction of the optical beam is tilted from the optic axis of the cornea by an angle, the resulting longitudinal modulus has contributions from both the tangential and axial elastic components. By analyzing the angle-dependent Brillouin frequency shifts, both the in-plane and on-axis longitudinal moduli are measured. Brillouin microscopy revealed that mechanical anisotropy is consistent with the structural properties of the cornea. Anisotropy was observed at different hydration conditions ranging from 72% to 96% in water content. This suggests that, unlike synthetic hydrogels for which Brillouin microscopy was found to lose sensitivity to mechanical changes at high water content (Scarcelli and Yun, 2018; Wu et al., 2018) Brillouin microscopy does not lose sensitivity to changes in solid mechanics of the cornea even for highly hydrated samples. High-resolution Brillouin microscopy demonstrated mechanical anisotropy at the level of single collagen fibrils depending on their orientations with respect to the illumination beam. It was shown that corneal anisotropy may be a critical factor to consider for accurate mechanical analysis of the cornea in clinical settings because the natural curvature of the cornea provides an inherent artifact in the angle at which the collagen fibrils are probed. These studies indicate a future direction in which Brillouin measurement of anisotropy can be used to decouple Brillouin shift dependence from hydration vs solid tissue.

5.4 Miniaturization and fiber-based instrumentation

Miniaturization of Brillouin spectroscopy and microscopy systems is another area of current and future development. Small footprint, robust packaging, and fiber-optic integration of Brillouin imaging equipment is necessary for the technology to go beyond research labs and enter industrial and clinical applications. Modern tandem Fabry–Perot interferometers take a foot print of $>1\ m^2$ and weigh close to 140 kg. VIPA-based spectrometers may be fit into a smaller space and have been commercialized.

So far the most common approach to Brillouin imaging measurements has been to combine a confocal microscope with a Brillouin spectrometer as an add-on. However, the use of a commercial bulky microscope, although straightforward to implement, restricts the range of applications. Replacing the large microscope by a narrow-diameter fiber-optic probe capable of transmitting the probe light and collecting the scattered light can help to shrink the size and complexity of the imaging system drastically while enabling expansion to medical and industrial applications where the measurements are expected to be taken on a spot, within the body, or remotely (e.g. within a chemical reactor).

Recent demonstrations of a dual and a hollow-core Brillouin fiber probes have shown promising paths for this exciting possibilities (Kabakova et al., 2017; Xiang et al., 2020). A prototype fiber probe fabricated by high-precision 3D printing was just over a centimeter in length and less than 2 mm in width. High imaging performance similar to a free-space Brillouin imaging system was achieved. This clearly demonstrated an exciting prospect for in situ measurements in challenging and otherwise difficult-to-reach environments in biomedical, material science and industrial applications.

6. Conclusion

This chapter introduced the recently-emerged field of research in Brillouin imaging and microscopy for micromechanical mapping of biological tissues and cells. This field has evolved from more traditional BLS spectroscopy to imaging, largely owing to the tremendous progress in measurement speed, functionality, and robustness of apparatus for detection and analysis of Brillouin scattered light. The continuing interests in mechanotransduction biology and biomechanics-based medical sciences, as well as biomaterials engineering, has been the major driving force in the development of Brillouin light scattering for biological and biomedical applications.

Despite many milestones that have been reached and the benefits of Brillouin imaging being demonstrated for the wide range of applications with scales from subcellular organelles to the whole organism, there are many outstanding challenges left for the researchers in this field to tackle. Although some of these topics have been reviewed in this chapter, this list is by no means complete. The field of Brillouin light scattering in biological systems is expected to continue to advance, grow in popularity, and become a fruitful scientific field that is here to stay for a long time.

References

Agus, D.B., Alexander, J.F., Arap, W., Ashili, S., Aslan, J.E., Austin, R.H., Backman, V., Bethel, K.J., Bonneau, R., Chen, W.C., Chen-Tanyolac, C., Choi, N.C., Curley, S.A., Dallas, M., Damania, D., Davies, P.C.W., Decuzzi, P., Dickinson, L., Estevez-Salmeron, L., Estrella, V., Ferrari, M., Fischbach, C., Foo, J., Fraley, S.I., Frantz, C., Fuhrmann, A., Gascard, P., Gatenby, R.A., Geng, Y., Gerecht, S., Gillies, R.J., Godin, B., Grady, W.M., Greenfield, A., Hemphill, C., Hempstead, B.L., Hielscher, A., Hillis, W.D., Holland, E.C., Ibrahim-Hashim, A., Jacks, T., Johnson, R.H., Joo, A., Katz, J.E., Kelbauskas, L., Kesselman, C., King, M.R., Konstantopoulos, K., Kraning-Rush, C.M., Kuhn, P., Kung, K., Kwee, B., Lakins, J.N., Lambert, G., Liao, D., Licht, J.D., Liphardt, J.T., Liu, L.Y., Lloyd, M.C., Lyubimova, A., Mallick, P., Marko, J., McCarty, O.J.T., Meldrum, D.R., Michor, F., Mumenthaler, S.M., Nandakumar, V., O'Halloran, T.V., Oh, S., Pasqualini, R., Paszek, M.J., Philips, K.G., Poultney, C.S., Rana, K., Reinhart-King, C.A., Ros, R., Semenza, G.L., Senechal, P., Shuler, M.L., Srinivasan, S., Staunton, J.R., Stypula, Y., Subramanian, H., Tlsty, T.D., Tormoen, G.W., Tseng, Y., van Oudenaarden, A., Verbridge, S.S., Wan, J.C., Weaver, V.M., Widom, J., Will, C., Wirtz, D., Wojtkowiak, J., Wu, P.H., Network, Physical Sciences–Oncology Centers, 2013. A physical sciences network characterization of non-tumorigenic and metastatic cells. Sci. Rep. 3, 1449. https://doi.org/10.1038/srep01449.

Akilbekova, D., Ogay, V., Yakupov, T., Sarsenova, M., Umbayev, B., Nurakhmetov, A., Tazhin, K., Yakovlev, V.V., Utegulov, Z.N., 2018. Brillouin spectroscopy and radiography for assessment of viscoelastic and regenerative properties of mammalian bones. Journal of Biomedical Optics 23 (9), 1–11. https://doi.org/10.1117/1.JBO.23.9.097004.

Altartouri, B., Bidhendi, A.J., Tani, T., Suzuki, J., Conrad, C., Chebli, Y., Liu, N., Karunakaran, C., Scarcelli, G., Geitmann, A., 2019. Pectin chemistry and cellulose crystallinity govern pavement cell morphogenesis in a multi-step mechanism. Plant Physiol. 181 (1), 127–141. https://doi.org/10.1104/pp.19.00303.

Ambekar, Y., Singh, M., Zhang, J.T., Nair, A., Aglyamov, S.R., Scarcelli, G., Larin, K.V., 2020a. Multimodal quantitative optical elastography of the crystalline lens with optical coherence elastography and Brillouin microscopy. Biomed. Opt. Express 11 (4), 2041–2051. https://doi.org/10.1364/boe.387361.

Ambekar, Y.S., Singh, M., Scarcelli, G., Rueda, E.M., Hall, B.M., Poche, R.A., Larin, K.V., 2020b. Characterization of retinal biomechanical properties using Brillouin microscopy. J. Biomed. Opt. 25 (9), 090502. https://doi.org/10.1117/1.jbo.25.9.090502.

Antonacci, G., Braakman, S., 2016. Biomechanics of subcellular structures by non-invasive Brillouin microscopy. Sci. Rep. 6. https://doi.org/10.1038/srep37217.

Antonacci, G., Pedrigi, R.M., Kondiboyina, A., Mehta, V.V., de Silva, R., Paterson, C., Krams, R., Torok, P., 2015. Quantification of plaque stiffness by Brillouin microscopy in experimental thin cap fibroatheroma. J. R. Soc. Interface 12, 20150843. https://doi.org/10.1098/rsif.2015.0843.

Antonacci, G., De Panfilis, S., Di Domenico, G., Delre, E., Ruocco, G., 2016. Breaking the contrast limit in single-pass Fabry-Perot spectrometers. Phys. Rev. Appl. 6 (5), 054020. https://doi.org/10.1103/PhysRevApplied.6.054020.

Antonacci, G., de Turris, V., Rosa, A., Ruocco, G., 2018. Background-deflection Brillouin microscopy reveals altered biomechanics of intracellular stress granules by ALS protein FUS. Commun. Biol. 1, 139. https://doi.org/10.1038/s42003-018-0148-x.

Bailey, M., Alunni-Cardinali, M., Correa, N., Caponi, S., Holsgrove, T., Barr, H., Stone, N., Winlove, C.P., Fioretto, D., Palombo, F., 2020. Viscoelastic properties of biopolymer hydrogels determined by Brillouin spectroscopy: a probe of tissue micromechanics. Sci. Adv. 6 (44). https://doi.org/10.1126/sciadv.abc1937.

Ballmann, C.W., Thompson, J.V., Traverso, A.J., Meng, Z., Scully, M.O., Yakovlev, V.V., 2015. Stimulated Brillouin scattering microscopic imaging. Sci. Rep. 5, 18139. https://doi.org/10.1038/srep18139.

Ballmann, C.W., Meng, Z., Traverso, A.J., Scully, M.O., Yakovlev, V.V., 2017. Impulsive Brillouin microscopy. Optica 4 (1), 124–128. https://doi.org/10.1364/OPTICA.4.000124. http://www.osapublishing.org/optica/abstract.cfm?URI=optica-4-1-124.

Bedborough, D.S., Jackson, D.A., 1976. Brillouin scattering study of gelatin gel using a double passed Fabry-Perot spectrometer. Polymer 17 (7), 573–576. https://doi.org/10.1016/0032-3861(76)90271-8.

Besner, S., Scarcelli, G., Pineda, R., Yun, S.-H., 2016. In vivo Brillouin analysis of the aging crystalline lens. Invest. Ophthalmol. Vis. Sci. 57 (13), 5093–5100. https://doi.org/10.1167/iovs.16-20143.

Binder, P.S., 2007. Analysis of ectasia after laser in situ keratomileusis: risk factors. J. Cataract Refract. Surg. 33 (9), 1530–1538. https://doi.org/10.1016/j.jcrs.2007.04.043.

Binder, P.S., Lindstrom, R.L., Stulting, R.D., Donnenfeld, E., Wu, H., McDonnell, P., Rabinowitz, Y., 2005. Keratoconus and corneal ectasia after LASIK. J. Refract. Surg. 21 (6), 749–752.

Cao, X., Moeendarbary, E., Isermann, P., Davidson, P.M., Wang, X., Chen, M.B., Burkart, A.K., Lammerding, J., Kamm, R.D., Shenoy, V.B., 2016. A chemomechanical model for nuclear morphology and stresses during cell transendothelial migration. Biophys. J. 111 (7), 1541–1552. https://doi.org/10.1016/j.bpj.2016.08.011.

Caponi, S., Fioretto, D., Mattarelli, M., 2020. On the actual spatial resolution of Brillouin imaging. Opt. Lett. 45 (5), 1063–1066. https://doi.org/10.1364/ol.385072.

Ching, S.C., Leung, P.T., Young, K., 1990. Spontaneous Brillouin scattering in a microdroplet. Phys. Rev. A 41 (9), 5026–5038. https://doi.org/10.1103/PhysRevA.41.5026.

Chiquet, M., 1999. Regulation of extracellular matrix gene expression by mechanical stress. Matrix Biol. 18 (5), 417–426.

Conrad, C., Gray, K., Stroka, K., Rizvi, I., Scarcelli, G., 2019. Mechanical characterization of 3D ovarian cancer nodules using Brillouin confocal microscopy. Cell. Mol. Bioeng. 12 (3), 215–226.

Conrad, C., Moore, K., Polacheck, W., Rizvi, I., Scarcelli, G., 2021. Mechanical modulation of tumor nodules under flow. IEEE Trans. Biomed. Eng. 69 (1), 294–301. https://doi.org/10.1109/TBME.2021.3092641.

Coppola, S., Schmidt, T., Ruocco, G., Antonacci, G., 2019. Quantifying cellular forces and biomechanical properties by correlative micropillar traction force and Brillouin microscopy. Biomed. Opt. Express 10 (5), 2202–2212. https://doi.org/10.1364/boe.10.002202.

Cross, S.E., Jin, Y.S., Rao, J., Gimzewski, J.K., 2007. Nanomechanical analysis of cells from cancer patients. Nat. Nanotechnol. 2 (12), 780–783. https://doi.org/10.1038/nnano.2007.388.

Cusack, S., Miller, A., 1979. Determination of the elastic constants of collagen by Brillouin light scattering. J. Mol. Biol. 135 (1), 39–51. https://doi.org/10.1016/0022-2836(79)90339-5.

Discher, D.E., Janmey, P., Wang, Y.L., 2005. Tissue cells feel and respond to the stiffness of their substrate. Science 310 (5751), 1139–1143.

Edrei, E., Gather, M.C., Scarcelli, G., 2017. Integration of spectral coronagraphy within VIPA-based spectrometers for high extinction Brillouin imaging. Opt. Express 25 (6), 6895–6903. https://doi.org/10.1364/oe.25.006895.

Elsayad, K., Werner, S., Gallemi, M., Kong, J.X., Guajardo, E.R.S., Zhang, L.J., Jaillais, Y., Greb, T., Belkhadir, Y., 2016. Mapping the subcellular mechanical properties of live cells in tissues with fluorescence emission-Brillouin imaging. Sci. Signal. 9 (435). https://doi.org/10.1126/scisignal.aaf6326.

Elsheikh, A., Alhasso, D., 2009. Mechanical anisotropy of porcine cornea and correlation with stromal microstructure. Exp. Eye Res. 88 (6), 1084–1091. https://doi.org/10.1016/j.exer.2009.01.010.

Eltony, A.M., Clouser, F., Shao, P., Pineda, R., Yun, S.H., 2020a. Brillouin microscopy visualizes centralized Corneal Edema in Fuchs endothelial dystrophy. Cornea 39 (2), 168–171. https://doi.org/10.1097/ico.0000000000002191.

Eltony, A.M., Shao, P., Yun, S.H., 2020b. Measuring mechanical anisotropy of the cornea with Brillouin microscopy. arXiv:2003.04344.

Engler, A.J., Sen, S., Sweeney, H.L., Discher, D.E., 2006a. Matrix elasticity directs stem cell lineage specification. Cell 126 (4), 677–689. https://doi.org/10.1016/j.cell.2006.06.044.

Engler, A.J., Sen, S., Sweeney, H.L., Discher, D.E., 2006b. Matrix elasticity directs stem cell lineage specification. Cell 126 (4), 677–689. https://doi.org/10.1016/j.cell.2006.06.044.

Faffe, D.S., Zin, W.A., 2009. Lung parenchymal mechanics in health and disease. Physiol. Rev. 89 (3), 759–775. https://doi.org/10.1152/physrev.00019.2007.

Fasciani, A., D'Annunzio, S., Poli, V., Fagnocchi, L., Beyes, S., Michelatti, D., Corazza, F., Antonelli, L., Gregoretti, F., Oliva, G., Belli, R., Peroni, D., Domenici, E., Zambrano, S., Intartaglia, D., Settembre, C., Conte, I., Testi, C., Vergyris, P., Ruocco, G., Zippo, A., 2020. MLL4-associated condensates counterbalance polycomb-mediated nuclear mechanical stress in Kabuki syndrome. Nat. Genet. 52 (12), 1397–1411. https://doi.org/10.1038/s41588-020-00724-8.

Fiore, A., Scarcelli, G., 2021. Multipass etalon cascade for high-resolution parallel spectroscopy. Opt. Lett. 46 (4), 781–784. https://doi.org/10.1364/ol.418090.

Fiore, A., Zhang, J.T., Shao, P., Yun, S.H., Scarcelli, G., 2016. High-extinction virtually imaged phased array-based Brillouin spectroscopy of turbid biological media. Appl. Phys. Lett. 108 (20), 203701. https://doi.org/10.1063/1.4948353.

Fukui, K., Takayanagi, S., Suga, D., Matsukawa, M., 2012. Measurement of wave velocity in cortical bone by micro-Brillouin scattering technique: effect of bone tissue properties. Jpn. J. Appl. Phys. 51, 07GF20. https://doi.org/10.1143/jjap.51.07gf20.

Gusev, V.E., Ruello, P., 2018. Advances in applications of time-domain Brillouin scattering for nanoscale imaging. Appl. Phys. Rev. 5 (3), 031101. https://doi.org/10.1063/1.5017241.

Harley, R., James, D., Miller, A., White, J.W., 1977. Phonons and the elastic moduli of collagen and muscle. Nature 267 (5608), 285–287. https://doi.org/10.1038/267285a0.

Hatami-Marbini, H., 2018. Influence of microstructure on stiffening effects of corneal cross-linking treatment. J. Refract. Surg. 34 (9), 622–627. https://doi.org/10.3928/1081597x-20180718-01.

Hatami-Marbini, H., Pachenari, M., 2020. Hydration related changes in tensile response of posterior porcine sclera. J. Mech. Behav. Biomed. Mater. 104, 103562. https://doi.org/10.1016/j.jmbbm.2019.103562.

Huang, S., Ingber, D.E., 1999. The structural and mechanical complexity of cell-growth control. Nat. Cell Biol. 1 (5), E131–E138.

Huang, S., Ingber, D.E., 2005. Cell tension, matrix mechanics, and cancer development. Cancer Cell 8 (3), 175–176. https://doi.org/10.1016/j.ccr.2005.08.009.

Ingber, D.E., 2002. Mechanical signalling and the cellular response to extracellular matrix in angiogenesis and cardiovascular physiology. Circ. Res. 91 (10), 877–887. https://doi.org/10.1161/01.res.0000039537.73816.e5.

Itoh, S., 1998. Very rapid nonscanning Brillouin spectroscopy using fixed etalons and multichannel detectors. Jpn. J. Appl. Phys. Part 1 37 (5B), 3134–3135. https://doi.org/10.1143/jjap.37.3134 (Itoh, S 18th Symposium on Ultrasonic Electronics (USE97) Nov 12–14, 1997 Chiba University, Chiba, Japan. Japan Soc Appl Physics Si).

Janmey, P.A., McCulloch, C.A., 2007. Cell mechanics: integrating cell responses to mechanical stimuli. Annu. Rev. Biomed. Eng. 9, 1–34. https://doi.org/10.1146/annurev.bioeng.9.060906.151927.

Jansen, K.A., Donato, D.M., Balcioglu, H.E., Schmidt, T., Danen, E.H.J., Koenderink, G.H., 2015. A guide to mechanobiology: where biology and physics meet. Biochim. Biophys. Acta (BBA) Mol. Cell Res. 1853 (11, Part B), 3043–3052, Mechanobiology. https://doi.org/10.1016/j.bbamcr.2015.05.007.

Johnson, D.L., 1982. Elastodynamics of gels. J. Chem. Phys. 77 (3), 1531–1539. https://doi.org/10.1063/1.443934.

Kabakova, I.V., Xiang, Y., Paterson, C., Török, P., 2017. Fiber-integrated Brillouin microspectroscopy: towards Brillouin endoscopy. J. Innov. Opt. Health Sci. 10 (06), 1742002. https://doi.org/10.1142/s1793545817420020.

Kim, K., Guck, J., 2020. The relative densities of cytoplasm and nuclear compartments are robust against strong perturbation. Biophys. J. 119 (10), 1946–1957. https://doi.org/10.1016/j.bpj.2020.08.044.

Kim, M., Besner, S., Ramier, A., Kwok, S.J.J., An, J., Scarcelli, G., Yun, S.H., 2016. Shear Brillouin light scattering microscope. Opt. Express 24 (1), 319–328. https://doi.org/10.1364/oe.24.000319.

Koski, K.J., Yarger, J.L., 2005. Brillouin imaging. Appl. Phys. Lett. 87 (6), 061903. https://doi.org/10.1063/1.1999857.

Koski, K.J., Akhenblit, P., McKiernan, K., Yarger, J.L., 2013. Non-invasive determination of the complete elastic moduli of spider silks. Nat. Mater. 12 (3), 262–267. https://doi.org/10.1038/nmat3549.

Krug, B., Koukourakis, N., Czarske, J.W., 2019. Impulsive stimulated Brillouin microscopy for non-contact, fast mechanical investigations of hydrogels. Opt. Express 27 (19), 26910–26923. https://doi.org/10.1364/oe.27.026910.

Kumar, S., Weaver, V.M., 2009. Mechanics, malignancy, and metastasis: the force journey of a tumor cell. Cancer Metastasis Rev. 28 (1–2), 113–127. https://doi.org/10.1007/s10555-008-9173-4.

Kwok, S.J.J., Kuznetsov, I.A., Kim, M., Choi, M., Scarcelli, G., Yun, S.H., 2016. Selective two-photon collagen crosslinking in situ measured by Brillouin microscopy. Optica 3 (5), 469–472. https://doi.org/10.1364/optica.3.000469.

La Cavera, S., Pérez-Cota, F., Smith, R.J., Clark, M., 2021. Phonon imaging in 3D with a fibre probe. Light Sci. Appl. 10 (1), 91. https://doi.org/10.1038/s41377-021-00532-7.

Levental, K.R., Yu, H.M., Kass, L., Lakins, J.N., Egeblad, M., Erler, J.T., Fong, S.F.T., Csiszar, K., Giaccia, A., Weninger, W., Yamauchi, M., Gasser, D.L., Weaver, V.M., 2009. Matrix crosslinking forces tumor progression by enhancing integrin signaling. Cell 139 (5), 891–906. https://doi.org/10.1016/j.cell.2009.10.027.

Lo, C.M., Wang, H.B., Dembo, M., Wang, Y.L., 2000. Cell movement is guided by the rigidity of the substrate. Biophys. J. 79 (1), 144–152.

Mathieu, V., Fukui, K., Matsukawa, M., Kawabe, M., Vayron, R., Soffer, E., Anagnostou, F., Haiat, G., 2011. Micro-Brillouin scattering measurements in mature and newly formed bone tissue surrounding an implant. Ultrasonics 133 (2), 021006. https://doi.org/10.1115/1.4003131. PMID: 21280878.

Matsukawa, M., Tsubota, R., Kawabe, M., Fukui, K., 2014. Application of a micro-Brillouin scattering technique to characterize bone in the GHz range. Ultrasonics 54 (5), 1155–1161. https://doi.org/10.1016/j.ultras.2013.09.016.

Mattana, S., Caponi, S., Tamagnini, F., Fioretto, D., Palombo, F., 2017. Viscoelasticity of amyloid plaques in transgenic mouse brain studied by Brillouin microspectroscopy and correlative Raman analysis. J. Innov. Opt. Health Sci. 10 (06), 1742001. https://doi.org/10.1142/s1793545817420019.

Mattana, S., Mattarelli, M., Urbanelli, L., Sagini, K., Emiliani, C., Serra, M.D., Fioretto, D., Caponi, S., 2018. Non-contact mechanical and chemical analysis of single living cells by microspectroscopic techniques. Light Sci. Appl. 7, 17139. https://doi.org/10.1038/lsa.2017.139.

McKee, C.T., Last, J.A., Russell, P., Murphy, C.J., 2011. Indentation versus tensile measurements of young's modulus for soft biological tissues. Tissue Eng. Part B Rev. 17 (3), 155–164. https://doi.org/10.1089/ten.teb.2010.0520.

Meng, Z.K., Traverso, A.J., Yakovlev, V.V., 2014. Background clean-up in Brillouin microspectroscopy of scattering medium. Opt. Express 22 (5), 5410–5415. https://doi.org/10.1364/OE.22.005410.

Moeendarbary, E., Valon, L., Fritzsche, M., Harris, A.R., Moulding, D.A., Thrasher, A.J., Stride, E., Mahadevan, L., Charras, G.T., 2013. The cytoplasm of living cells behaves as a poroelastic material. Nat. Mater. 12 (3), 253–261. https://doi.org/10.1038/nmat3517.

Mofrad, M.R.K., Kamm, R.D., 2006. Cytoskeletal Mechanics. Cambridge University Press, New York.

Nikolic, M., Scarcelli, G., 2019. Long-term Brillouin imaging of live cells with reduced absorption-mediated damage at 660 nm wavelength. Biomed. Opt. Express 10 (4), 1567–1580. https://doi.org/10.1364/boe.10.001567. Nikolic, Milos Scarcelli, Giuliano.

Nikolic, M., Conrad, C., Zhang, J.T., Scarcelli, G., 2018. Noninvasive imaging: Brillouin confocal microscopy. In: Dong, C., Zahir, N., Konstantopoulos, K. (Eds.), Biomechanics in Oncology. Advances in Experimental Medicine and Biology, 1092, pp. 351–364.

Palombo, F., Winlove, C.P., Edginton, R.S., Green, E., Stone, N., Caponi, S., Madami, M., Fioretto, D., 2014. Biomechanics of fibrous proteins of the extracellular matrix studied by Brillouin scattering. J. R. Soc. Interface 11 (101), 20140739. https://doi.org/10.1098/rsif.2014.0739.

Payne, S.L., 2005. Lysyl oxidase regulates Breast Cancer cell migration and Adhesion through a hydrogen peroxide-mediated mechanism. Cancer Res. 65, 11429–11436. https://doi.org/10.1158/0008-5472.can-05-1274.

Pelham, R.J., Wang, Y.L., 1997. Cell locomotion and focal adhesions are regulated by substrate flexibility. Proc. Natl. Acad. Sci. U. S. A. 94 (25), 13661–13665.

Popescu, A., Walther, T., 2005. On the potential of faraday anomalous dispersion optical filters as high-resolution edge filters. Laser Phys. 15 (1), 55–60.

Prevedel, R., Antonacci, G., 2019. Brillouin microscopy: an emerging tool for mechanobiology. Nat. Methods. https://doi.org/10.1038/s41592-019-0543-3.

Prevedel, R., Diz-Munoz, A., Ruocco, G., Antonacci, G., 2019. Brillouin microscopy: an emerging tool for mechanobiology. Nat. Methods 16 (10), 969–977. https://doi.org/10.1038/s41592-019-0543-3.

Provenzano, P., Inman, D., Eliceiri, K., Keely, P., 2009. Matrix density-induced mechanoregulation of breast cell phenotype, signaling and gene expression through a FAK-ERK linkage. Oncogene 28 (49), 4326–4369. https://doi.org/10.1038/onc.2009.299.

Rabinowitz, Y., 2006. Ectasia after laser in situ keratomileusis. Curr. Opin. Ophthalmol. 17 (5), 421–427. https://doi.org/10.1097/01.icu.0000243015.51886.3a.

Randall, J.T., Vaughan, J.M., 1982. The measurement and interpretation of Brillouin scattering in the lens of the eye. 214 (1197), 449–470. https://doi.org/10.1098/rspb.1982.0021.

Randall, J.T., Vaughan, J.M., Buckingham, A.D., Series, G.W., Pike, E.R., Powles, J.G., 1979. Brillouin scattering in systems of biological significance. 293 (1402), 341–348. https://doi.org/10.1098/rsta.1979.0101.

Randleman, J.B., Su, J.P., Scarcelli, G., 2017. Biomechanical changes after LASIK flap creation combined with rapid cross-linking measured with Brillouin microscopy. J. Refract. Surg. 33 (6), 408–414. https://doi.org/10.3928/1081597x-20170421-01.

Remer, I., Bilenca, A., 2016a. Background-free Brillouin spectroscopy in scattering media at 780 nm via stimulated Brillouin scattering. Opt. Lett. 41 (5), 926–929. https://doi.org/10.1364/OL.41.000926.

Remer, I., Bilenca, A., 2016b. High-speed stimulated Brillouin scattering spectroscopy at 780 nm. APL Photonics 1 (6), 061301. https://doi.org/10.1063/1.4953620.

Remer, I., Shaashoua, R., Shemesh, N., Ben-Zvi, A., Bilenca, A., 2020. High-sensitivity and high-specificity biomechanical imaging by stimulated Brillouin scattering microscopy. Nat. Methods 17 (9), 913–916. https://doi.org/10.1038/s41592-020-0882-0.

Remmerbach, T.W., Wottawah, F., Dietrich, J., Lincoln, B., Wittekind, C., Guck, J., 2009. Oral cancer diagnosis by mechanical phenotyping. Cancer Res. 69 (5), 1728–1732. https://doi.org/10.1158/0008-5472.can-08-4073.

Roberts, A.B., Zhang, J.T., Singh, V.R., Nikolic, M., Moeendarbary, E., Kamm, R.D., So, P.T.C., Scarcelli, G., 2021. Tumor cell nuclei soften during transendothelial migration. J. Biomech. 121, 110400. https://doi.org/10.1016/j.jbiomech.2021.110400.

Roy, A.S., Dupps Jr, W.J., 2011. Patient-specific computational modeling of keratoconus progression and differential responses to collagen cross-linking. Invest. Ophthalmol. Vis. Sci. 52 (12), 9174–9187. http://www.ncbi.nlm.nih.gov/pubmed/22039252. http://www.ncbi.nlm.nih.gov/pmc/articles/PMC3253542/pdf/z7g9174.pdf.

Ryu, S., Martino, N., Kwok, S.J.J., Bernstein, L., Yun, S.-H., 2021. Label-free histological imaging of tissues using Brillouin light scattering contrast. Biomed. Opt. Express 12 (3), 1437–1448. https://doi.org/10.1364/BOE.414474.

Sandercock, J.R., 1976. Simple stabilization scheme for maintenance of mirror alignment in a scanning February-Perot interferometer. J. Phys. E Sci. Instrum. 9 (7), 566–569. https://doi.org/10.1088/0022-3735/9/7/016.

Sandercock, J.R., 1978. Light scattering from surface acoustic phonons in metals and semiconductors. Solid State Commun. 26, 547.

Scarcelli, G., Yun, S.H., 2008. Confocal Brillouin microscopy for three-dimensional mechanical imaging. Nat. Photonics 2 (1), 39–43. https://doi.org/10.1038/nphoton.2007.250.

Scarcelli, G., Yun, S., 2011. Multistage VIPA etalons for high-extinction parallel Brillouin spectroscopy. Opt. Express 19 (11), 10913–10922. https://doi.org/10.1364/OE.19.010913.

Scarcelli, G., Yun, S.H., 2018. Reply to 'water content, not stiffness, dominates Brillouin spectroscopy measurements in hydrated materials'. Nat. Methods 15 (8), 562–563. https://doi.org/10.1038/s41592-018-0075-2.

Scarcelli, G., Kim, P., Yun, S.H., 2011. In vivo measurement of age-related stiffening in the crystalline lens by Brillouin optical microscopy. Biophys. J. 101 (6), 1539–1545. https://doi.org/10.1016/j.bpj.2011.08.008.

Scarcelli, G., Pineda, R., Yun, S.H., 2012. Brillouin optical microscopy for corneal biomechanics. Invest. Ophthalmol. Vis. Sci. 53 (1), 185–190. https://doi.org/10.1167/iovs.11-8281.

Scarcelli, G., Kling, S., Quijano, E., Pineda, R., Marcos, S., Yun, S.H., 2013. Brillouin microscopy of collagen crosslinking: noncontact depth-dependent analysis of corneal elastic modulus. Invest. Ophthalmol. Vis. Sci. 54 (2), 1418–1425. https://doi.org/10.1167/iovs.12-11387.

Scarcelli, G., Besner, S., Pineda, R., Yun, S.H., 2014. Biomechanical characterization of keratoconus corneas ex vivo with Brillouin microscopy. Invest. Ophthalmol. Vis. Sci. 55 (7), 4490–4495. https://doi.org/10.1167/iovs.14-14450.

Scarcelli, G., Besner, S., Pineda, R., Kalout, P., Yun, S.H., 2015a. In vivo biomechanical mapping of normal and keratoconus corneas. JAMA Ophthalmol. 133 (4), 480–482. https://doi.org/10.1001/jamaophthalmol.2014.5641.

Scarcelli, G., Polacheck, W.J., Nia, H.T., Patel, K., Grodzinsky, A.J., Kamm, R.D., Yun, S.H., 2015b. Noncontact three-dimensional mapping of intracellular hydromechanical properties by Brillouin microscopy. Nat. Methods 12 (12), 1132–1134. https://doi.org/10.1038/nmeth.3616.

Scarcelli G., Hutchins R., Zanini G., 2022. Full-field brillouin microscopy systems and methods, US Patent App. 17/393, 532.

Schlüßler, R., Möllmert, S., Abuhattum, S., Cojoc, G., Müller, P., Kim, K., Möckel, C., Zimmermann, C., Czarske, J., Guck, J., 2018. Mechanical mapping of spinal cord growth and repair in living Zebrafish Larvae by Brillouin imaging. Biophys. J. 115 (5), 911–923. https://doi.org/10.1016/j.bpj.2018.07.027.

Schlüßler, R., Kim, K., Nötzel, M., Taubenberger, A., Abuhattum, S., Beck, T., Müller, P., Maharana, S., Cojoc, G., Girardo, S., Hermann, A., Alberti, S., Guck, J., 2020. Combined fluorescence, optical diffraction tomography and Brillouin microscopy. https://doi.org/10.1101/2020.10.30.361808.

Seiler, T.G., Shao, P., Frueh, B.E., Yun, S.H., Seiler, T., 2018. The influence of hydration on different mechanical moduli of the cornea. Graefes Arch. Clin. Exp. Ophthalmol. 256 (9), 1653–1660. https://doi.org/10.1007/s00417-018-4069-7.

Shao, P., Seiler, T.G., Eltony, A.M., Ramier, A., Kwok, S.J.J., Scarcelli, G., II, R.P., Yun, S.-H., 2018. Effects of corneal hydration on Brillouin microscopy in vivoeffect of corneal hydration on Brillouin frequency. Invest. Ophthalmol. Vis. Sci. 59 (7), 3020–3027. https://doi.org/10.1167/iovs.18-24228.

Shao, P., Eltony, A.M., Seiler, T.G., Tavakol, B., Pineda, R., Koller, T., Seiler, T., Yun, S.H., 2019. Spatially-resolved Brillouin spectroscopy reveals biomechanical abnormalities in mild to advanced keratoconus in vivo. Sci. Rep. 9, 7467. https://doi.org/10.1038/s41598-019-43811-5.

Shirasaki, M., 1996. Large angular dispersion by a virtually imaged phased array and its application to a wavelength demultiplexer. Opt. Lett. 21 (5), 366–368. https://doi.org/10.1364/OL.21.000366.

Singh, M., Li, J., Han, Z., Wu, C., Aglyamov, S., Twa, M.D., Larin, K.V., 2016. Investigating elastic anisotropy of the porcine cornea as a function of intraocular pressure with optical coherence elastography. J. Refract. Surg. 32 (8), 562–567. https://doi.org/10.3928/1081597X-20160520-01.

Stephen, M.A., Krainak, M.A., Fahey, M.E., 2015. Lateral-transfer recirculating etalon spectrometer. Opt. Express 23 (23), 30020–30027. https://doi.org/10.1364/oe.23.030020.

Tse, H.T.K., Gossett, D.R., Moon, Y.S., Masaeli, M., Sohsman, M., Ying, Y., Mislick, K., Adams, R.P., Rao, J.Y., Di Carlo, D., 2013. Quantitative diagnosis of malignant pleural effusions by single-cell mechanophenotyping. Sci. Transl. Med. 5 (212), 212ra163. https://doi.org/10.1126/scitranslmed.3006559.

Tzima, E., Irani-Tehrani, M., Kiosses, W.B., Dejana, E., Schultz, D.A., Engelhardt, B., Cao, G., DeLisser, H., Schwartz, M.A., 2005. A mechanosensory complex that mediates the endothelial cell response to fluid shear stress. Nature 437 (7057), 426–431. https://doi.org/10.1038/nature03952. http://www.nature.com/nature/journal/v437/n7057/suppinfo/nature03952_S1.html.

Weaver, V.M., 2017. Cell and tissue mechanics: the new cell biology frontier. Mol. Biol. Cell 28 (14), 1815–1818. https://doi.org/10.1091/mbc.e17-05-0320. PMID: 28684606.

Webb, J.N., Su, J.P., Scarcelli, G., 2017. Mechanical outcome of accelerated corneal crosslinking evaluated by Brillouin microscopy. J. Cataract Refract. Surg. 43 (11), 1458–1463. https://doi.org/10.1016/j.jcrs.2017.07.037.

Webb, J.N., Langille, E., Hafezi, F., Randleman, J.B., Scarcelli, G., 2019. Biomechanical impact of localized corneal cross-linking beyond the irradiated treatment area. J. Refract. Surg. 35 (4), 253–260. https://doi.org/10.3928/1081597x-20190304-01.

Webb, J.N., Zhang, H.Y., Roy, A.S., Randleman, J.B., Scarcelli, G., 2020. Detecting mechanical anisotropy of the cornea using Brillouin microscopy. Transl. Vis. Sci. Technol. 9 (7), 26. https://doi.org/10.1167/tvst.9.7.26.

Weber, I.P., Yun, S.H., Scarcelli, G., Franze, K., 2017. The role of cell body density in ruminant retina mechanics assessed by atomic force and Brillouin microscopy. Phys. Biol. 14 (6), 065006. https://doi.org/10.1088/1478-3975/aa6d18.

Willits, J.T., Weiner, A.M., Cundiff, S.T., 2012. Line-by-line pulse shaping with spectral resolution below 890 mHz. Opt. Express 20 (3), 3110–3117. https://doi.org/10.1364/oe.20.003110.

Wisniewski, E.O., Mistriotis, P., Bera, K., Law, R.A., Zhang, J., Nikolić, M., Weiger, M., Parlani, M., Tuntithavornwat, S., Afthinos, A., Zhao, R., Wirtz, D., Kalab, P., Scarcelli, G., Friedl, P., Konstantopoulos, K., 2020. Dorsoventral polarity directs cell responses to migration track geometries. Sci. Adv. 6 (31), eaba6505.

Wolfenson, H., Yang, B., Sheetz, M.P., 2019. Steps in mechanotransduction pathways that control cell morphology. Annu. Rev. Physiol. 81 (1), 585–605. https://doi.org/10.1146/annurev-physiol-021317-121245. PMID: 30403543.

Wolff, J., 1892. Das Gesetz Der Transformation Der Knochen. Hirschwald.

Wollensak, G., 2006. Crosslinking treatment of progressive keratoconus: new hope. Curr. Opin. Ophthalmol. 17 (4), 356–360.

Wu, P.-J., Kabakova, I.V., Ruberti, J.W., Sherwood, J.M., Dunlop, I.E., Paterson, C., Török, P., Overby, D.R., 2018. Water content, not stiffness, dominates Brillouin spectroscopy measurements in hydrated materials. Nat. Methods 15 (8), 561–562. https://doi.org/10.1038/s41592-018-0076-1.

Wu, P.-J., Masouleh, M.I., Dini, D., Paterson, C., Török, P., Overby, D.R., Kabakova, I.V., 2019. Detection of proteoglycan loss from articular cartilage using Brillouin microscopy, with applications to osteoarthritis. Biomed. Opt. Express 10 (5), 2457–2466. https://doi.org/10.1364/BOE.10.002457.

Xiang, Y., Basirun, C., Chou, J., Warkiani, M.E., Török, P., Wang, Y., Gao, S., Kabakova, I.V., 2020. Background-free fibre optic Brillouin probe for remote mapping of micromechanics. Biomed. Opt. Express 11 (11), 6687–6698. https://doi.org/10.1364/BOE.404535.

Zhang, J.T., Scarcelli, G., 2021. Mapping mechanical properties of biological materials via an add-on Brillouin module to confocal microscopes. Nat. Protoc. 16, 1251–1275. https://doi.org/10.1038/s41596-020-00457-2.

Zhang, J.T., Fiore, A., Yun, S.H., Kim, H., Scarcelli, G., 2016. Line-scanning Brillouin microscopy for rapid non-invasive mechanical imaging. Sci. Rep. 6, 35398. https://doi.org/10.1038/srep35398.

Zhang, J., Nou, X.A., Kim, H., Scarcelli, G., 2017. Brillouin flow cytometry for label-free mechanical phenotyping of the nucleus. Lab Chip 17 (4), 663–670. https://doi.org/10.1039/C6LC01443G.

Zhang, H.Y., Roozbahani, M., Piccinini, A.L., Golan, O., Hafezi, F., Scarcelli, G., Randleman, J.B., 2019. Depth-dependent reduction of biomechanical efficacy of contact lens-assisted corneal cross-linking analyzed by Brillouin microscopy. J. Refract. Surg. 35 (11), 721–728. https://doi.org/10.3928/1081597x-20191004-01.

Zhang, H.Y., Roozbahani, M., Piccinini, A.L., Hafezi, F., Scarcelli, G., Randleman, J.B., 2020a. Brillouin microscopic depth-dependent analysis of corneal crosslinking performed over or under the LASIK flap. J. Cataract Refract. Surg. 46 (11), 1543–1547. https://doi.org/10.1097/j.jcrs.0000000000000294.

Zhang, J.T., Alisafaei, F., Nikolic, M., Nou, X.F.A., Kim, H., Shenoy, V.B., Scarcelli, G., 2020b. Nuclear mechanics within intact cells is regulated by cytoskeletal network and internal nanostructures. Small 16 (18). https://doi.org/10.1002/smll.201907688.

Zhao, P., Vanderwal, J.J., 1997. Brillouin scattering study of gelatin gel. Polym. Gels Netw. 5 (1), 23–36. https://doi.org/10.1016/S0966-7822(96)00029-9.

Index

Note: Page numbers followed by "*f*" indicate figures.

K

L

M

N

W

Y

Z

FSC
www.fsc.org
MIX
Papier | Fördert gute Waldnutzung
FSC® C083411

Kolibri 360 GmbH
Ferdinand-Jühlke-Straße 7
99095 Erfurt, Deutschland
produktsicherheit@kolibri360.de